AF388953

Machine Learning for Biomedical Application

Machine Learning for Biomedical Application

Editors

Michał Strzelecki
Pawel Badura

MDPI • Basel • Beijing • Wuhan • Barcelona • Belgrade • Manchester • Tokyo • Cluj • Tianjin

Editors
Michał Strzelecki
Lodz University of Technology
Poland

Pawel Badura
Silesian University of Technology
Poland

Editorial Office
MDPI
St. Alban-Anlage 66
4052 Basel, Switzerland

This is a reprint of articles from the Special Issue published online in the open access journal *Applied Sciences* (ISSN 2076-3417) (available at: https://www.mdpi.com/journal/applsci/special_issues/Machine_Learning_Biomedical).

For citation purposes, cite each article independently as indicated on the article page online and as indicated below:

LastName, A.A.; LastName, B.B.; LastName, C.C. Article Title. *Journal Name* **Year**, *Volume Number*, Page Range.

ISBN 978-3-0365-3445-9 (Hbk)
ISBN 978-3-0365-3446-6 (PDF)

Contents

About the Editors

Michał Strzelecki Professor Michał Strzelecki has been employed at the Institute of Electronics, Lodz University of Technology (TUL) since 1988, and he had been a Full Professor since 2007. His scientific interests include the processing and analysis of biomedical signals and images, data analysis methods, and artificial intelligence. In particular, he is interested in issues related to image texture analysis and the applications and hardware implementations of artificial neural networks. He also works on the development of software aimed to support medical imaging diagnostics. Within these interests, Prof. Strzelecki conducts extensive cooperation with domestic and foreign medical universities on the construction and development of systems supporting the diagnostic process.

Pawel Badura Dr. Pawel Badura, PhD, DSc, has been with the Faculty of Biomedical Engineering, Silesian University of Technology, Poland, since 2007. He is an associate professor in the Department of Medical Informatics and Artificial Intelligence, which defines his scientific interests. Dr. Badura is experienced in medical image and signal analysis, computer-aided diagnosis systems, and artificial intelligence for biomedical applications. In recent years, he has worked with machine learning and deep learning applied to various fields of biomedical engineering, e.g., image analysis, acoustic analysis, or classification.

applied sciences

Editorial

Machine Learning for Biomedical Application

Michał Strzelecki [1,*] and Pawel Badura [2]

1 Institute of Electronics, Lodz University of Technology, Żeromskiego 116, 90-924 Lodz, Poland
2 Faculty of Biomedical Engineering, Silesian University of Technology, Roosevelta 40, 41-800 Zabrze, Poland; pawel.badura@polsl.pl
* Correspondence: michal.strzelecki@p.lodz.pl; Tel.: +48-42-631-26-31

Citation: Strzelecki, M.; Badura, P. Machine Learning for Biomedical Application. *Appl. Sci.* **2022**, 12, 2022. https://doi.org/10.3390/app12042022

Received: 26 January 2022
Accepted: 8 February 2022
Published: 15 February 2022

Publisher's Note: MDPI stays neutral with regard to jurisdictional claims in published maps and institutional affiliations.

The tremendous development of technology also affects medical science, including imaging diagnostics. Computer tomographs enable the non-invasive visualization of internal human organs and tissues without the need for a surgical operation. This leads to the research and development of new, more efficient, and reliable diagnostic and therapeutic procedures. Medical imaging, such as biomedical signal acquisition, plays an increasingly important role not only in diagnostics, but also in therapy, monitoring its effects, and in rehabilitation. On the other hand, the increasing data sets generated by medical diagnostic devices make it difficult to explore and analyze data in a non-automatic way—as doctors have done so far. Hence, recently, more and more attention has been paid to the design and development of automatic data analysis systems, the application of which covers various issues in the field of healthcare. Such systems can dynamically adapt to changing conditions, and thus facilitate the analysis and solving of complex problems. Such systems often implement machine learning methods.

Machine learning (ML) is a subset of artificial intelligence (AI). Algorithms are trained to find patterns and correlations in large data sets, and to make the best decisions as well as predictions based on the results of such analysis. Machine learning systems become more effective over time, and the more data they have access to, the more accurate they are. Nowadays, deep learning methods are also often used in medical imaging [1]. Deep learning is a part of machine learning. It is based on complex artificial neural networks. The learning process is deep because the structure of artificial neural networks consists of many input, output, and hidden layers, which are often interconnected. Deep networks achieve much better results in terms of the recognition, classification, and prediction of medical data compared to classical machine learning algorithms.

Thanks to ML technology, including DL, health care workers, including doctors, can cope with complex problems that would be difficult, time-consuming, and ineffective to solve on their own. This Special Issue includes 10 publications that discuss the use of broadly understood machine learning methods for processing and analyzing biomedical signals and images coming from many medical modalities. The use of these methods allows a better understanding of the principles of the human body functioning at various levels (cellular, anatomical, and physiological) by providing additional, quantitative, reliable data extracted from medical data.

Ihsanto [2] proposes an algorithm developed for automated electrocardiogram (ECG) classification. ECG is a popular biosignal in heart disease diagnostics. However, it is non-stationary; thus, the implementation of classic signal analysis techniques (such as time-based analysis feature extraction and classification) is rather difficult. Thus, a machine learning approach based on the ensemble of depthwise separable convolutional (DSC) neural networks for the classification of cardiac arrhythmia ECG beats was proposed. This method reduces the standard path of ECG analysis (QRS detection, preprocessing, feature extraction, and classification) to two steps only, i.e., QRS detection and classification. Since feature extraction was combined with classification, no ECG preprocessing was required. To reduce the computational cost and maintain method reliability, All Convolutional Network

(ACN), Batch Normalization (BN), and ensemble convolutional neural networks were implemented. The developed ensemble of deep networks was validated using the MIT-BIH arrhythmia database. The obtained classification results (16 class problem) resulted in sensitivity (Sn), specificity (Sp), and positive predictivity (Pp), and accuracy (Acc) equal to 99.03%, 99.94%, 99.03%, and 99.88%, respectively. It was demonstrated that presented classification quality measures outperformed other state-of-the-art methods.

Biomedical signals are often used for the design and development of human–machine interfaces, which is emerging branch biomedical engineering. Borowska-Terka [3] proposes such a system dedicated to persons with disabilities, which that is a hands-free head-gesture-controlled interface. It can help, for example, paralyzed people to send messages or the visually impaired to handle travel aids. The system contains a small stereovision rig with a built-in inertial measurement unit (IMU). To recognize head movements, two methods are considered. In the first approach, for various time window sizes of the signals recorded from a three-axis accelerometer and a three-axis gyroscope, selected statistical parameters were calculated. In the second technique, the direct analysis of signal samples recorded from the IMU was performed. Next, the accuracies of 16 different data classifiers for distinguishing the head movements: pitch, roll, yaw, and immobility were evaluated. The highest accuracies were obtained for the direct classification of unprocessed samples of IMU signals and with the use of SVM classifier (95% correct recognitions), while the random forests classifier reached 93%. Such results indicate that a person with physical or sensory disability can efficiently communicate with other people or manage applications using simple head gesture sequences.

MRI is one of most common imaging modalities used in diagnostics and treatment planning. Klepaczko [4] shows an application of dynamic contrast-enhanced magnetic resonance imaging (DCE-MRI) for visualizing and quantifying kidney perfusion, which is one of the most important indicators of an organ's state. In clinical practice, kidney function is assessed by measuring the glomerular filtration rate (GFR). Estimating the GFR based on DCE-MRI data requires the application of an organ-specific pharmacokinetic (PK) model, but the determination of the model parameters is sensitive to the determination of the arterial input function (AIF). Thus, a multi-layer perceptron network was proposed for PK model parameter determination. As a reference method, the trust-region reflective algorithm was used. The efficiency of the proposed approach was tested for 20 data sets, collected for 10 healthy volunteers whose image-derived GFR scores were compared with ground-truth blood test values. The achieved mean difference between the image-derived and ground-truth GFR values was 2.35 mL/min/1.73 m^2, which is comparable to the result obtained for the reference estimation method (-5.80 mL/min/1.73 m^2). It was demonstrated that the implemented neural networks ensure agreement with ground-truth measurements at a comparable level. The advantages of using a neural network are twofold. First, it can estimate a GFR value without the need to determine the AIF for each individual patient. Second, a reliable estimate can be obtained, without the need to manually set up either the initial parameter values or the constraints thereof. After further validation and more exhaustive patient testing, the proposed approach can be implemented in clinical practice.

Another important application of ML to fundus blood vessel image segmentation is presented in [5]. Such analysis is important in the diagnosis and treatment of several diseases, such as hypertension, coronary heart disease, and diabetes. The analysis of such images is rather complicated, and classic algorithms suffer from relatively low segmentation accuracy. Thus, an improved U-shaped neural network (MRU-NET) segmentation method for retinal vessels was proposed. After the application of the image enhancement algorithm, the image contrast is improved. Applied random segmentation divides the image into smaller image blocks that help to reduce the complexity of the U-shaped neural network model. Next, the residual learning is introduced into the encoder and decoder to improve the efficiency of feature analysis. Finally, a feature balancing module is implemented followed by the feature fusion module that is introduced between the encoder and decoder to extract image features with different granularities. The developed segmentation technique was tested on the DRIVE and STARE datasets. It was demonstrated that the obtained accuracies (96.11% and 96.62% for DRIVE and STARE, respectively) and sensitivities (86.13% and 78.87%, respectively) outperform other state-of-the-art methods presented in the literature.

A computer tool dedicated to the comprehensive analysis of lung changes in computed tomography (CT) images is described in [6]. The proposed system enables the analysis of the correlation between the radiation dose delivered during radiotherapy and the density changes in lungs caused by the fibrosis. The input data, including patient dose, are extracted from the CT images coded in DICOM format. The convolution neural networks are used for CT processing. Next, the selected slices are segmented and registered by the developed algorithms. The results of the analysis are visualized graphically, enabling, for example, the presentation of dose distribution maps in the lungs. It is expected that, thanks to the developed application, it will be possible to demonstrate the statistically significant impact of low doses on lung function for a large number of patients.

Some areas of computer-aided diagnosis benefit from worldwide competitions for research teams mainly due to the access to extensive and well-annotated datasets. The study of Sage and Badura [7] addresses the detection of five subtypes of intracranial hemorrhage (ICH) in head CT. The authors trained and validated their tools using a public database provided by the Radiological Society of North America for the international ICH detection competition. The set used in [7] consists of over 372k images from ten thousand patients. Five separate models of a dedicated double-branch architecture were trained to detect every ICH subtype. Each ResNet-50-based branch analyzes two artificial RGB images: one reflecting the original CT slice in three different intensity windows and the other combining the image with its neighbors in the series. The authors report the results produced by each separate branch, but the best results came from another framework. Features extracted by ResNet-50 cores are concatenated and passed to a classifier: either an SVM or a random forest. Depending on the ICH type, the F1 score reaches from 75.3 to 96.6%, surpassing other studies in the field thus far.

The ResNet-50 architecture is also the core of a method proposed by Song et al. [8] for the automated detection of cephalometric landmarks in head X-ray images. Again, the study takes advantage of a public database for an international ISBI Grand Challenge in dental X-ray image analysis. However, the authors support the investigation with their own dataset this time. The analysis is performed within patches extracted from the X-ray image at a registration-based coarse landmark localization stage. The ResNet-50 architecture is replicated and trained nineteen times to address each of nineteen defined landmarks separately. The fully connected layer works in a regression mode to estimate the x and y coordinates of the landmark with a two-argument mean squared error (MSE) loss function. The authors assess their methodology using a radial detection error and a successful detection rate with an assumed tolerance. Despite a relatively small training dataset, the method can detect most landmarks accurately, especially in the ISBI dataset images. The comparison with some state-of-the-art methods favors the current study in most evaluation metrics. The authors indicate that their approach still has room for improvement in terms of computational time, mainly in the initial automated registration.

Mazur-Milecka et al. [9] applied deep learning models to segment laboratory rodents in thermal images. Their study aimed at the efficient and non-intrusive monitoring of animals' activity and physiological changes, also in low light conditions. The image data were captured at 60 frames per second by a thermal camera over a cage with two rats. The authors implemented and investigated two approaches: a two-stage detect-then-segment Mask R-CNN and a single-stage TensorMask network. Various training schemes and parameter setups were used, including the transfer learning of models pre-trained on public datasets. The TensorMask model produced the most accurate results when pre-trained using visible light images and then trained with thermal sequences (mean average precision over 90%). The authors also verified possible benefits from inputting alternative images created by rescaling the original thermal intensities in various automatically adjusted ranges of animal temperature. However, they did not find any such modifications improving the segmentation accuracy. In general, the single-stage methods performed better than two-stage ones if pre-trained; the opposite was true in the case of models trained from scratch.

Huang et al. [10] designed a prediction system for analyzing the risk of suffering amyotrophic lateral sclerosis (ALS) based on combinatorial comorbidity patterns. They employed medical history data gathered in electronic medical records (EMR) to support the identification of high-risk ALS subjects in the early stage. One of the main contributions of the study is the weighted Jaccard index (WJI) defined to balance the impact of particular comorbidities in predicting the output risk. Comprehensive patient-group-wise distributions were determined over hundreds of individual- and mid-level diseases characterized over the EMR database. This resulted in the determination of WJI weights. Four machine learning models were trained and validated: logistic regression, random forest, SVM, and XGBoost. The first two were considered most efficient based on seven canonic classification assessment measures, including accuracy exceeding 80%. In general, the individual-level disease classification turned out to be more efficient than the mid-level analysis, and the WJI proved its advantage over the pure Jaccard index in the ALS risk prediction.

Finally, Loh et al. [11] delivered a systematic review of the automated detection of sleep stages using deep learning models from the last decade. The authors summarize 36 studies from 2013 to 2020 that employ various deep models to analyze polysomnogram (PSG) recordings. After providing some medical and machine learning background and introducing five or six sleep stages, depending on the assumed standard, they analyze their detection and classification from different points of view. First, the models and architectures are introduced (convolutional or recurrent neural networks, long short-term memory, autoencoders, and hybrid models). Second, the available databases are described. Then, deep learning approaches addressing the sleep stage analysis are presented and compared. Finally, the authors discuss the topic in detail and draw conclusions. Even though electroencephalography (EEG) seems to be the most widely used signal in sleep stage detection, the current review states that it may not be able to work efficiently enough: automated systems should also involve other PSG recordings, e.g., electrooculography (EOG) or electromyography (EMG).

The variety of topics covered by the articles in this Special Issue is yet another proof of enormous opportunities for biomedical engineers and scientists familiar with artificial intelligence, machine, and deep learning. These techniques respond to the emerging needs of computer-aided diagnosis and therapy. Their development follows the explosion of computing power; hence, the enormity of diagnostic, signal, or image data has a chance to be processed, pre-analyzed, and presented to the physician within a reasonable time. From predictive models based on classification and regression models to automated multi-scale feature extraction from 2D, 3D, and even higher-dimensional data structures, contextual analysis, segmentation, this is the future of the software branch of biomedical engineering. The tools themselves are becoming increasingly reliable through the development of the so-called explainable artificial intelligence (XAI). All these issues guarantee that the broad biomedical application of machine learning is far from exhausted.

Funding: This research received no external funding.

Conflicts of Interest: The authors declare no conflict of interest.

References

1. Lundervold, A.S.; Lundervold, A. An overview of deep learning in medical imaging focusing on MRI. *Zeitschrift fur Medizinische Physik.* **2019**, *29*. [CrossRef] [PubMed]
2. Ihsanto, E.; Ramli, K.; Sudiana, D.; Gunawan, T. An Efficient Algorithm for Cardiac Arrhythmia Classification Using Ensem-ble of Depthwise Separable Convolutional Neural Networks. *Appl. Sci.* **2020**, *10*, 483. [CrossRef]
3. Borowska-Terka, A.; Strumillo, P. Person Independent Recognition of Head Gestures from Parametrised and Raw Signals Recorded from Inertial Measurement Unit. *Appl. Sci.* **2020**, *10*, 4213. [CrossRef]

4. Klepaczko, A.; Strzelecki, M.; Kociołek, M.; Eikefjord, E.; Lundervold, A. A Multi-Layer Perceptron Network for Perfusion Parameter Estimation in DCE-MRI Studies of the Healthy Kidney. *Appl. Sci.* **2020**, *10*, 5525. [CrossRef]
5. Ding, H.; Cui, X.; Chen, L.; Zhao, K. MRU-NET: A U-Shaped Network for Retinal Vessel Segmentation. *Appl. Sci.* **2020**, *10*, 6823. [CrossRef]
6. Konkol, M.; Śniatała, K.; Śniatała, P.; Wilk, S.; Baczyńska, B.; Milecki, P. Computer Tools to Analyze Lung CT Changes after Radiotherapy. *Appl. Sci.* **2021**, *11*, 1582. [CrossRef]
7. Sage, A.; Badura, P. Intracranial Hemorrhage Detection in Head CT Using Double-Branch Convolutional Neural Network, Support Vector Machine, and Random Forest. *Appl. Sci.* **2020**, *10*, 7577. [CrossRef]
8. Song, Y.; Qiao, X.; Iwamoto, Y.; Chen, Y.-W. Automatic Cephalometric Landmark Detection on X-ray Images Using a Deep-Learning Method. *Appl. Sci.* **2020**, *10*, 2547. [CrossRef]
9. Mazur-Milecka, M.; Kocejko, T.; Ruminski, J. Deep Instance Segmentation of Laboratory Animals in Thermal Images. *Appl. Sci.* **2020**, *10*, 5979. [CrossRef]
10. Huang, C.-H.; Yip, B.-S.; Taniar, D.; Hwang, C.-S.; Pai, T.-W. Comorbidity Pattern Analysis for Predicting Amyotrophic Lateral Sclerosis. *Appl. Sci.* **2021**, *11*, 1289. [CrossRef]
11. Loh, H.W.; Ooi, C.P.; Vicnesh, J.; Oh, S.L.; Faust, O.; Gertych, A.; Acharya, U.R. Automated Detection of Sleep Stages Using Deep Learning Techniques: A Systematic Review of the Last Decade (2010–2020). *Appl. Sci.* **2020**, *10*, 8963. [CrossRef]

*applied
sciences*

Review

Automated Detection of Sleep Stages Using Deep Learning Techniques: A Systematic Review of the Last Decade (2010–2020)

Hui Wen Loh [1], Chui Ping Ooi [1], Jahmunah Vicnesh [2], Shu Lih Oh [2], Oliver Faust [3], Arkadiusz Gertych [4,5,*] and U. Rajendra Acharya [1,2,6,7,8,*]

[1] School of Science and Technology, Singapore University of Social Sciences, Singapore 599494, Singapore; hwloh002@suss.edu.sg (H.W.L.); cpooi@suss.edu.sg (C.P.O.)

[2] School of Engineering, Ngee Ann Polytechnic, Singapore 599489, Singapore; e0145834@u.nus.edu (J.V.); shulih@hotmail.com (S.L.O.)

[3] Department of Engineering and Mathematics, Sheffield Hallam University, Sheffield S1 1WB, UK; oliver.faust@gmail.com

[4] Department of Surgery, Department of Pathology and Laboratory Medicine, Cedars-Sinai Medical Center, Los Angeles, CA 90040, USA

[5] Faculty of Biomedical Engineering, Silesian University of Technology, Roosevelta 40, 41-800 Zabrze, Poland

[6] Department of Bioinformatics and Medical Engineering, Asia University, Taichung 41354, Taiwan

[7] International Research Organization for Advanced Science and Technology (IROAST), Kumamoto University, Kumamoto 860-8555, Japan

[8] School of Management and Enterprise, University of Southern Queensland, Darling Heights, QLD 4350, Australia

* Correspondence: agertych@gmail.com (A.G.); aru@np.edu.sg (U.R.A.)

Received: 3 November 2020; Accepted: 8 December 2020; Published: 15 December 2020

Abstract: Sleep is vital for one's general well-being, but it is often neglected, which has led to an increase in sleep disorders worldwide. Indicators of sleep disorders, such as sleep interruptions, extreme daytime drowsiness, or snoring, can be detected with sleep analysis. However, sleep analysis relies on visuals conducted by experts, and is susceptible to inter- and intra-observer variabilities. One way to overcome these limitations is to support experts with a programmed diagnostic tool (PDT) based on artificial intelligence for timely detection of sleep disturbances. Artificial intelligence technology, such as deep learning (DL), ensures that data are fully utilized with low to no information loss during training. This paper provides a comprehensive review of 36 studies, published between March 2013 and August 2020, which employed DL models to analyze overnight polysomnogram (PSG) recordings for the classification of sleep stages. Our analysis shows that more than half of the studies employed convolutional neural networks (CNNs) on electroencephalography (EEG) recordings for sleep stage classification and achieved high performance. Our study also underscores that CNN models, particularly one-dimensional CNN models, are advantageous in yielding higher accuracies for classification. More importantly, we noticed that EEG alone is not sufficient to achieve robust classification results. Future automated detection systems should consider other PSG recordings, such as electroencephalogram (EEG), electrooculogram (EOG), and electromyogram (EMG) signals, along with input from human experts, to achieve the required sleep stage classification robustness. Hence, for DL methods to be fully realized as a practical PDT for sleep stage scoring in clinical applications, inclusion of other PSG recordings, besides EEG recordings, is necessary. In this respect, our report includes methods published in the last decade, underscoring the use of DL models with other PSG recordings, for scoring of sleep stages.

Keywords: sleep disorder; obstructive sleep disorder; overnight polysomnogram; EEG; EMG; ECG; HRV signals; deep learning

1. Introduction

Sleep is crucial for the maintenance and regulation of various biological functions at a molecular level [1], which helps humans to restore physical and mental wellbeing and proper brain function during the day [2]. There are two primary types of sleep: non-rapid eye movement (NREM) and rapid eye movement (REM) sleep. NREM sleep comprises four stages, after which, it continues into the REM sleep stage. NREM and REM sleep stages are connected and cyclically alternated through the sleep process wherein unbalanced cycling or the absence of sleep stages give rise to sleep disorders [3]. Unfortunately, sleep disorders, which lead to poor sleep quality, is often neglected [4]. Stranges et al. [4] highlighted that sleep-related problems is a looming global health issue. In their study, datasets from the World Health Organization (WHO) and International Network for the Demographic Evaluation of Populations and Their Health (INDEPTH) were used to investigate the prevalence of sleep problems in low-income countries. It was reported that 16.6% of the adult population, which amounts to approximately 150 million, have sleep problems and current trends indicate that this figure will increase to 260 million by 2030.

To date, it is mandatory that sleep stage scoring is done manually by human experts [5,6]. However, human experts have limited capacity to handle slow changes in background electroencephalography (EEG) and learn the different rules to score sleep stages for various polysomnogram (PSG) recordings [6]. Furthermore, evaluations by human experts are prone to inter- and intra-observer variabilities that can negatively affect the quality of sleep stage scoring [7]. Another important factor affecting sleep stage scoring is the patient convenience and the cost of diagnosis. As such, a sleep lab is a highly controlled environment that requires dedicated facilities and highly trained personnel. Hence, sleep labs tend to be in urban centers and patients must travel there to spend one or multiple nights in the facility. These factors make sleep labs inconvenient for patients and the cost per diagnosis is high. Other diagnostic methods, such as portable monitoring devices for sleep stages, exhibit some advantages, such as enhancing access to patients, low cost, and user-friendliness. However, these advantages are outweighed by several disadvantages, such as having diagnostic limitations, failure of device, reliability concerns, and underestimating the apnea/hypopnea index, amongst others [8]. To improve the situation requires a fundamental change in the sleep stage scoring process. We need machines to replace the labor carried out by human experts. This can only be done with systems that understand sleep stages in much of the same way as human experts do. Deep learning (DL) is hailed as a method to mechanize knowledge work, such as sleep stage scoring. However, before we join and adopt this technology, it is prudent to investigate both capabilities and limitations of current DL methods.

This paper aims to capture both capabilities and limitations of current DL methods in sleep stage classification. It is intended to provide deep cohesive information for experts to consolidate and extend their knowledge in the field. This knowledge might also be of interest to policy makers and healthcare administrators because DL technologies are going to shape future sleep stage scoring systems. This review paper summarizes the various DL models employed in the last 10 years and their performances as sleep stage classification systems. This information is valuable for those who plan to use established techniques to address a related problem. This review will also help establish the distinctiveness of a study, because any claim for novelty requires an overview of established methods. In this paper, we focused the review process on DL techniques, because during our practice in the field we found that, among the various artificial intelligence technologies, DL is the most suitable to be developed into a decision support tool for sleep stage scoring. In the foreseeable future, all studies on the topic would either include DL or mention DL-based techniques as a reference point.

To support our claim that DL technology will benefit sleep stage scoring, we have structured the remainder of the manuscript as follows. Sections 3 and 4 describe programmed diagnostic tools (PDTs) and various DL tools, respectively. Section 5 describes the guidelines for sleep stage classification and the publicly available databases with sleep recordings to train and evaluate DL models. Section 6 discusses the key findings of automated sleep stage classification studies based on different DL models. In Section 7, we elaborate on the potential future direction of sleep analysis. Section 8 concludes the paper by highlighting our review findings, which includes a discussion of the best DL models and PSG signals employed for automated sleep stage classification.

2. Medical Background

The discovery of obstructive sleep apnea (OSA) in 1965 is lauded as the greatest progress in the history of sleep medicine [9]. For many years, OSA was regarded as the occasional closure of the upper airway, thus, early treatments, such as tracheostomy, prioritized primarily on reducing the obstruction to airway [10]. However, recent studies show that OSA is linked to the risk of cardiovascular disease and death [11], emphasizing the need to consider other factors for treatment options.

The emergence of sleep disorders, such as insomnia, OSA, and various other sleep-related disorders further contribute to poor sleep quality [12,13]. Some sleep disorders manifest themselves in sleep interruptions, such as early morning awaking, and the lack or absence of restful sleep [14]. In OSA, this is more severe with symptoms such as extreme daytime drowsiness, snoring and repeated occurrences of interruptions to the respiratory airflow during sleep, which stems from the collapse of the upper airway in the throat [15]. These in turn affect the cardiovascular physiology, causing cardiovascular diseases, such as stroke, angina, and heart failure [16]. OSA has also been linked to higher morbidity and mortality rates [17] and low quality of vital scores [18]. Some studies had also showed that a lack of sleep increases fatigue during the day, which decreases the performance of individuals at work and threatens their occupational safety [19,20].

Overnight polysomnogram (PSG) is currently the "gold standard" to measure multiple physiologic parameters of sleep, and it is used to score sleep stages [6]. These recordings include electroencephalograms (EEGs), electrooculograms (EOGs), electromyograms (EMGs), electrocardiogram (ECGs), respiratory efforts, airflow, and blood oxygenation [5]. According to the American Academy of Sleep Medicine guidelines [21], sleep should be scored by segmenting these PSG recordings into 30-s fragments, also known as epochs. Each epoch is then scored and categorized based on the sleep stages that appear most often. For example, an epoch with one characteristic of Sleep Stage 1 and two characteristics of Sleep Stage 2 will be classified as Sleep Stage 2.

Technology has been shown to help improve convenience and bring down healthcare costs. In this case, such technology could take the form of a cost-effective programmed diagnostic tool (PDT) for automated detection of sleep disorders. Some studies had demonstrated the ability of PDTs to perform as well as experts in sleep stage scoring [22–24], and even outperforming the experts in the detection of microstructures of sleep, such as arousal and cyclic alternating pattern. These highlight PDT versatility and the potential to first augment (and perhaps replace) human experts in the analysis of sleep recordings [6,25].

3. Programmed Diagnostic Tools (PDTs) for Polysomnogram (PSG) Analysis

PDTs are based on artificial intelligence, namely conventional machine learning and DL techniques.

Unfortunately, machine learning is often subjected to the curse of dimensionality or p(features) >> n(samples), a common problem that surfaces when training machine learning models with high dimensional data, such as PSG recordings, which cause the model to overfit [26]. Hence, data reduction step such as feature extraction is helpful for traditional machine learning algorithms to prevent overfitting problem when handling high dimensional PSG recordings [27]. Often, the feature extraction from PSG recordings is done manually through experience and acquired skills of human observers [27,28]. The extracted features are fed to standard machine learning models, such as the support vector machine or random forest classifiers, to classify PSG recordings into respective sleep stages. However, this explicit feature engineering involves converting PSG recordings to a low dimensional vector which can result in information loss [27]. Furthermore, the aforementioned machine learning techniques are not ideally suited to handle high dimensional data because of the lack of depth to capture relevant relationships between the covariates in large volumes of data. As a result, standard machine learning is limited in its ability to reliably classify PSG recordings with high precision and accuracy.

On the other hand, DL models can be trained with raw PSG recordings without the need for information reduction [27]. DL techniques attempt to train the model, such that they learn how to make sense of data on their own by extracting features automatically from PSG signals. The extracted knowledge determines the inference. When a large data volume is available, DL models often outperform machine learning models because they can utilize all the available information and make accurate predictions [29]. In addition, DL models learn features that are useful to make inference and neglect those that are not. Hence, DL techniques are more suitable than traditional machine learning techniques when dealing with high-dimensional PSG signals and are thus considered state-of-art methods for automated sleep stage classification.

4. DL Models

In contrast to other programming languages (such as MATLAB), the Python programming language includes a plethora of libraries that can facilitate the development of DL models with greater ease. Figure 1 shows that over 75% of the automated sleep stages classification studies employed DL tools from Python (TensorFlow, Theano, Keras, PyTorch, Lasagne). Keras is a high-level Application Programming Interface (API) for Python that can use either TensorFlow or Theano as its backend. The programming process in TensorFlow and Theano is also simplified, such that the cognitive load for human to build a DL model is significantly reduced.

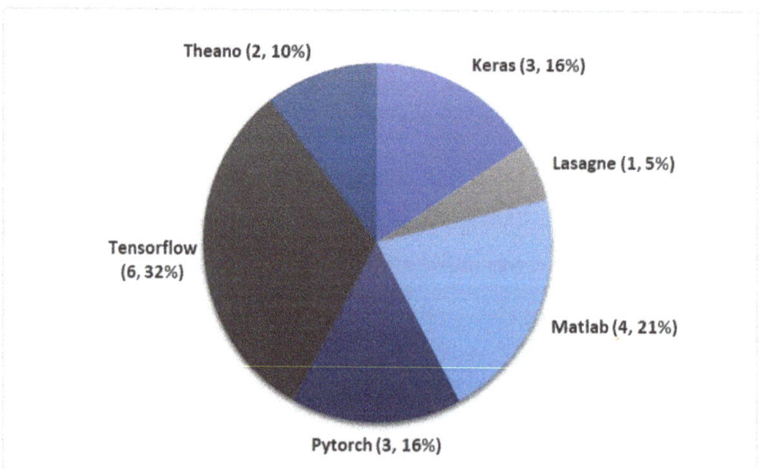

Figure 1. Pie chart representation of the number of times (19 in total) and percentage each deep learning (DL) tools were used to develop DL models in automated sleep stage classification studies as listed in Tables 1–5.

4.1. Convolutional Neural Network (CNN)

The first CNN was created to mimic the human visual system. In the visual cortex, simple and complex cortical cells break down visual information into simpler representations, so that it is easier for the brain to perceive and classify the image [30,31]. In an analysis of multiple light recordings, a typical

CNN model comprises three main layers: convolutional, pooling, and fully connected, as depicted in Figure 2. In this model, an input data is first broken down by learned filters at the convolutional layer to extract important features, and feature maps are created as outputs. These feature maps contain different kinds of information about the data characteristics [32]. The pooling layer follows immediately after the convolutional layer, and it is responsible for reducing the feature map dimension. By doing so, the feature map complexity is reduced, and the visual information is broken down further. Another desired effect of this architecture is the reduction of overfitting [33]. Multiple convolutional and pooling layers can be incorporated in a model to make it "deeper" and increase its recognition ability to classify complex images. After a series of convolutional and pooling layers, the resulting feature map is flattened into a single list of vectors before it is fed to the fully connected layers, which establishes connections between output and input via learnable weights [32]. The superiority of this architecture was first demonstrated in image recognition and classification by Krizhevsky et al. [34], where their proposed CNN model was awarded the top five award in an image classification competition, known as ImageNet Large Scale Visual Recognition Competition (ILSVRC-2012).

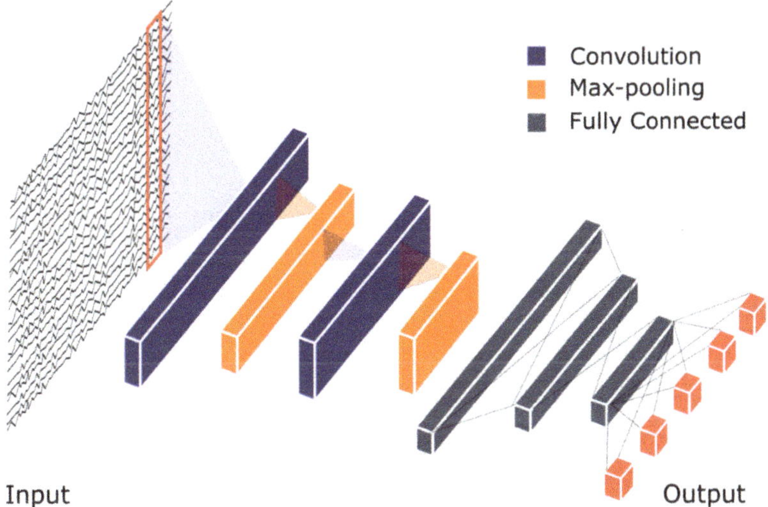

Figure 2. Basic Convolutional Neural Network (CNN) model architecture for polysomnogram (PSG) recording analysis. The CNN output model is represented by five red boxes indicating five prediction classes for example W (Wakefulness), S1–S3 (Non-REM sleep Stages 1–3) and REM (rapid eye movement) sleep.

4.2. Recurrent Neural Network (RNN) and Long Short-Term Memory (LSTM)

Both RNN and LSTM models are designed for sequential data processing, such as speech [35], [36], text [37], and handwriting recognition [38]. These models attempt to recognize a pattern in the sequence. Previously, Kumar et al. [39] proposed LSTM model to analyze brain EEG for brain-computer interface (BCI) systems. Their model achieved a low misclassification rate of 3.09% and 2.07% for two publicly available BCI datasets. A recent study by Kim et al. [40] demonstrated the effectiveness of the LSTM and RNN in analyzing bio signals. The LSTM-based deep RNN model, proposed in their study, achieved exemplary performance: 100% accuracy, precision, and sensitivity in personal authentication based on ECG signals. This makes them suitable for classifying bio signals like PSG recordings, which have distinct patterns in different sleep stages.

It needs to be noted that very early versions of RNNs were incapable of learning long-term dependencies, because these RNNs were unable to form connections between old and new data when a large information gap existed between them [41]. This resulted in a phenomenon known as vanishing gradient problem, where the error signals vanished during backpropagation, which eventually led to a model breakdown. Hochreiter and Schmidhuber [42] developed LSTM to solve the vanishing gradient problem, but Ger et al. [43] showed that LSTM was still not able to efficiently learn sequences that were very long or continuous. The reason for this failure was that the internal values of the memory cell in LSTM models had grown out of bound from the continuous input stream, despite the fact that the LSTM model was programmed to reset itself when faced with this kind of problem. As a remedy, "forget gates" were introduced to LSTMs, which remove data that were no longer relevant in memory cells, thereby forgetting and resetting information in memory cells at appropriate times [43]. Useful information, on the other hand, was continuously backpropagated, thus, allowing these models to memorize relevant information and recognize patterns in long-term dependencies. The architecture of the LSTM model is shown in Figure 3.

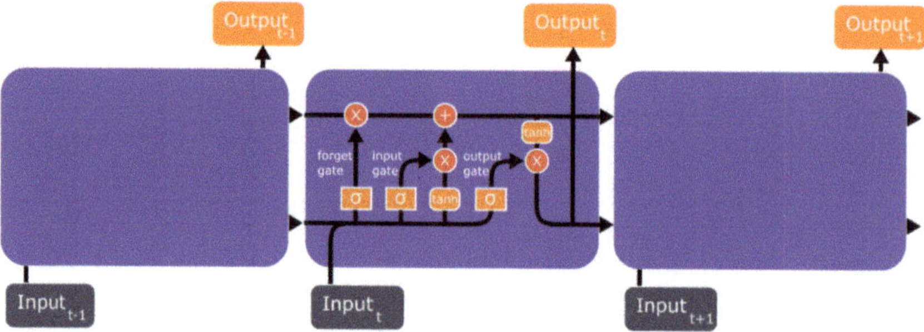

Figure 3. Basic Long Short-Term Memory (LSTM) model architecture.

However, high computational complexity is the downside of RNN and LSTM, and they need a large memory bandwidth to train [44,45]. As such, hardware designers often experience difficulty when dealing with RNN or LSTM models, because these models occupy a large amount of cloud space, which is not scalable. Therefore, reducing the computational complexity of RNN and LSTM models must be considered for real-time or mobile applications.

4.3. Autoencoders (AEs)

Rumelhart et al. [46] were the first to propose Autoencoders (AEs), a DL technique that is specialized in dimensionality reduction and denoising of data. The key player in an autoencoder operation is the latent (hidden) representation, h, as shown in Figure 4. It plays the role of a bottleneck, which retains only those features that are necessary to reconstruct the input data on the AE output [47]. This latent representation (features) is often used in data classification tasks. This implies that AEs and CNNs have common characteristics and attempt to extract and learn only important features. While CNNs make use of convolutional layers to concentrate the extracted features and improve recognition ability, AEs make use of latent representation (h) to compress the data received from the encoder unit. This step retains salient features and removes irrelevant data. Thus, the operation of AEs includes denoising and reducing the data dimension while extracting features simultaneously. This reduces the computational complexity which in certain classification tasks makes AE models easier to train [48]. However, AEs are also associated with disadvantages, such as poor data compressibility and the inability to train a model effectively for certain tasks. For instance, the latent representation (h) will fail to capture salient features if errors are present in the encoder unit [47]. This is because h cannot get rid of errors; instead, it will compute the average of the input data rather than retain salient features for the decoder unit. The goal of AEs is to reconstruct the input data as shown in Figure 4, hence, the encoder units of AEs have to ensure that there are minimum errors before feeding the input data into the latent representation, h.

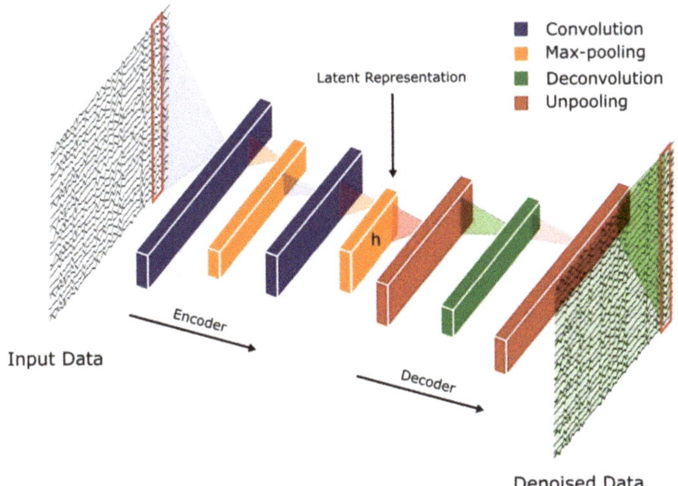

Figure 4. Basic autoencoder architecture.

4.4. Hybrid Models

Hybrid models are based on either CNN–RNN or CNN–LSTM models. The idea of creating such hybrids is to combine the advantages associated with both CNNs and RNN/LSTMs, in terms of feature extraction and pattern recognition ability in sequential data [49,50]. In these hybrid models, the convolutional layers are at the frontline of the model to extract important features from PSG signals, while RNN or LSTM layers would attempt to recognize patterns in feature maps received from the convolutional layers.

5. Sleep Stages Classification Using DL Models

5.1. Different Stages of Sleep

According to Rechtschaffen and Kales (R and K) [51], humans can experience six discrete stages during sleep: (1) wakefulness (W), (2) rapid eye movement (REM) sleep, and (3) four stages of non-REM (NREM) sleep (S1 to S4) [52]. Based on the sleep electroencephalogram (EEG) characteristics, W occurs when the brain is most active, which is represented by high frequency of alpha rhythms. In the NREM sleep, these alpha rhythms eventually diminish when entering the S1 wherein theta rhythm dominates instead. In the S2, sleep spindles and occasional K-complex waveform will appear. The K-complex waveform usually lasts for approximately 1 to 2 s. The S3 sleep occurs when low frequency delta rhythms appear intermittently and eventually, they dominate in the S4 sleep. Finally, REM sleep usually follows after the S4 sleep. In the REM sleep, theta rhythms resurface again, but unlike in the S1 sleep, theta rhythms are accompanied with EEG flattening [52]. Following the guidelines from American Academy of Sleep Medicine (AASM), the S3 and S4 sleep stages can be merged into one sleep stage S3, because of the similarity in their characteristics [21]. Since the delta rhythms are the slowest EEG waves, S3 and S4 sleep stages are known as Slow Wave Sleep (SWS) or the deep sleep. Thus, most sleep classification studies are based on five: W, S1, S2, S3, and REM sleep stages, instead of six (Figure 5).

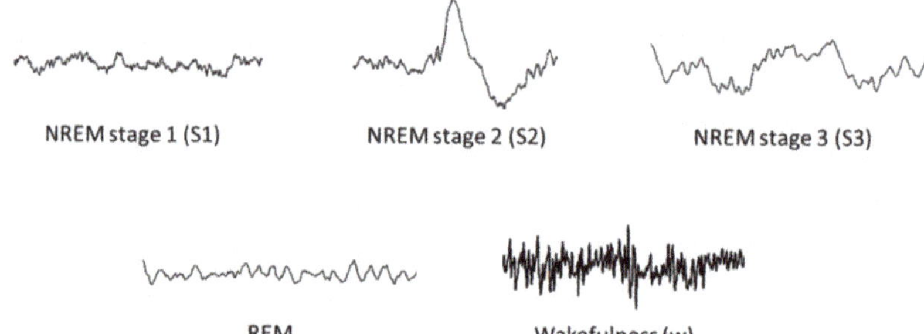

Figure 5. Examples of electroencephalography (EEG) signals in different sleep stages.

5.2. Sleep Databases

Eight main sleep databases have been used for automated sleep stage classifications. Five of the databases are free to download from PhysioNet [53], namely the Sleep-EDF [54], the expanded Sleep-EDF [54], the St. Vincent's University Hospital/University College Dublin Sleep Apnea Database (UCD) [53], the Sleep Heart Health Study (SHHS) [55,56], and the Massachusetts Institute of Technology-Beth Israel Hospital (MIT-BIH) [57] database. The ISRUC-Sleep datasets [58] can be downloaded from the official websites. Permission is required to obtain the sleep datasets from the Montreal Archive of Sleep Studies (MASS) [59].

The PSG recordings, in most of the sleep databases, are scored according to R and K rules [51], wherein scoring is done based on wakefulness, NREM sleep and REM sleep. NREM sleep is then subdivided into four stages (S1 to S4). Exceptions are ISRUC and MASS which follow the AASM guideline and partition the recordings into five sleep stages instead of six [21].

5.3. DL Techniques Used in Automatic Sleep Stage Classification

The development of a program diagnostic tool (PDT) for automatic sleep stage classification using DL techniques is shown in Figure 6. First, PSG recordings have to be pre-processed to achieve standardization or normalization. Depending on the requirement and architecture of the proposed DL model, additional steps to convert the PSG recordings into the right input format is required; for example, converting one-dimensional (1D) signals into a two-dimensional (2D) format to train 2D-CNN models. Subsequently, the pre-processed signals are split into training, validation, and testing sets. The training set is used to train the model, the validation set is to fine-tune the model, and the testing set is used to evaluate the model's performance. A well-trained model can accurately classify PSG recordings into the five sleep stages.

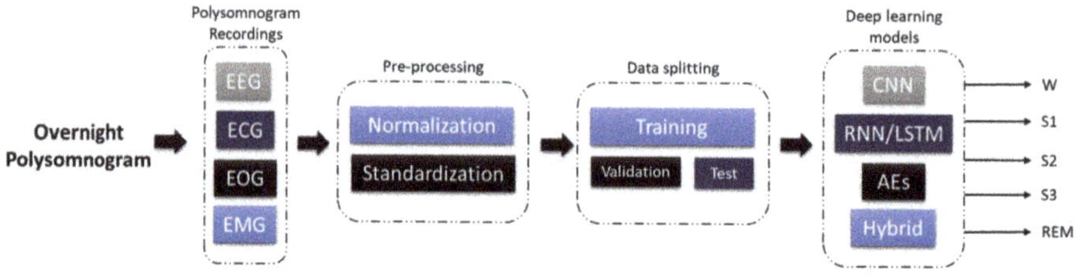

Figure 6. Programmed diagnostic tool (PDT) block diagram with DL for automated sleep stage classification.

Figure 7 illustrates the number of times each sleep database had been used by studies for automated sleep stage classification using DL techniques, from 2010 to 2020. The DL methods and accuracy obtained from the respective sleep databases are summarized as follows: Sleep-EDF (Table 1), expanded Sleep-EDF (Table 2), MASS (Table 3), MIT-BIH, and SHHS (Table 4), and studies that used the remaining two sleep databases (ISRUC and UCD) and private datasets are listed in Table 5. With the exception of three studies [60–62], which classified sleep into four stages, all automated sleep stage classification studies, in Tables 1–5, followed the AASM guidelines [21] and classified sleep into five stages. In studies with sleep databases following the R and K rules [51], (i.e., Sleep-EDF, expanded Sleep-EDF, UCD, SHHS, and MIT-BIH), the S3 and S4 stages were often combined manually before pre-processing the PSG signals.

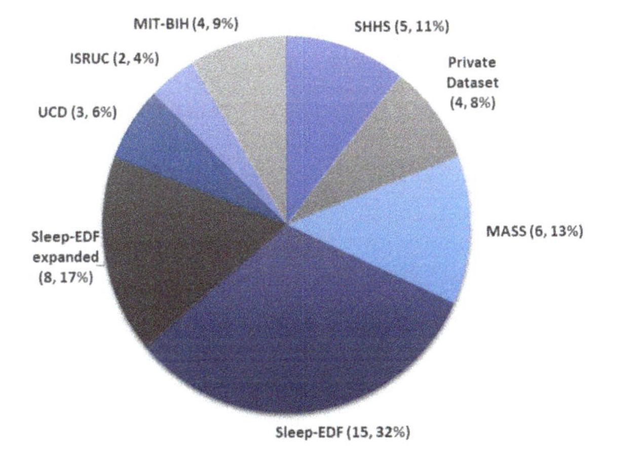

Figure 7. Pie chart representation of the frequency in which each sleep database was used in automated sleep stage classification studies. The total number of studies was 47, as listed in Tables 1–5. * Summary statistics: using various databases for sleep stage classification.

Table 1. Summary of automated sleep stage classification approaches with DL applied to PSG recordings in Sleep-EDF dataset.

Author	Signals	Samples	Approach	Tools/Programming Languages	Accuracy (%)
Zhu et al. [63] 2020	EEG	15,188	attention CNN	–	93.7
Qureshi et al. [64] 2019	EEG	41,900	CNN	–	92.5
Yildirim et al. [65] 2019	EEG	15,188	1D CNN	Keras	90.8
Hsu et al. [66] 2013	EEG	2880	Elman RNN	–	87.2
Michielli et al. [67] 2019	EEG	10,280	RNN-LSTM	MATLAB	86.7
Wei et al. [68] 2017	EEG	–	CNN	–	84.5
Mousavi et al. [69] 2019	EEG	42,308	CNN-BiRNN	TensorFlow	84.3
Seo et al. [70] 2020	EEG	42,308	CRNN	PyTorch	83.9
Zhang et al. [71] 2020	EEG	–	CNN	–	83.6
Supratak et al. [72] 2017	EEG	41,950	CNN-BiLSTM	TensorFlow	82.0
Phan et al. [73] 2019	EEG	–	Multi-task CNN	TensorFlow	81.9
Vilamala et al. [74] 2017	EEG	–	CNN	–	81.3
Phan et al. [75] 2018	EEG	–	1-max CNN	–	79.8
Phan et al. [76] 2018	EEG	–	Attentional RNN	–	79.1
Yildirim et al. [65] 2019	EOG	15,188	1D-CNN	Keras	89.8
Yildirim et al. [65] 2019	EEG + EOG	15,188	1D-CNN	Keras	91.2
Xu et al. [77] 2020	PSG signals	–	DNN	–	86.1
Phan et al. [73] 2019	EEG + EOG	–	Multi-task CNN	TensorFlow	82.3

Table 2. Summary of automated sleep stage classification approaches with DL applied to PSG recordings in Expanded Sleep-EDF dataset.

Author	Signals	Samples	Approach	Tools/Programming Languages	Accuracy (%)
Wang et al. [78] 2018	EEG	–	C-CNN	–	–
Wang et al. [78] 2018	EEG	–	RNN-biLSTM	–	–
Fernandez-Blanco et al. [79] 2020	EEG	–	CNN	–	92.7
Yildirim et al. [65] 2019	EEG	127,512	1D-CNN	Keras	90.5
Jadhav et al. [80] 2020	EEG	62,177	CNN	–	83.3
Zhu et al. [63] 2020	EEG	42,269	attention CNN	–	82.8
Mousavi et al. [69] 2019	EEG	222,479	1D-CNN	TensorFlow	80.0
Tsinalis et al. [81] 2016	EEG	–	2D-CNN	Lasagne + Theano	74.0
Yildirim et al. [65] 2019	EOG	127,512	1D-CNN	Keras	88.8
Yildirim et al. [65] 2019	EEG + EOG	127,512	1D-CNN	Keras	91.0
Sokolovsky et al. [82] 2019	EEG + EOG	–	CNN	TensorFlow + Keras	81.0

Table 3. Summary of automated sleep stages classification approaches with DL applied to PSG recordings in Montreal Archive of Sleep Studies (MASS) dataset.

Author	Signals	Samples	Approach	Tools/Programming Languages	Accuracy (%)
Seo et al. [70] 2020	EEG	57,395	CRNN	PyTorch	86.5
Supratak et al. [72] 2017	EEG	58,600	CNN-BiLSTM	TensorFlow	86.2
Phan et al. [73] 2019	EEG	–	Multi-task CNN	TensorFlow	78.6
Dong et al. [83] 2018	EOG F4	–	MNN RNN-LSTM	Theano	85.9
Dong et al. [83] 2018	EOG Fp2	–	MNN RNN-LSTM	Theano	83.4
Chambon et al. [84] 2018	EEG/EOG + EMG	–	2D-CNN	Keras	–
Phan et al. [85] 2019	EEG + EOG + EMG	–	Hierarchical RNN	TensorFlow	87.1
Phan et al. [73] 2019	EEG + EOG + EMG	–	Multi-task CNN	TensorFlow	83.6
Phan et al. [73] 2019	EEG + EOG	–	Multi-task CNN	TensorFlow	82.5

Table 4. Summary of automated sleep stage classification approaches with DL applied to PSG recordings in Sleep Heart Health Study (SHHS) and Massachusetts Institute of Technology-Beth Israel Hospital (MIT-BIH) datasets.

Database	Author	Signals	Samples	Approach	Tools/Programming Languages	Accuracy (%)
MIT-BIH	Zhang et al. [86] 2020	EEG	–	Orthogonal CNN	–	87.6
	Zhang et al. [87] 2018	EEG	–	CUCNN	MATLAB	87.2
	Sors et al. [88] 2018	EEG	5793	CNN	–	87.0
SHHS	Seo et al. [70] 2020	EEG	5,421,338	CRNN	PyTorch	86.7
	Fernández-Varela et al. [89] 2019	EEG + EOG + EMG	1,209,971	1D-CNN	–	78.0
	Zhang et al. [90] 2019	EEG + EOG + EMG	5793	CNN-LSTM	–	–
SHHS	Li et al. [60] 2018	ECG HRV	400,547	CNN	MATLAB	65.9
MIT-BIH	Li et al. [60] 2018	ECG HRV	2829	CNN	MATLAB	75.4
	Tripathy et al. [61] 2018	EEG + HRV	7500	DNN Autoencoder	MATLAB	73.7

Table 5. Summary of automated sleep stage classification approaches with DL applied to PSG recordings in ISRUC, Massachusetts General Hospital (MGH), and University College Dublin Sleep Apnea Database (UCD) datasets.

Database	Author	Signals	Samples	Approach	Tools/Programming Languages	Accuracy (%)
ISRUC	Cui et al. [91] 2018	EEG	–	CNN	–	92.2
	Yang et al. [92] 2018	EEG	–	CNN-LSTM	–	–
	Zhang et al. [86] 2020	EEG	–	Orthogonal CNN	–	88.4
UCD	Zhang et al. [87] 2018	EEG	–	CUCNN	MATLAB	87.0
	Yuan et al. [93] 2019	Multivariate PSG signals	287,840	Hybrid CNN	PyTorch	74.2
	Zhang et al. [71] 2020	EEG	264,736	CNN	–	96.0
	Biswal et al. [94] 2018	PSG signals	10,000	RCNN	PyTorch	87.5
	Biswal et al. [95] 2017	EEG	10,000	RCNN	TensorFlow	85.7
	Radha et al. [62] 2019	ECG HRV	541,214	LSTM	–	Class = 4 77.0

* The accuracy scores in Tables 1–5 are based on AASM guidelines, five class classification [21].

Figure 8 shows the number of times PSG recordings such as EEG, EOG, EMG, and ECG signals were used for sleep stage classification studies. It is not surprising that EEG signal was the most popular input for DL models. The characteristic waves and description of each sleep stages are often based on EEG characteristics (i.e., alpha waves, theta waves, delta waves, etc.); Figure 5.

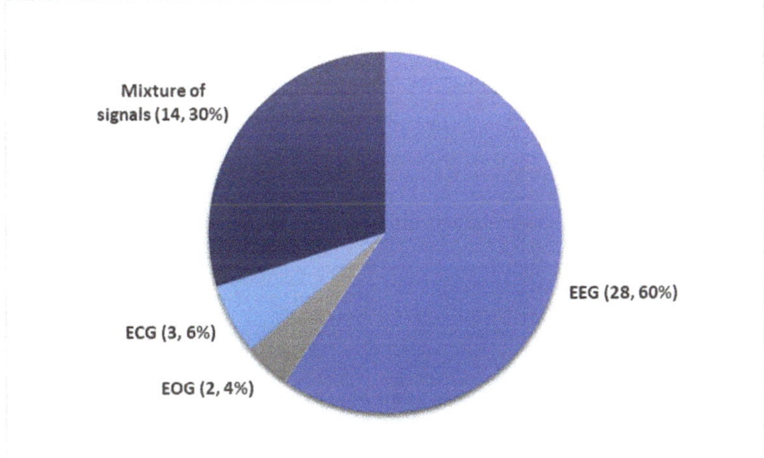

Figure 8. Different subsets of PSG recordings used to train DL models for automated sleep stage classification as listed in Tables 1–5. Of the 36 studies, the mixture of signals (electrooculogram (EOG), electromyogram (EMG), and electroencephalography (EEG)) was employed 14 times while EEG signals were used 28 times. Only a small fraction (five studies) employed ECG or EOG time series. * Summary statistics: using EEG versus EEG + additional signals.

Nonetheless, other signals within the PSG recordings are indispensable, because they provide additional information on biological aspects of sleep that may not be manifested in EEG recordings. Since REM sleep is characterized by the movement of eyes and loss in muscle tone of the body core, EOG, and EMG signals may provide key information to separate the REM sleep stage from the other stages. It was shown that some of the REM sleep stages could be overlooked in single-channel EEG input [27]. Therefore, a combination of signals, comprising of EOG, EMG, and EEG, are second in terms of frequency of use after single-channel EEG inputs (Figure 8).

Although ECG is an important sleep parameter [96], it is not common to use raw ECG signals as a direct input for DL models. As seen in Table 4, heart rate variability (HRV) parameters derived from ECG signals, were used to train the DL models instead. There are only three studies that employed HRV parameters, and these studies classified sleep into four stages instead of five: wakefulness (W), light sleep (S1 and S2), deep sleep (S3 and S4), and REM sleep. Li et al. [60] proposed a 3-layer CNN model. They used a cardiorespiratory coupling (CRC) spectrogram, which was derived from ECG and HRV. Besides alternations in physiological signals, there are other changes in body system changes in some individuals such as cardiovascular [97], respiratory [98], or blood flow in the brain [99]. Hence, the CRC picks up the cardiovascular and respiratory changes. Their model achieved an overall accuracy of 65.9% and 75.4% for SHHS and MIT-BIH respectively, as seen in Table 4. Tripathy et al. [61] combined EEG and HRV features as input to an AE model. During testing, the model achieved an overall accuracy of 73.7%. Radha et al. [62] published the only study that was based on ECG signals from a private dataset that was collected as part of the European Union SIESTA project [100] as shown in Table 5. Likewise, they converted ECG signals into HRV and used the HRV features to train an LSTM model, which achieved an accuracy of 77.0%.

6. Discussion

Even though CNNs are primarily used in image classification, they can also be successfully applied to 1D PSG recordings. Most of the automated sleep stage classification studies rely on the CNN approach (Figure 9a). However, in order to convert 1D signals to 2D images, the input signals need to be reshaped so that the 2D convolutional layer is able to read the data. There are various 1D to 2D transformation methods, such as spectrogram [101], time-frequency representations, which can be established via Hilbert–Huang transform [86], or bispectrum algorithms [102]. To date, there are eight studies that included 2D-convolutional layers in their models [60,74,81,84,86,90,93,95]. However, the process of converting 1D signals to 2D signal representations should be carried out with caution due to potential loss of useful information during the conversion step [103].

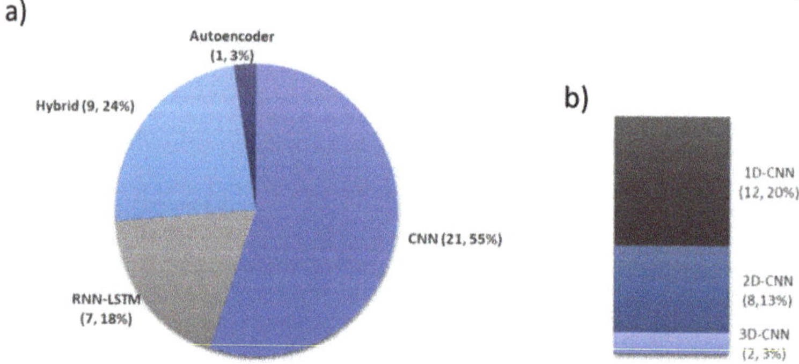

Figure 9. Selection of various DL techniques for automated sleep stage classification based on 36 studies. (**a**) Number of times and percentage of DL models, (**b**) Various CNN-based models. * Summary statistics: using autoencoders versus hybrid versus LSTM versus CNN models (one-dimensional (1D), 2D, 3D).

One-dimensional (1D)-CNNs were specifically designed to process 1D signals [104]. Unlike traditional 2D-CNNs, which require the input data to be in a matrix format, and 1D-CNNs can run with a simple array, hence, significantly reducing the computational complexity. In addition, 2D-CNNs require deeper model architecture to learn 1D signals. In contrast, 1D-CNN can easily learn 1D signals with shallow model architecture. This means that training of 1D-CNN models with 1D signals is simpler, easier, faster and, therefore, more efficient. This also highlights the 1D-CNN models' compatibility with near real-time processing and deployment in mobile applications, which can potentially be used to track and recognize sleep patterns at home [104]. The popularity of 1D-CNNs for the analysis of 1D PSG signals is demonstrated in this review. Almost all of the studies that proposed CNN-based models had employed 1D convolutional layers, as seen in Figure 9b. Furthermore, studies that proposed 1D-CNN models had higher performance compared to those with 2D-CNN. The highest accuracy score obtained by a 1D-CNN model was 96% [71]. Conversely, none of the eight studies that proposed 2D-CNN models surpassed an accuracy score of 90%.

Two studies included a 3D convolutional layer in their proposed CNN models. Phan et al. [73] used 3D filters to process a combination of signals, namely EEG, EOG, and EMG. Similar to signal pre-processing for a 2D-CNN, these three signals were converted to a 2D time-frequency image before arranging them as a 3D input. As a result, a higher accuracy was obtained with a combination of input signals when compared to a single channel input. Jadhav et al. [80] converted their EEG input into 2D-Continuous Wavelet Transform (CWT) images. The CWT images came with three main colors, red–green–blue (RGB), which provided the third-dimensional input. Hence, their convolutional layer in their proposed model had 3D filters to read the CWT images in terms of width, height, and color of the image.

From Table 4, it is evident that only Tripathy et al. [61] proposed AE models for automated sleep stage classification. In the study by Zhang et al. [87], AE was used solely as a dimensionality reduction tool to pre-process the EEG time-frequency distribution.

6.1. Proposed CNN-Based Models

From Figure 10, it can be observed that the performance of the CNN-based models had improved over the years. To date, the model by Zhang et al. [71] achieved the best overall accuracy (96%) for sleep stage classification. However, this accuracy was achieved using a private clinical dataset. When they evaluated the performance of their model using the Sleep-EDF dataset, an overall accuracy score of 86.4% was achieved. This was lower than the accuracy obtained by Zhu et al. [63] (93.7%) who used single channel EEG signals from the same database. The unique feature of the model proposed by Zhu et al. was the attention mechanism that they incorporated into the CNN's learning framework. This attention mechanism improved the feature extraction performance of the model through intra- and inter-epoch feature learning. However, the same model achieved a lower accuracy of 82.8% when it was tested on EEG signals from the expanded Sleep-EDF database. Another unique CNN model was proposed by Cui et al. [91] wherein fine-grained methods were used to assist the CNN model to find the best time segment in the EEG signals. Fine-grained methods construct time series from the EEG signals. Basically, if the time window in fine-grained method is set to 3, every 3 time-steps along the EEG signals will be combined together as one-time segment, hence reducing the complexity of EEG signals. The layers in their proposed CNN were shallow-7 layers, including 2 convolutional layers. Yet, they were able to achieve a high overall accuracy of 92%.

A more versatile and consistent CNN model was proposed by Yildirim et al. [65], which was a 19-layer 1D-CNN model with 10 convolutional layers. It achieved an accuracy higher than 90% in both Sleep-EDF dataset and its expanded version, as seen in Figure 10. The peak accuracy was achieved (91.2%) when a mixture of PSG signals was used as input (EEG and EOG), but when single channel EOG signals were used, the accuracy of the model decreased to below 90% (88.8% and 89.8%), (Figure 11).

Both Zhu et al. [63] and Cui et al. [91] showed that CNN models with a small number of layers can achieve a high classification performance of sleep stages by improving the feature extraction ability through additional tools, such as attention mechanism or fine-graining. On the other hand, Yildirim et al. [65] showed that a deeper CNN model can achieve high accuracy classifications across different inputs of PSG recordings (EEG, EOG, EEG + EOG).

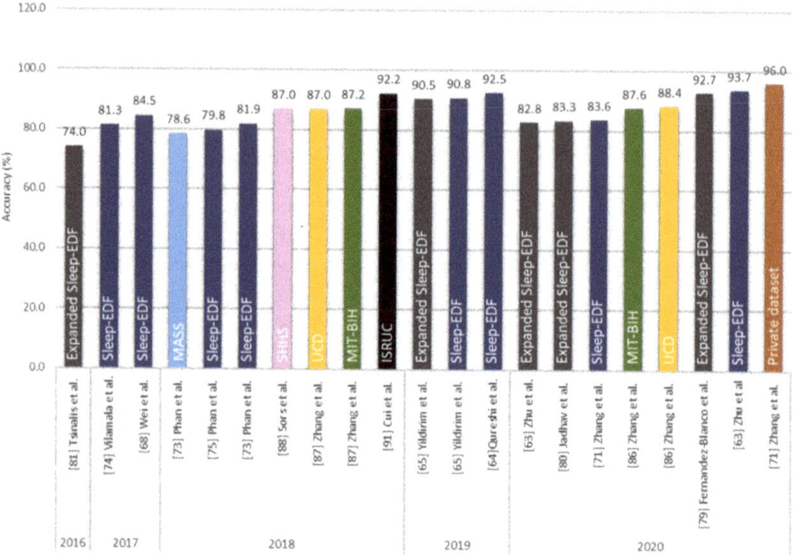

Figure 10. Performance of CNN-based models analyzing only EEG signals. Sleep datasets are presented in different colors. * Summary statistics: Various sleep databases used to develop CNN models for sleep stage detection.

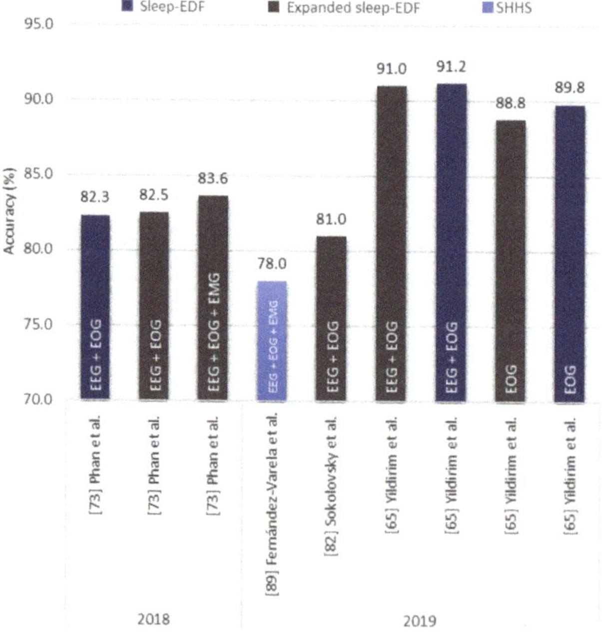

Figure 11. Sleep stage classification accuracy of CNN-based models based on stand-alone EOG signals or a mixture of EEG, EOG, and/or EMG signals. Various sleep datasets are represented by different colors and the type of signals described in the bar chart. * Summary statistics: Using EEG+ other signals/EOG in 2018–2019.

6.2. Proposed RNN/LSTM-Based Models

Contrary to CNN models, very little automated sleep stage classification studies were done using RNN/LSTM models. The best performance was observed in a study by Hsu et al. [66] back in 2013, wherein a 4-layer RNN model was used. They had adopted the structure of an Elman network and had successfully classified various sleep stages with an overall accuracy score of 87.2%, as seen in Figure 12. A similar accuracy of 86.7% was achieved by Michielli et al. [67], where a cascaded RNN network was proposed, with 2 LSTM units. Two other studies explored a mixture of signals to train RNN-based models: Dong et al. [83] proposed a mixed neural network where a RNN model was substituted for a multi-layer perception and an LSTM model was exchanged for an RNN model. The proposed final model achieved accuracies of 85.9% and 83.4% using F4-EOG and Fp2-EOG inputs, respectively. Both F4 and Fp2 were single channel EEG signals recorded at different electrode placements. Hence, F4-EOG and Fp2-EOG inputs were considered a mixture of signals (EEG + EOG). Subsequently, Phan et al. [85] proposed an end-to-end hierarchical RNN model, known as SeqSleepNet, which consisted of an attention-based recurrent model and filter bank layers. Short-time Fourier transform was used to convert multiple PSG recordings (EEG, EOG, and EMG) into power spectra signals which were then fed to train the proposed model. The model achieved a high accuracy score of 87.1%, which was the highest amongst the RNN/LSTM-based models.

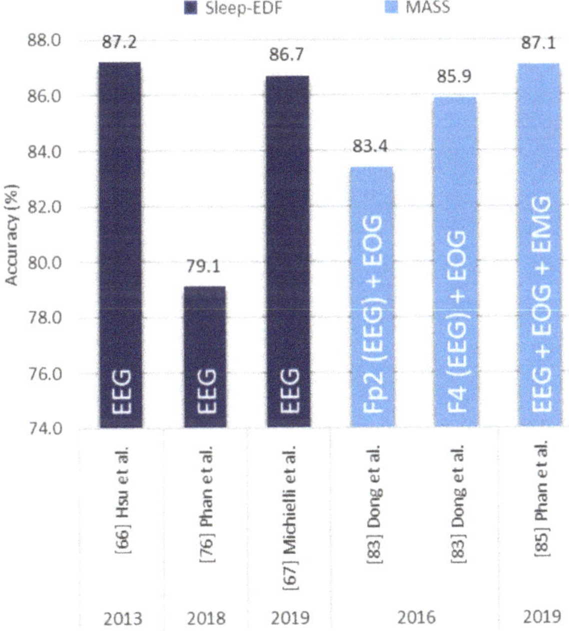

Figure 12. Sleep stage classification accuracy of RNN/LSTM-based models based on stand-alone EEG signals or a mixture of EEG, EOG, and ECG signals. Various sleep datasets are represented by different colors and the type of signals described in the bar chart. * Summary statistics: Using RNN versus LSTM models for EEG/EEG + other signals from 2013–2019.

6.3. Proposed Hybrid Models

There are limited studies on the employment of hybrid models. The best performing hybrid model for EEG signals was proposed by Seo et al. [70]. Their IITNet model consisted of CNN layers and two bidirectional LSTM layers. The CNN layers were responsible for extracting representative features in each epoch and producing sequential feature maps, which were analyzed by the bidirectional LSTM layers to capture temporal sleep stage information [72]. Figure 13 shows that this model achieved an overall accuracy score of 86.7% (SHHS database), compared to 83.9% when applied to Sleep-EDF database. On the other hand, Mousavi et al. [69] proposed SleepEEGNet, a CNN-RNN model with bidirectional RNN units. The difference between this model and Seo et al.'s model [70] was in the architecture of bidirectional RNN units which resembled AEs in the former model. Comparing within the same Sleep-EDF databases, this model achieved a higher overall accuracy score of 84.3%, as seen in Figure 13. However, their proposed model achieved a lower accuracy score of 80.0%, with EEG signals from the Expanded Sleep-EDF database.

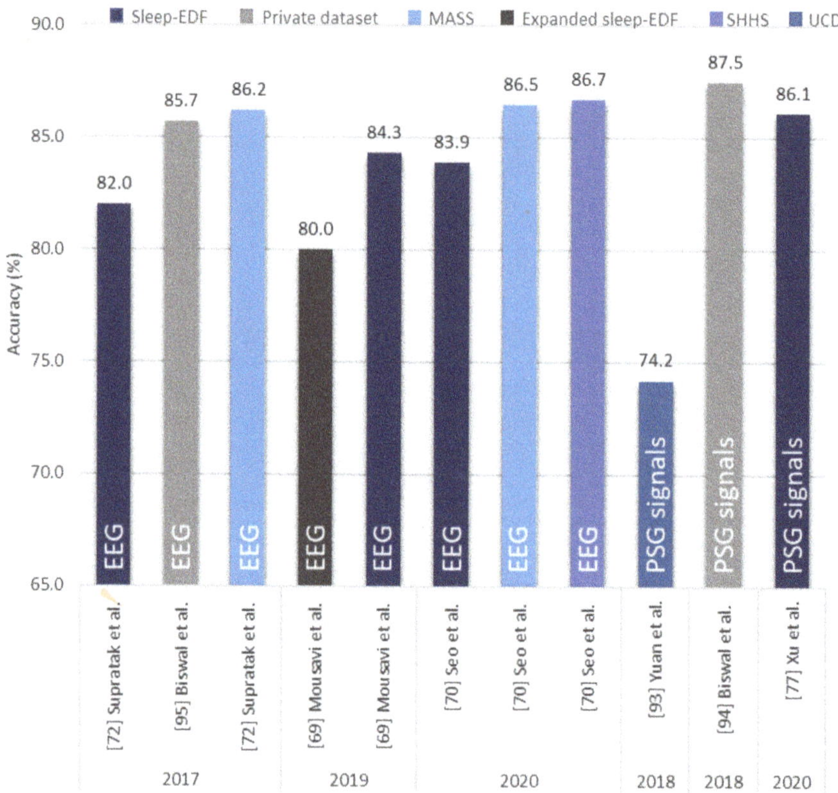

Figure 13. Performance of proposed hybrid models using EEG signals or a mixture of signals. Various sleep datasets are represented by different colors and the type of signals are described in the bar chart. * Summary statistics: different datasets used to build hybrid models using EEG/PSG signals. * The accuracy scores in Figures 9–12 are based on AASM guidelines and pertain to the five class classification [21].

When a mixture of PSG signals were taken into consideration, the CNN-RNN based model proposed by Biswal et al. [94] outperformed all other models. However, in their study they used a private sleep database with recordings from the Massachusetts General Hospital (MGH) sleep laboratory to train and test the model. They also trained their model with the MGH dataset and then tested it on the SHHS dataset. With that setup, they obtained an overall accuracy of 77.7%. This score is lower than that obtained by the CNN-based model proposed by Fernández-Varela et al. [89] who employed a mixture of signals from the SHHS database to train and test their model.

The number of studies employing DL methods for automated sleep stage classification has increased in the past few years (Figure 14), most of the studies incorporated CNN models. The CNN models became popular after the publication by Krizhevsky et al. [34] in 2015. However, the number of studies relying on this architecture started to decline after 2018. This implies that CNN-based approaches have perhaps reached their peak ability and peak performance in classifying sleep stages. This decline is also likely due to the fact that best CNN based architectures are now less competitive when compared to the other DL models. Conversely, the research on RNN/LSTM and hybrid models remained stagnant from 2018 to 2019, and no further studies were carried out using AE models after 2018. This suggests that more attention should be paid to improve the performances of these models, particularly to RNN/LSTM, and hybrid models in automated sleep stage classifications.

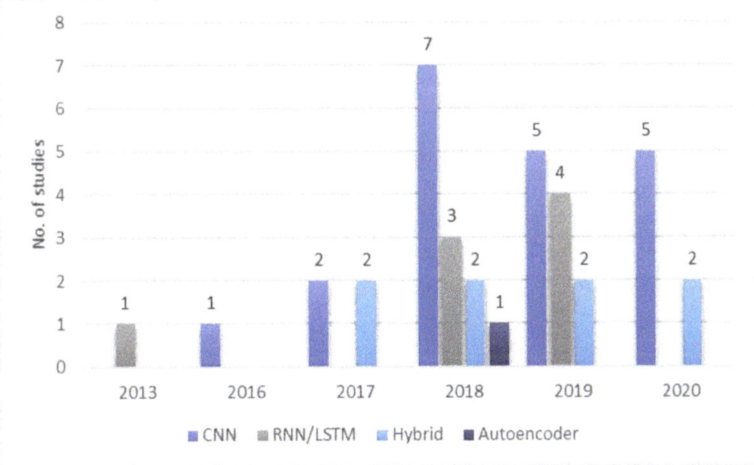

Figure 14. Number of studies published between 2013 and 2020 describing implementation of various DL models for PSG recording analysis. * Summary statistics: year of the analyses for various deep learning models used in sleep classification.

When assessing polysomnographic recordings, clinical experts rely on a combination of EEG, EOG, and EMG signals before they determine the sleep stage for each sleep phase [63]. In order to be on par with the clinical experts, an ideal DL model should effectively classify sleep stages based on a mixture of signals. At present, the majority of automated sleep stage classification studies demonstrate a high performance with single-EEG channel, but only a small fraction (25%) evaluated the performance of their approaches on a mixture of signals (Figure 8).

In summary, research on a mixture of PSG signals should be the main focus in the automated detection of sleep stages. The RNN/LSTM and hybrid models have yet to reach their peak performance in classifying sleep stages. Further research on these models to evaluate their performance across different databases could decrease the bias in these methods and help to identify architectures capable of processing mixed PSG signals. A model with optimal architecture could be employed in various applications and platforms, such as mobile, point-of-care monitoring devices.

This review underscores the advantages (key points discussed) as follows:

1. Numerous studies (15 from Figure 10) employed CNN models with EEG signals, and that CNN models are effective in recognizing characteristic features of sleep EEG.
2. One-dimensional (1D)-CNN models were used more often than 2D- and 3D-CNN models. From Figure 9b, 12, 8, and 2 1D, 2D, and 3D models were used, respectively.
3. Most studies (60% from Figure 8) used EEG signals and achieved high classification accuracy.

4. EEG signals were mainly used in studies that explored a mixture of PSG signals. In other words, EEG could be a reference signal when considering mixture of PSG signals to train and evaluate newly proposed models.

The limitations of this review are as follows:

1. It is difficult to compare various models and identify the best performing approach, because the majority of studies used data from only one sleep database to train and test the model.
2. There is a lack of studies that utilized other PSG recordings, such as EOG, EMG, or ECG signals. Studies that used these PSG recordings also did not perform equally well as those that used only EEG signals. Hence, this limits the implementation of these PSG recordings in real world applications for automated sleep stages classification.

7. Future Work

We anticipate that more public databases with large data records will become available to conduct studies on sleep stage classification and the development of more accurate approaches. When constructing a DL model, the ability to implement it in cloud-based processing systems should be considered, (Figure 15). Extracted EEG or PSG signals could then be sent to the model, for processing in the cloud and results of the analysis returned to the clinician. The analysis could be performed on-line or off-line with the goal to reduce expert workload who otherwise would have to review hours of recordings manually. Subsequently, verified results could be sent to the patient's mobile phone. With the DL model deployed in the cloud, any device with access to internet, such as mobile applications, could use it anytime and anywhere. For example, Patanaik et al. [105] developed a cloud-based framework that could classify sleep stages using real-time data with 90% accuracy. Their model performed better than expert scorers (82%). This type of framework is suitable for wearable sleep devices, such as the Kokoon and Dreem headband that reads real-time data [105].

Figure 15. Block diagram of cloud-based sleep stage classification system using EEG signals.

Furthermore, future studies should focus on using various EEG signals to detect sleep stages. Currently, the most common electrode placing to record EEG signals are Fpz-Cz, Pz-Oz, C4-A1, and C3-A2.

Another area of development and research is the microstructure of sleep, such as Cyclic Alternating Pattern (CAP) and arousal. Their duration is shorter than a half-minute epoch, hence, often undetected by humans in conventional sleep stage scoring [6]. Since quantifying microstructures by humans has poor reproducibility and is prone to errors, it would be desirable to develop automated approaches to address this deficiency [6]. However, CAP sleep database (CAPSLPDB) is currently the only database that provide PSG recordings for development of tools for CAP detection [106]. Hence, more public databases designed for microstructure of sleep assessment should be available to allow advancement of PDT in the detection of CAP.

8. Conclusions

Sleep disorders are a pressing global issue and the most dangerous sleep disorder is the obstructive sleep apnea, which can lead to cardiovascular diseases, if left untreated. Hence, efficient and accurate diagnostic tools are required for early interventions. In this work, we reviewed 36 studies that employed programmed diagnostic tools with the DL models as the backbone, analyzing overnight polysomnogram recordings to classify sleep stages. Presently, CNN models can offer higher performance in classifying sleep stages, especially with EEG signals. Hence, they are consistently and favorably used by researchers to classify sleep stages as compared to the other machine learning models and physiological signals. Moreover, employing 1D-CNN models is advantageous, because they yield high classification results on EEG signals. However, EEG signals alone may not be sufficient to achieve robust classifications. To achieve robustness and high accuracy one could develop a system that takes advantage of an automated processing and human experts in the interpretation of EEG, EOG, and EMG signals together when classifying sleep stages. Therefore, in this review, we highlighted that future studies should focus on classifying sleep stages using all or a combination of these signals. Furthermore, other DL models, such as RNN/LSTM and hybrid models, should also be explored as their full potential has yet to be realized. Future studies could focus on the compatibility and applicability of the DL models in mobile and real time applications. Lastly, more research in developing DL models to detect sleep microstructures is required, as these are often undetected in sleep stage scoring.

Author Contributions: All authors contributed to this article. The idea for the article was provided by H.W.L. and U.R.A.; H.W.L. drafted the paper. S.L.O. provided some of the diagrams. C.P.O., O.F., A.G., J.V., and U.R.A. edited the paper and provided suggestions to improve the quality of the paper. All authors have read and agreed to the published version of the manuscript.

Funding: This research received no external funding.

Conflicts of Interest: The authors declare that they have no conflict of interest.

References

1. Laposky, A.; Bass, J.; Kohsaka, A.; Turek, F.W. Sleep and circadian rhythms: Key components in the regulation of energy metabolism. *FEBS Lett.* **2007**, *582*, 142–151. [CrossRef] [PubMed]
2. Cho, J.W.; Duffy, J.F. Sleep, sleep disorders, and sexual dysfunction. *World J. Men's Health* **2019**, *37*, 261–275. [CrossRef] [PubMed]
3. Institute of Medicine (US), Committee on Sleep Medicine and Research. *Sleep Disorders and Sleep Deprivation: An Unmet Public Health Problem*; Colten, H.R., Altevogt, B.M., Eds.; National Academies Press: Washington, DC, USA, 2006.
4. Stranges, S.; Tigbe, W.; Gómez-Olivé, F.X.; Thorogood, M.; Kandala, N.-B. Sleep problems: An emerging global epidemic? Findings from the INDEPTH WHO-SAGE study among more than 40,000 older adults from 8 countries across Africa and Asia. *Sleep* **2012**, *35*, 1173–1181. [CrossRef] [PubMed]
5. Schulz, H. Rethinking sleep analysis. *J. Clin. Sleep Med.* **2008**, *4*, 99–103. [CrossRef]
6. Spriggs, W.H. *Essentials of Polysomnography*; Jones & Bartlett Learning: Burlington, MA, USA, 2014.
7. Silber, M.H.; Ancoli-Israel, S.; Bonnet, M.H.; Chokroverty, S.; Grigg-Damberger, M.M.; Hirshkowitz, M.; Kapen, S.; A Keenan, S.; Kryger, M.H.; Penzel, T.; et al. The visual scoring of sleep in adults. *J. Clin. Sleep Med.* **2007**, *3*, 121–131. [CrossRef]
8. Corral, J.; Pepin, J.-L.; Barbé, F. Ambulatory monitoring in the diagnosis and management of obstructive sleep apnoea syndrome. *Eur. Respir. Rev.* **2013**, *22*, 312–324. [CrossRef]
9. Jung, R.; Kuhlo, W. Neurophysiological studies of abnormal night sleep and the Pickwickian syndrome. *Prog. Brain Res.* **1965**, *18*, 140–159. [CrossRef]
10. Bahammam, A. Obstructive sleep apnea: From simple upper airway obstruction to systemic inflammation. *Ann. Saudi Med.* **2011**, *31*, 1–2. [CrossRef]
11. Marshall, N.S.; Wong, K.K.H.; Liu, P.Y.; Cullen, S.R.J.; Knuiman, M.; Grunstein, R.R. Sleep apnea as an independent risk factor for all-cause mortality: The Busselton health study. *Sleep* **2008**, *31*, 1079–1085. [CrossRef]
12. Hirotsu, C.; Tufik, S.; Andersen, M.L. Interactions between sleep, stress, and metabolism: From physiological to pathological conditions. *Sleep Sci.* **2015**, *8*, 143–152. [CrossRef]
13. Schilling, C.; Schredl, M.; Strobl, P.; Deuschle, M. Restless legs syndrome: Evidence for nocturnal hypothalamic-pituitary-adrenal system activation. *Mov. Disord.* **2010**, *25*, 1047–1052. [CrossRef] [PubMed]

14. Hungin, A.P.S.; Close, H. Sleep disturbances and health problems: Sleep matters. *Br. J. Gen. Pract.* **2010**, *60*, 319–320. [CrossRef]

15. Hudgel, D.W. The role of upper airway anatomy and physiology in obstructive sleep. *Clin. Chest Med.* **1992**, *13*, 383–398. [PubMed]

16. Shahar, E.; Whitney, C.W.; Redline, S.; Lee, E.T.; Newman, A.B.; Nieto, F.J.; O'Connor, G.; Boland, L.L.; Schwartz, J.E.; Samet, J.M. Sleep-disordered Breathing and Cardiovascular Disease. *Am. J. Respir. Crit. Care Med.* **2001**, *163*, 19–25. [CrossRef] [PubMed]

17. Mohsenin, V. Obstructive sleep apnea and hypertension: A critical review. *Curr. Hypertens. Rep.* **2014**, *16*, 482. [CrossRef] [PubMed]

18. Balachandran, J.S.; Patel, S.R. Obstructive sleep apnea. *Ann. Intern. Med.* **2014**, *161*. [CrossRef]

19. Williamson, A.; Lombardi, D.A.; Folkard, S.; Stutts, J.; Courtney, T.K.; Connor, J.L. The link between fatigue and safety. *Accid. Anal. Prev.* **2011**, *43*, 498–515. [CrossRef]

20. Léger, D.; Guilleminault, C.; Bader, G.; Lévy, E.; Paillard, M. Medical and socio-professional impact of insomnia. *Sleep* **2002**, *25*, 625–629. [CrossRef]

21. Iber, C.; Ancoli-Israel, S.; Chesson, A.L.; Quan, S.F. *The AASM Manual for the Scoring of Sleep and Associated Events: Rules, Terminology and Technical Specification*; American Academy of Sleep Medicine: Darien, IL, USA, 2007.

22. Svetnik, V.; Ma, J.; Soper, K.A.; Doran, S.; Renger, J.J.; Deacon, S.; Koblan, K.S. Evaluation of automated and semi-automated scoring of polysomnographic recordings from a clinical trial using zolpidem in the treatment of insomnia. *Sleep* **2007**, *30*, 1562–1574. [CrossRef]

23. Pittman, M.S.D.; Macdonald, R.M.M.; Fogel, R.B.; Malhotra, A.; Todros, K.; Levy, B.; Geva, D.A.B.; White, D.P. Assessment of automated scoring of polysomnographic recordings in a population with suspected sleep-disordered breathing. *Sleep* **2004**, *27*, 1394–1403. [CrossRef]

24. Anderer, P.; Gruber, G.; Parapatics, S.; Woertz, M.; Miazhynskaia, T.; Klösch, G.; Saletu, B.; Zeitlhofer, J.; Barbanoj, M.J.; Danker-Hopfe, H.; et al. An E-health solution for automatic sleep classification according to Rechtschaffen and Kales: Validation study of the Somnolyzer 24 × 7 utilizing the siesta database. *Neuropsychobiology* **2005**, *51*, 115–133. [CrossRef] [PubMed]

25. Acharya, U.R.; Bhat, S.; Faust, O.; Adeli, H.; Chua, E.C.-P.; Lim, W.J.E.; Koh, J.E.W. Nonlinear dynamics measures for automated EEG-based sleep stage detection. *Eur. Neurol.* **2015**, *74*, 268–287. [CrossRef] [PubMed]

26. Mirza, B.; Wang, W.; Wang, J.; Choi, H.; Chung, N.C.; Ping, P. Machine learning and integrative analysis of biomedical big data. *Genes* **2019**, *10*, 87. [CrossRef] [PubMed]

27. Faust, O.; Razaghi, H.; Barika, R.; Ciaccio, E.J.; Acharya, U.R. A review of automated sleep stage scoring based on physiological signals for the new millennia. *Comput. Methods Programs Biomed.* **2019**, *176*, 81–91. [CrossRef]

28. Shoeibi, A.; Ghassemi, N.; Khodatars, M.; Jafari, M.; Hussain, S.; Alizadehsani, R.; Moridian, P.; Khosravi, A.; Hosseini-Nejad, H.; Rouhani, M.; et al. Epileptic Seizure Detection Using Deep Learning Techniques: A Review. *arXiv* **2020**, arXiv:2007.01276.

29. Faust, O.; Hagiwara, Y.; Hong, T.J.; Lih, O.S.; Acharya, U.R. Deep learning for healthcare applications based on physiological signals: A review. *Comput. Methods Programs Biomed.* **2018**, *161*, 1–13. [CrossRef]

30. Silva, D.B.; Cruz, P.P.; Molina, A.; Molina, A.M. Are the long–short term memory and convolution neural networks really based on biological systems? *ICT Express* **2018**, *4*, 100–106. [CrossRef]

31. Hubel, D.H.; Wiesel, T.N. Receptive fields, binocular interaction and functional architecture in the cat's visual cortex. *J. Physiol.* **1962**, *160*, 106–154. [CrossRef]

32. Yamashita, R.; Nishio, M.; Do, R.K.G.; Togashi, K. Convolutional neural networks: An overview and application in radiology. *Insights Imaging* **2018**, *9*, 611–629. [CrossRef]

33. Tabian, I.; Fu, H.; Khodaei, Z.S. A convolutional neural network for impact detection and characterization of complex composite structures. *Sensors* **2019**, *19*, 4933. [CrossRef]

34. Krizhevsky, A.; Sutskever, I.; Hinton, G.E. Imagenet classification with deep convolutional neural networks. *Commun. ACM* **2017**, *60*, 84–90. [CrossRef]

35. Goehring, T.; Keshavarzi, M.; Carlyon, R.P.; Moore, B.C.J. Using recurrent neural networks to improve the perception of speech in non-stationary noise by people with cochlear implants. *J. Acoust. Soc. Am.* **2019**, *146*, 705. [CrossRef] [PubMed]

36. Coto-Jiménez, M. Improving post-filtering of artificial speech using pre-trained LSTM neural networks. *Biomimetics* **2019**, *4*, 39. [CrossRef] [PubMed]

37. Lyu, C.; Chen, B.; Ren, Y.; Ji, D. Long short-term memory RNN for biomedical named entity recognition. *BMC Bioinform.* **2017**, *18*, 462. [CrossRef] [PubMed]

38. Graves, A.; Liwicki, M.; Fernández, S.; Bertolami, R.; Bunke, H.; Schmidhuber, J. A Novel connectionist system for unconstrained handwriting recognition. *IEEE Trans. Pattern Anal. Mach. Intell.* **2008**, *31*, 855–868. [CrossRef] [PubMed]

39. Kumar, S.; Sharma, A.; Tsunoda, T. Brain wave classification using long short-term memory network based OPTICAL predictor. *Sci. Rep.* **2019**, *9*, 1–13. [CrossRef]

40. Kim, B.-H.; Pyun, J.-Y. ECG identification for personal authentication using LSTM-based deep recurrent neural networks. *Sensors* **2020**, *20*, 3069. [CrossRef]

41. Yu, Y.; Si, X.; Hu, C.; Zhang, J.-X. A Review of recurrent neural networks: LSTM cells and network architectures. *Neural Comput.* **2019**, *31*, 1235–1270. [CrossRef]

42. Hochreiter, S.; Schmidhuber, J. Long short-term memory. *Neural Comput.* **1997**, *9*, 1735–1780. [CrossRef]

43. Gers, F.A.; Schmidhuber, J.; Cummins, F. Learning to forget: Continual prediction with LSTM. In Proceedings of the 9th International Conference on Artificial Neural Networks—ICANN'99, Edinburgh, UK, 7–10 September 1999.

44. Masuko, T. Computational cost reduction of long short-term memory based on simultaneous compression of input and hidden state. In Proceedings of the 2017 IEEE Automatic Speech Recognition and Understanding Workshop, Okinawa, Japan, 16–20 December 2017; pp. 126–133. [CrossRef]

45. Dash, S.; Acharya, B.R.; Mittal, M.; Abraham, A.; Kelemen, A. (Eds.) *Deep Learning Techniques for Biomedical and Health Informatics*; Springer International Publishing: Cham, Switzerland, 2020.

46. Rumelhart, D.E.; McClelland, J.L. Learning internal representations by error propagation. In *Parallel Distributed Processing: Explorations in the Microstructure of Cognition: Foundations*; MIT Press: Cambridge, MA, USA, 1987; pp. 318–362.

47. Voulodimos, A.; Doulamis, N.; Doulamis, A.; Protopapadakis, E. Deep learning for computer vision: A brief review. *Comput. Intell. Neurosci.* **2018**, *2018*, 1–13. [CrossRef]

48. Testolin, A.; Diamant, R. Combining denoising autoencoders and dynamic programming for acoustic detection and tracking of underwater moving targets. *Sensors* **2020**, *20*, 2945. [CrossRef] [PubMed]

49. Trabelsi, A.; Chaabane, M.; Ben-Hur, A. Comprehensive evaluation of deep learning architectures for prediction of DNA/RNA sequence binding specificities. *Bioinformatics* **2019**, *35*, i269–i277. [CrossRef] [PubMed]

50. Long, H.; Liao, B.; Xu, X.; Yang, J. A hybrid deep learning model for predicting protein hydroxylation sites. *Int. J. Mol. Sci.* **2018**, *19*, 2817. [CrossRef] [PubMed]

51. Hori, T.; Sugita, Y.; Koga, E.; Shirakawa, S.; Inoue, K.; Uchida, S.; Kuwahara, H.; Kousaka, M.; Kobayashi, T.; Tsuji, Y.; et al. Proposed sments and amendments to 'A Manual of Standardized Terminology, Techniques and Scoring System for Sleep Stages of Human Subjects', the Rechtschaffen & Kales (1968) standard. *Psychiatry Clin. Neurosci.* **2001**, *55*, 305–310. [CrossRef]

52. Carley, D.W.; Farabi, S.S. Physiology of sleep. *Diabetes Spectr.* **2016**, *29*, 5–9. [CrossRef] [PubMed]

53. Goldberger, A.L.; Amaral, L.A.N.; Glass, L.; Hausdorff, J.M.; Ivanov, P.C.; Mark, R.G.; Mietus, J.E.; Moody, G.B.; Peng, C.-K.; Stanley, H.E. PhysioBank, PhysioToolkit, and PhysioNet. *Circulation* **2000**, *101*, e215–e220. [CrossRef]

54. Kemp, B.; Zwinderman, A.; Tuk, B.; Kamphuisen, H.; Oberye, J. Analysis of a sleep-dependent neuronal feedback loop: The slow-wave microcontinuity of the EEG. *IEEE Trans. Biomed. Eng.* **2000**, *47*, 1185–1194. [CrossRef]

55. Zhang, G.-Q.; Cui, L.; Mueller, R.; Tao, S.; Kim, M.; Rueschman, M.; Mariani, S.; Mobley, D.R.; Redline, S. The national sleep research resource: Towards a sleep data commons. *J. Am. Med. Inform. Assoc.* **2018**, *25*, 1351–1358. [CrossRef]

56. Quan, S.F.; Howard, B.V.; Iber, C.; Kiley, J.P.; Nieto, F.J.; O'Connor, G.; Rapoport, D.M.; Redline, S.; Robbins, J.; Samet, J.M.; et al. The sleep heart health study: Design, rationale, and methods. *Sleep* **1997**, *20*, 1077–1085. [CrossRef]

57. Ichimaru, Y.; Moody, G. Development of the polysomnographic database on CD-ROM. *Psychiatry Clin. Neurosci.* **1999**, *53*, 175–177. [CrossRef]

58. Khalighi, S.; Sousa, T.; Santos, J.M.; Nunes, U. ISRUC-Sleep: A comprehensive public dataset for sleep researchers. *Comput. Methods Programs Biomed.* **2016**, *124*, 180–192. [CrossRef] [PubMed]

59. O'Reilly, C.; Gosselin, N.; Carrier, J.; Nielsen, T. Montreal archive of sleep studies: An open-access resource for instrument benchmarking and exploratory research. *J. Sleep Res.* **2014**, *23*, 628–635. [CrossRef] [PubMed]

60. Li, Q.; Li, Q.C.; Liu, C.; Shashikumar, S.P.; Nemati, S.; Clifford, G.D. Deep learning in the cross-time frequency domain for sleep staging from a single-lead electrocardiogram. *Physiol. Meas.* **2018**, *39*, 124005. [CrossRef] [PubMed]

61. Tripathy, R.; Acharya, U.R. Use of features from RR-time series and EEG signals for automated classification of sleep stages in deep neural network framework. *Biocybern. Biomed. Eng.* **2018**, *38*, 890–902. [CrossRef]

62. Radha, M.; Fonseca, P.; Moreau, A.; Ross, M.; Cerny, A.; Anderer, P.; Long, X.; Aarts, R.M. Sleep stage classification from heart-rate variability using long short-term memory neural networks. *Sci. Rep.* **2019**, *9*, 1–11. [CrossRef]

63. Zhu, T.; Luo, W.; Yu, F. Convolution-and attention-based neural network for automated sleep stage classification. *Int. J. Environ. Res. Public Health* **2020**, *17*, 4152. [CrossRef]

64. Qureshi, S.; Karrila, S.; Vanichayobon, S. GACNN SleepTuneNet: A genetic algorithm designing the convolutional neuralnetwork architecture for optimal classification of sleep stages from a single EEG channel. *Turk. J. Electr. Eng. Comput. Sci.* **2019**, *27*, 4203–4219. [CrossRef]

65. Yıldırım, Ö.; Baloglu, U.B.; Acharya, U.R. A deep learning model for automated sleep stages classification using PSG signals. *Int. J. Environ. Res. Public Health* **2019**, *16*, 599. [CrossRef]

66. Hsu, Y.-L.; Yang, Y.-T.; Wang, J.-S.; Hsu, C.-Y. Automatic sleep stage recurrent neural classifier using energy features of EEG signals. *Neurocomputing* **2013**, *104*, 105–114. [CrossRef]

67. Michielli, N.; Acharya, U.R.; Molinari, F. Cascaded LSTM recurrent neural network for automated sleep stage classification using single-channel EEG signals. *Comput. Biol. Med.* **2019**, *106*, 71–81. [CrossRef]

68. Wei, L.; Lin, Y.; Wang, J.; Ma, Y. Time-Frequency Convolutional Neural Network for Automatic Sleep Stage Classification Based on Single-Channel EEG. In Proceedings of the 2017 IEEE 29th International Conference on Tools with Artificial Intelligence, Boston, MA, USA, 6–8 November 2017; IEEE: Piscataway, NJ, USA, 2017; pp. 88–95.

69. Mousavi, S.; Afghah, F.; Acharya, U.R. SleepEEGNet: Automated sleep stage scoring with sequence to sequence deep learning approach. *PLoS ONE* **2019**, *14*, e0216456. [CrossRef] [PubMed]

70. Seo, H.; Back, S.; Lee, S.; Park, D.; Kim, T.; Lee, K. Intra- and inter-epoch temporal context network (IITNet) using sub-epoch features for automatic sleep scoring on raw single-channel EEG. *Biomed. Signal Process. Control* **2020**, *61*, 102037. [CrossRef]

71. Zhang, X.; Xu, M.; Li, Y.; Su, M.; Xu, Z.; Wang, C.; Kang, D.; Li, H.; Mu, X.; Ding, X.; et al. Automated multi-model deep neural network for sleep stage scoring with unfiltered clinical data. *Sleep Breath.* **2020**, *24*, 581–590. [CrossRef] [PubMed]

72. Supratak, A.; Dong, H.; Wu, C.; Guo, Y. DeepSleepNet: A model for automatic sleep stage scoring based on raw single-channel EEG. *IEEE Trans. Neural Syst. Rehabil. Eng.* **2017**, *25*, 1998–2008. [CrossRef] [PubMed]

73. Phan, H.; Andreotti, F.; Cooray, N.; Chén, O.Y.; de Vos, M. Joint classification and prediction CNN framework for automatic sleep stage classification. *IEEE Trans. Biomed. Eng.* **2019**, *66*, 1285–1296. [CrossRef]

74. Vilamala, A.; Madsen, K.H.; Hansen, L.K. Deep convolutional neural networks for interpretable analysis of EEG sleep stage scoring. In Proceedings of the 2017 IEEE 27th International Workshop on Machine Learning for Signal Processing (MLSP), Tokyo, Japan, 25–28 September 2017; pp. 1–6. [CrossRef]

75. Phan, H.; Andreotti, F.; Cooray, N.; Chen, O.Y.; de Vos, M. DNN filter bank improves 1-max pooling CNN for single-channel EEG automatic sleep stage classification. In Proceedings of the 2018 40th Annual International Conference of the IEEE Engineering in Medicine and Biology Society (EMBC), Honolulu, HI, USA, 18–21 July 2018; pp. 453–456. [CrossRef]

76. Phan, H.; Andreotti, F.; Cooray, N.; Chen, O.Y.; De Vos, M. Automatic Sleep Stage Classification Using Single-Channel EEG: Learning Sequential Features with Attention-Based Recurrent Neural Networks. In Proceedings of the 2018 40th Annual International Conference of the IEEE Engineering in Medicine and Biology Society (EMBC), Honolulu, HI, USA, 18–21 July 2018.

77. Xu, M.; Wang, X.; Zhangt, X.; Bin, G.; Jia, Z.; Chen, K. Computation-Efficient Multi-Model Deep Neural Network for Sleep Stage Classification. In Proceedings of the ASSE '20: 2020 Asia Service Sciences and Software Engineering Conference, Nagoya, Japan, 13–15 May 2020; Association for Computing Machinery: New York, NY, USA, 2020; pp. 1–8.

78. Wang, Y.; Wu, D. Deep Learning for Sleep Stage Classification. In Proceedings of the 2018 Chinese Automation Congress (CAC), Xi'an, China, 30 November–2 December 2018; IEEE: Piscataway, NJ, USA, 2018; pp. 3833–3838.

79. Fernandez-Blanco, E.; Rivero, D.; Pazos, A. Convolutional neural networks for sleep stage scoring on a two-channel EEG signal. *Soft Comput.* **2019**, *24*, 4067–4079. [CrossRef]

80. Jadhav, P.; Rajguru, G.; Datta, D.; Mukhopadhyay, S. Automatic sleep stage classification using time-frequency images of CWT and transfer learning using convolution neural network. *Biocybern. Biomed. Eng.* **2020**, *40*, 494–504. [CrossRef]

81. Tsinalis, O.; Matthews, P.M.; Guo, Y.; Zafeiriou, S. *Automatic Sleep Stage Scoring with Single-Channel EEG Using Convolutional Neural Networks*; Imperial College London: London, UK, 2016.

82. Sokolovsky, M.; Guerrero, F.; Paisarnsrisomsuk, S.; Ruiz, C.; Alvarez, S.A. Deep learning for automated feature discovery and classification of sleep stages. *IEEE/ACM Trans. Comput. Biol. Bioinform.* **2019**, *17*. [CrossRef]

83. Dong, H.; Supratak, A.; Pan, W.; Wu, C.; Matthews, P.M.; Guo, Y. Mixed neural network approach for temporal sleep stage classification. *IEEE Trans. Neural Syst. Rehabil. Eng.* **2018**, *26*, 324–333. [CrossRef]

84. Chambon, S.; Galtier, M.N.; Arnal, P.J.; Wainrib, G.; Gramfort, A. A deep learning architecture for temporal sleep stage classification using multivariate and multimodal time series. *IEEE Trans. Neural Syst. Rehabil. Eng.* **2018**, *26*, 758–769. [CrossRef]

85. Phan, H.; Andreotti, F.; Cooray, N.; Chén, O.Y.; de Vos, M. SeqSleepNet: End-to-end hierarchical recurrent neural network for sequence-to-sequence automatic sleep staging. *IEEE Trans. Neural Syst. Rehabil. Eng.* **2019**, *27*, 400–410. [CrossRef] [PubMed]

86. Zhang, J.; Yao, R.; Ge, W.; Gao, J. Orthogonal convolutional neural networks for automatic sleep stage classification based on single-channel EEG. *Comput. Methods Programs Biomed.* **2020**, *183*, 105089. [CrossRef] [PubMed]

87. Zhang, J.; Wu, Y. Complex-valued unsupervised convolutional neural networks for sleep stage classification. *Comput. Methods Programs Biomed.* **2018**, *164*, 181–191. [CrossRef] [PubMed]

88. Sors, A.; Bonnet, S.; Mirek, S.; Vercueil, L.; Payen, J.-F. A convolutional neural network for sleep stage scoring from raw single-channel EEG. *Biomed. Signal Process. Control* **2018**, *42*, 107–114. [CrossRef]

89. Fernández-Varela, I.; Hernández-Pereira, E.; Alvarez-Estevez, D.; Moret-Bonillo, V. A Convolutional Network for Sleep Stages Classification. *arXiv* **2019**, arXiv:1902.05748v1. [CrossRef]

90. Zhang, L.; Fabbri, D.; Upender, R.; Kent, D.T. Automated sleep stage scoring of the Sleep Heart Health Study using deep neural networks. *Sleep* **2019**, *42*. [CrossRef]

91. Cui, Z.; Zheng, X.; Shao, X.; Cui, L. Automatic sleep stage classification based on convolutional neural network and fine-grained segments. *Complexity* **2018**, *2018*, 1–13. [CrossRef]

92. Yang, Y.; Zheng, X.; Yuan, F. A Study on Automatic Sleep Stage Classification Based on CNN-LSTM. In Proceedings of the ICCSE'18: The 3rd International Conference on Crowd Science and Engineering, Singapore, 28–31 July 2018; Association for Computing Machinery: New York, NY, USA, 2018; pp. 1–5.

93. Yuan, Y.; Jia, K.; Ma, F.; Xun, G.; Wang, Y.; Su, L.; Zhang, A. A hybrid self-attention deep learning framework for multivariate sleep stage classification. *BMC Bioinform.* **2019**, *20*, 1–10. [CrossRef]

94. Biswal, S.; Sun, H.; Goparaju, B.; Westover, M.B.; Sun, J.; Bianchi, M.T. Expert-level sleep scoring with deep neural networks. *J. Am. Med. Inform. Assoc.* **2018**, *25*, 1643–1650. [CrossRef]

95. Biswal, S.; Kulas, J.; Sun, H.; Goparaju, B.; Westover, M.B.; Bianchi, M.T.; Sun, J. SLEEPNET: Automated Sleep Staging System via Deep Learning. *arXiv* **2017**, arXiv:1707.08262.

96. Hoshide, S.; Kario, K. Sleep Duration as a risk factor for cardiovascular disease—A review of the recent literature. *Curr. Cardiol. Rev.* **2010**, *6*, 54–61. [CrossRef]

97. Woods, S.L.; Froelicher, E.S.S.; Motzer, S.U.; Bridges, S.J. *Cardiac Nursing*, 5th ed.; Lippincott Williams and Wilkins: London, UK, 2005.

98. Krieger, J. Breathing during sleep in normal subjects. *Clin. Chest Med.* **1985**, *6*, 577–594. [PubMed]

99. Madsen, P.L.; Schmidt, J.F.; Wildschiodtz, G.; Friberg, L.; Holm, S.; Vorstrup, S.; Lassen, N.A. Cerebral O_2 metabolism and cerebral blood flow in humans during deep and rapid-eye-movement sleep. *J. Appl. Physiol.* **1991**, *70*, 2597–2601. [CrossRef] [PubMed]

100. Klosh, G.; Kemp, B.; Penzel, T.; Schlogl, A.; Rappelsberger, P.; Trenker, E.; Gruber, G.; Zeithofer, J.; Saletu, B.; Herrmann, W.; et al. The SIESTA project polygraphic and clinical database. *IEEE Eng. Med. Boil. Mag.* **2001**, *20*, 51–57. [CrossRef] [PubMed]

101. Yıldırım, Ö.; Talo, M.; Ay, B.; Baloglu, U.B.; Aydin, G.; Acharya, U.R. Automated detection of diabetic subject using pre-trained 2D-CNN models with frequency spectrum images extracted from heart rate signals. *Comput. Biol. Med.* **2019**, *113*, 103387. [CrossRef] [PubMed]

102. Pham, T.-H.; Vicnesh, J.; Koh, J.E.; Oh, S.L.; Arunkumar, N.; Abdulhay, E.; Ciaccio, E.J.; Acharya, U.R. Autism spectrum disorder diagnostic system using HOS bispectrum with EEG Signals. *Int. J. Environ. Res. Public Health* **2020**, *17*, 971. [CrossRef] [PubMed]

103. Khan, S.A.; Kim, J.-M. Automated bearing fault diagnosis using 2D analysis of vibration acceleration signals under variable speed conditions. *Shock. Vib.* **2016**, *2016*, 1–11. [CrossRef]

104. Kiranyaz, S.; Avci, O.; Abdeljaber, O.; Ince, T.; Gabbouj, M.; Inman, D.J. 1D convolutional neural networks and applications: A survey. *Mech. Syst. Signal Process.* **2021**, *151*. [CrossRef]

105. Patanaik, A.; Ong, J.L.; Gooley, J.J.; Ancoli-Israel, S.; Chee, M.W.L. An end-to-end framework for real-time automatic sleep stage classification. *Sleep* **2018**, *41*. [CrossRef]

106. Terzano, M.G.; Parrino, L.; Sherieri, A.; Chervin, R.; Chokroverty, S.; Guilleminault, C.; Hirshkowitz, M.; Mahowald, M.; Moldofsky, H.; Rosa, A.; et al. Atlas, rules, and recording techniques for the scoring of cyclic alternating pattern (CAP) in human sleep. *Sleep Med.* **2001**, *2*, 537–553. [CrossRef]

Publisher's Note: MDPI stays neutral with regard to jurisdictional claims in published maps and institutional affiliations.

Article

An Efficient Algorithm for Cardiac Arrhythmia Classification Using Ensemble of Depthwise Separable Convolutional Neural Networks

Eko Ihsanto [1], Kalamullah Ramli [1,*], Dodi Sudiana [1] and Teddy Surya Gunawan [2]

[1]   Department of Electrical Engineering, Universitas Indonesia, Depok, Jawa Barat 16424, Indonesia; eko.ihsanto@gmail.com (E.I.); dodi.sudiana@ui.ac.id (D.S.)

[2]   Department of Electrical and Computer Engineering, International Islamic University Malaysia, Kuala Lumpur 53100, Malaysia; tsgunawan@iium.edu.my

*   Correspondence: kalamullah.ramli@ui.ac.id

Received: 30 November 2019; Accepted: 3 January 2020; Published: 9 January 2020

Abstract: Many algorithms have been developed for automated electrocardiogram (ECG) classification. Due to the non-stationary nature of the ECG signal, it is rather challenging to use traditional handcraft methods, such as time-based analysis of feature extraction and classification, to pave the way for machine learning implementation. This paper proposed a novel method, i.e., the ensemble of depthwise separable convolutional (DSC) neural networks for the classification of cardiac arrhythmia ECG beats. Using our proposed method, the four stages of ECG classification, i.e., QRS detection, preprocessing, feature extraction, and classification, were reduced to two steps only, i.e., QRS detection and classification. No preprocessing method was required while feature extraction was combined with classification. Moreover, to reduce the computational cost while maintaining its accuracy, several techniques were implemented, including All Convolutional Network (ACN), Batch Normalization (BN), and ensemble convolutional neural networks. The performance of the proposed ensemble CNNs were evaluated using the MIT-BIH arrythmia database. In the training phase, around 22% of the 110,057 beats data extracted from 48 records were utilized. Using only these 22% labeled training data, our proposed algorithm was able to classify the remaining 78% of the database into 16 classes. Furthermore, the sensitivity (S_n), specificity (S_p), and positive predictivity (P_p), and accuracy (Acc) are 99.03%, 99.94%, 99.03%, and 99.88%, respectively. The proposed algorithm required around 180 µs, which is suitable for real time application. These results showed that our proposed method outperformed other state of the art methods.

Keywords: depthwise separable convolution (DSC); all convolutional network (ACN); batch normalization (BN); ensemble convolutional neural network (ECNN); electrocardiogram (ECG); MIT-BIH database

---

1. Introduction

ECG signals can be easily acquired by putting one's finger on the sensor for about 30 s [1]. There are at least two types of important information contained in the ECG signal, including those related to health or biomedical [2–4] and those related to the person identification or biometrics [5–7]. Due to its convenience, many ECG classification algorithms have been developed, including handcraft [4,8,9] and machine learning [10–15] methods. The handcraft method is rather difficult to utilize on non-stationary signals, such as ECG, while machine learning methods normally require high computational resources. Due to its high accuracy, the machine learning method is preferable compared to the handcraft method. However, an efficient algorithm is required to minimize the computational requirements while still maintaining its high accuracy.

Many researches have been conducted on the implementation of handcraft techniques, including the extraction of time-based ECG features using Fourier [8] and wavelet [4,9] transforms. Both Fourier and wavelet transform can be used for ECG beats detection (QRS detection), as well as feature extraction, such as R-peak, RR-interval, T-wave region, and QT-zone. After QRS detection, amplitude and duration-based ECG features can also be measured using weighted diagnostic distortion (WDD) [16]. The feature extraction stage is usually followed by a classification stage with various methods such as vector quantization [17], random forest [18–20], k-nearest neighbor (kNN) [10,20,21], support vector machine (SVM) [10,13,18,20], multi-layer perceptron (MLP) [22–24] and convolutional neural network (CNN) [25].

If the feature extraction and classification stages are done separately, SVM can be used and optimized using particle swarm optimization (PSO) [10]. As presented in [10], after supervised training was conducted using 500 beats, the model can classify 40,438 test beats into five classes with accuracy of 89.72%, outperforming the other methods such as kNN or radial basis function (RBF). Nevertheless, the quality of this classification can still be improved in terms of increasing the number of classes and/or accuracy. For example, discrete orthogonal Stockwell transform (DOST) could be used during feature extraction followed by principal component analysis (PCA) to reduce feature dimensions [13]. As shown in [13], after supervised training was conducted on 23,996 beats, the remainder 86,113 test beats could be classified into 16 classes with better accuracy of 98.82%.

Another promising method for improving efficiency is by combining both feature extraction and classification stages using MLP [26] and CNN [27]. For example, in [27], a neural network model containing three layers of CNN and two layers of MLP was proposed. The input of this model is a raw ECG beat signal containing 64 or 128 samples centered on the R-peak. While the number of ECG beats used for training is kept at minimum at 245 and the testing beats is set to 100,144, it can achieve accuracy of 95.14% to classify five classes. In [28], autoencoder was utilized with a rather good result but it needs to fairly evaluate the performance with and without denoising. Moreover, the deep networks configuration could be further optimized to reduce computational time.

Although many researches have been conducted, an efficient algorithm for cardiac arrhythmia classification is still required. Therefore, the objective of this paper was to simplify the overall process to lower the computational cost, while maintaining high accuracy. A neural network model presented in [27] was adopted and modified to now classify 16 classes as in [13]. The performance of our proposed algorithms was evaluated using number of classes, prediction stages, and accuracy.

2. 1D Convolutional Neural Network and Its Enhancement

In this paper, we use two stages of ECG classification, namely beat segmentation and classification. For classification purposes, CNN could be used as stated in [27,29]. Although CNN hyperparameters such as number of filters, filter size, padding type, activation type, pooling, backpropagation, still have to be done intuitively or by trial and error, there are still some techniques that can be used to reduce the amount of trial and error attempts to achieve the best results. Further enhancement to the CNN could be done using All Convolutional Network (ACN) [30], Batch Normalization (BN) [31]. Depthwise Separable Convolution (DSC) [32], ensemble CNN [33], which are further elaborated in this section. Out of various methods for CNN enhancement, DSC has the greatest effect on decreasing training time, while ensemble CNN enables further improvement on the classification rate.

ACN is used to replace pooling layer with stride during convolution. Pooling or stride is used for downsampling as CNN output parameters are less than the input parameter. In our case, the CNN input parameter is set to 256, while the output parameter is set to 16. Normally, the last or output layer of CNN uses the SoftMax activation function to determine the output class based on its highest probability. To reduce the number of layers, pooling layer could be replaced with stride during convolution [30].

Input normalization is required to solve internal covariate shift, i.e., the change in the distribution of network activations due to the change in network parameters during training. Without normalization, this will slow down the training iteration or even stop the iteration before reaching adequate accuracy. To solve this issue, BN can be conducted for each training mini-batch [31]. To reduce computational cost, BN could be conducted during the convolution process before nonlinear activation, such as rectified linear unit (ReLU). On the other hand, DSC could be used to reduce the number of parameters and floating points multiplication operation, with negligible performance degradation [34]. DSC could be performed on one layer or a group of layers of CNNs.

Another thing that must be considered in designing the CNN model is the implementation of the Flatten layer before the Fully Connected (FC) layer. Even though Flatten technically can be replaced by the AveragePool layer, these two techniques differ in terms of execution time and the final results obtained. The Flatten process does not require any further calculations, it only changes the arrangement of parameters in the last layer, while AveragePool must perform arithmetic operations to get the average value of each group in the last layer according to its position. The next effect of Flatten causes more neurons connected to FC, in comparison to the number of neurons produced by AveragePool which is obviously related to the number of arithmetic operations at the FC layer. It should be noted that the number of neurons in the Flatten layer represents all local features on the last layer without having to be combined in the average value.

2.1. 1-D CNNs

As described in [27,29], during the forward propagation, the input map of the next layer neuron will be obtained by the cumulation of the final output maps of the previous layer neurons convolved with their individual kernels as follows:

$$x_k^l = \beta_k^l + \sum_{i=1}^{N_{l-1}} conv1D\left(\omega_{ik}^{l-1}, s_i^{l-1}\right) \tag{1}$$

where $conv1D(\cdot,\cdot)$ is 1-D convolution, x_k^l is the input, β_k^l is the bias of the k-th neuron at layer l, s_i^{l-1} is the output of the i-th neuron at layer $l-1$, ω_{ik}^{l-1} is the kernel (weight) from the i-th neuron at layer $l-1$ to the k-th neuron at layer l. Let $l=1$ and $l=L$ be the input and output layers, respectively. The inter backpropagation delta error of the output s_k^l can be expressed as follows:

$$\Delta s_k^l = \sum_{i=1}^{N_{l+1}} conv1Dz\left(\Delta_i^{l+1}, rev\left(\omega_{ik}^l\right)\right) \tag{2}$$

where $rev(\cdot)$ flips the array, and $conv1Dz(\cdot,\cdot)$ performs full convolution in 1-D with $K-1$ zero padding. Lastly, the weight and bias sensitivities can be expressed as follows:

$$\frac{\partial E}{\partial \omega_{ik}^l} = conv1D\left(s_k^l, \Delta_i^{l+1}\right) \tag{3}$$

$$\frac{\partial E}{\partial \beta_k^l} = \sum_n \Delta_k^l(n) \tag{4}$$

2.2. All Convolutional Network

All Convolutional Network (ACN) [30] is utilized by removing the max pooling layer and replacing it with convolutional stride. As a result, the computational cost will be reduced, as it removes several layers as well as reducing the floating point operation on the convolution operation. For example, conv stride = 1 continued with max pooling 2 could be replaced with one layer, i.e., conv stride = 2 with an almost similar result.

Let f denote a feature map produced by some layer of a CNN, while N is the number of filters in this layer. Then p-norm subsampling with pooling size m (or half-length $\frac{m}{2}$) and stride r applied to the feature map f is a 3-dimensional array $s(f)$ with the following entries:

$$s_{i,j,u}(f) = \left(\sum_{h=-\lfloor \frac{m}{2} \rfloor}^{\lfloor \frac{m}{2} \rfloor} \sum_{w=-\lfloor \frac{m}{2} \rfloor}^{\lfloor \frac{m}{2} \rfloor} \left| f_g(h,w,i,j,u) \right|^p \right)^{\frac{1}{p}} \tag{5}$$

where $g(h,w,i,j,u) = (r \cdot i + h, r \cdot j + w, u)$ is the function mapping from positions in s to positions in f representing the stride, p is the order of the p-norm ($p \to \infty$ represents max pooling). If $r > m$, pooling regions do not overlap. The standard definition of a convolution layer c applied to the feature map f is given as:

$$c_{i,j,o}(f) = \sigma \left(\sum_{h=-\lfloor \frac{m}{2} \rfloor}^{\lfloor \frac{m}{2} \rfloor} \sum_{w=-\lfloor \frac{m}{2} \rfloor}^{\lfloor \frac{m}{2} \rfloor} \sum_{u=1}^{N} \theta_{h,w,u,o} \cdot f_{g(h,w,i,j,u)} \right) \tag{6}$$

where θ are the convolutional weights, $\sigma(\cdot)$ is the activation function, typically a rectified linear activation ReLU $\sigma(x) = max(x, 0)$, and $o \in [1, M]$ is the number of output features of the convolutional layer. From the two equations, it is evident that the computational cost will be reduced.

2.3. Batch Normalization

Batch Normalization (BN) is intended to avoid non-linear saturation or internal covariate shifts causing a faster learning process [31]. The Batch Normalization allows much higher learning rates and reduces dependence on the initialization process. It can also act as a regularizer to reduce the generalization error and to avoid overfitting without the implementation of dropout layer. To reduce computation, the Batch Normalization is carried out directly after the convolution and before ReLU. ReLU before BN can mess up the calculations due to the non-linear nature of ReLU. Suppose we have network activation as follows:

$$z = g(\omega u + \beta) \tag{7}$$

where ω and β are learned parameters of the model, and $g(\cdot)$ is the nonlinear activation function such as sigmoid or ReLU. Since we normalize $\omega u + \beta$, the bias β can be ignored since its effect will be cancelled by the subsequent mean subtraction. Therefore, Equation (7) is replaced with:

$$z = g(BN(\omega u)) \tag{8}$$

where the BN transform is applied indenpendently to each dimension of $x = \omega u$, with a separate pair of learned parameters.

2.4. Depthwise Separable Convolution

In addition to the implementation of BN, the convolution part can also be further optimized, for example by using depthwise separable convolution (DSC) to reduce the computational costs by reducing the number of arithmetic operations while preserving the same final results [32,34,35]. This technique is applied with changes in filter sizes 3, 1, and 3, respectively. In addition to DSC, the size of the stride in the third convolution should be set to be greater than one, for example 2, 3, or 4, allowing the downsampling process to be implemented within the convolution process, instead of on a special layer such as the MaxPool layer [30]. DSC [34] splits the convolution into two calculation stages, i.e., depthwise and pointwise, as follows:

$$Conv(\omega, y)_{(i,j)} = \sum_{k.l.m}^{K.L.M} \omega_{(k.l.m)} \cdot y_{(i+k,j+l,m)} \tag{9}$$

$$PointwiseConv(\omega, y)_{(i,j)} = \sum_{m}^{M} \omega_m \cdot y_{(i,j,m)} \tag{10}$$

$$DepthwiseConv(\omega, y)_{(i,j)} = \sum_{k,l}^{K,L} \omega_{(k,l)} \odot y_{(i+k,j+l)} \tag{11}$$

$$SepConv(\omega_p, \omega_d, y)_{(i,j)} = PointwiseConv_{(i,j)}\left(\omega_p, DepthwiseConv_{(i,j)}(\omega_d, y)\right) \tag{12}$$

3. Proposed Ensemble of Depthwise Separable Convolutional Neural Networks

Traditional methods of ECG classification are varied in stages, including four, three, and two. The four stages are beat detection morphology feature extraction, feature dimension reduction, and classification, as stated in [13,36]. The three stages are beat detection, feature detection, and classification [10]. Moreover, the two stages of classification are beat detection and classification [27,37], in which the feature extraction stage is combined with classification. In [38], one stage was used for arrhythmia detection using 34 layer CNNs. However, it cannot be directly compared to our proposed algorithm as they used a different database with 12 heart arrhythmias, sinus rhythm, and noise for a total output of 14 classes. This paper proposes a two-stage ECG beat detection, as one patient might experience a normal beat and another arrhythmia beats as described in the MIT-BIH database. Furthermore, as will be explained in Section 5, the beat detection and segmentation require minimum computational time while improving the classification process. In this section, beat detection and segmentation, and ensemble of Depthwise Separable CNNs consists of around 34,719 train parameters with 21 layer CNNs are explained.

3.1. Beat Detection and Segmentation

In this paper, beat detection, QRS detection, or R-peak detection are performed based on analysis of gradient, amplitude, and duration of the ECG signals similar to [39]. R-peak detection of ECG signals can be done through wavelet transforms with a detection accuracy of more than 99% [39,40]. R-peak detection was performed on 48 records of the MIT-BIH database. With a 360 Hz sampling rate, each record contains 650,000 samples or a duration of 30 min [41]. Each sample is a conversion of a range of 10 mV using an 11-bit ADC [41].

Figure 1 shows examples of ECG pieces of 3000 samples or 8.33 s from the 2nd, 18th, and 36th records. The RR intervals on ECG chunk from the 2nd and 36th files tend to be uniform, while the RR interval on the ECG chunk from the 18th file looks more diverse. Hence, the R-peak detection and the RR interval measurement affect the beat segmentation process.

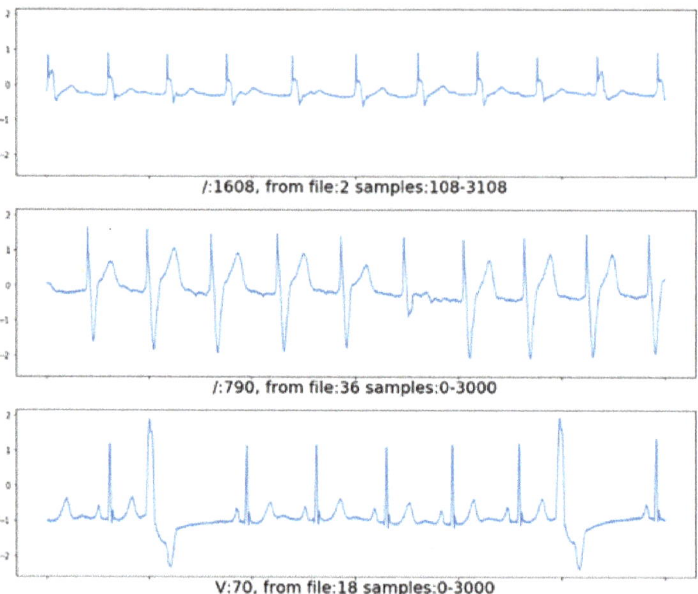

Figure 1. The ECG chunk contains 3000 samples from 3 files, i.e., the 2nd, 18th, and 36th files. The vertical axis is the voltage in mV, while the horizontal axis is time with a scale of 500 samples or 1388 s. The text on the horizontal axis contains the type of beat, and its R-peak position, for example, /: 608, means that the beat is Paced Beat (see Table 1) with R-peak position at the 1608th sample from 650,000 samples in the 2nd file.

In this paper, after R-peak detection, beat segmentation starts from $\frac{1}{4}L_i$ to $\frac{3}{4}R_i$, in which L_i is the RR-interval right before the detected R-peak, while R_i is the RR-interval right after the detected R-peak. This automatic segmentation window is necessary to ensure that there is only a single beat or R-peak in each segment. The maximum number of samples taken for each segment is 256 points which is equivalent to a duration of 256/360 s or 711 ms. The MIT-BIH database used a sampling rate of 360 Hz, so the 256 sample segment size will be more than adequate as the typical RR interval is around 500 ms [42].

Figure 2 shows the example of our proposed automatic beat segmentation. In our segmentation method, if the RR-interval is too short, or there is more than one R-peak at the segment, then zero-paddings are performed to keep the segment size to 256 samples with only one R-peak. If required, this segment size could be downsampled to 128 or even 64 samples to further reduce the computational cost. However, there will be a slight decrease in the classification accuracy. More experiments are conducted on the sample sizes and its accuracy in the next section.

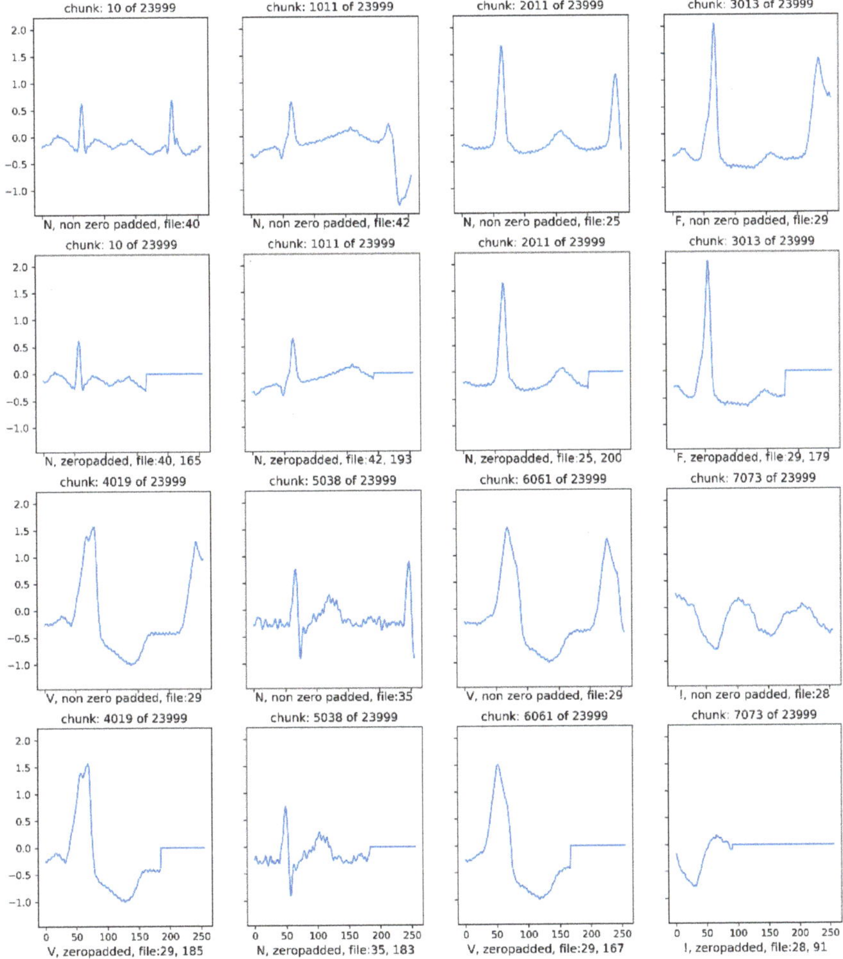

Figure 2. Samples of 16 chunks out of 23,999 beats training data. The 10th chunk, our top left chunk, contains the Normal beat taken from the 40th file. The first and third row were the original signals, while the second and fourth row were the zero-padded signals.

3.2. Ensemble CNNs

Each segmented ECG beat can be replicated into three beat sizes, i.e., 64, 128, and 256 samples. The 64 and 128 samples were the downsampled version of the original 256 beats segmented automatically as explained in Section 3.1. In total, we have three CNN configurations with a difference only in layer 1 (input size), i.e., 64, 128, and 256. After the training and testing phase, the three outputs from three CNNs were ensembled using the averaging method.

Using ensemble CNN by averaging, it can improve further the accuracy by reducing the variance [33]. In our proposed algorithm, we calculate the average of all tensor SoftMax from 3 CNNs as shown in Equation (13) and Figure 3. Let $l = L$ denote the last layer of the CNN model, and s_k^L is the output of the k-th neuron at the output layer, and let $m = 1, \ldots, M$ denotes the number of CNNs to be ensembled. The final output of the ensemble CNN using averaging can be calculated as follows:

$$\overline{P}(s) = \frac{1}{N} \sum_{m=1}^{M} s_k^L[m] \tag{13}$$

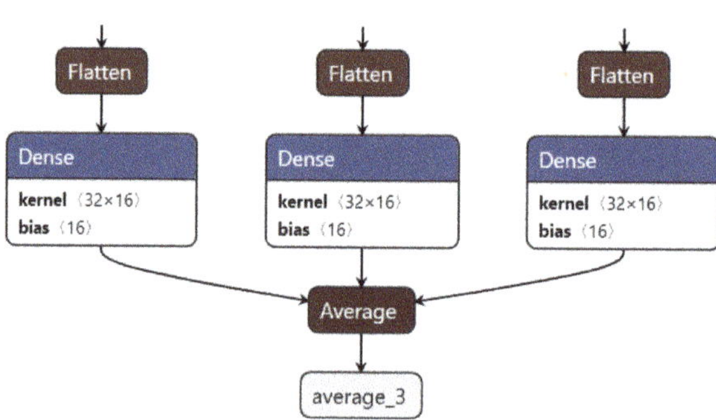

Figure 3. Ensemble CNN.

4. Implementation and Experimental Setup

This section discusses the ECG datasets from the MIT-BIH arrhythmia database, the computational platform used for simulation, and the proposed ensemble CNN algorithm.

4.1. MIT-BIH Arrhythmia Database and Computing Platform

In this paper, ECG datasets from the MIT-BIH arrhythmia database [41] are used for the performance evaluation of our proposed ensemble CNN algorithms. This database is the widely used database for testing classification performance. It contains 48 records, in which each contain two- channel ECG signals for 30 min duration selected from 24 h recording of 47 patients. The database consists of 19 types of ECG beats, in which 16 of them are related to cardiac arrhythmia.

From the total MIT-BIH database, we only used around 110,157 ECG beats out of 112,647 total beats which were categorized into 16 classes as shown in Table 1. These beat samples were then further divided into 23,999 beats for training and 86,158 beats for testing. As described in Figure 2, the total of 110,157 beats was also replicated with zero padding. Therefore, there are two sets of experimental data, with and without zero padding.

Table 1. Sixteen classes of cardiac arrhythmia ECG beats from the MIT-BIH Database and its amount of Training and Testing beats.

No	Symbol	Annotation Description	Total	Training	Testing
1	N	Normal beat	75,052	11,257	63,795
2	L	Left bundle branch block beat	8075	2826	5249
3	R	Right bundle branch block beat	7259	2540	4719
4	V	Premature ventricular contraction	7130	2495	4635
5	/	Paced beat	7028	2459	4569
6	A	Atrial premature contraction	2546	891	1655
7	f	Fusion of paced and normal beat	982	491	491
8	F	Fusion of ventricular and normal beat	803	401	402
9	!	Ventricular flutter wave	472	236	236
10	j	Nodal (junctional) escape beat	229	114	115
11	x	Non-conducted P-wave	193	96	97
12	a	Aberrated atrial premature beat	150	75	75
13	E	Ventricular escape beat	106	53	53
14	J	Nodal (junctional) premature beat	83	41	42
15	e	Atrial escape beat	16	8	8
16	Q	Unclassifiable beat	33	16	17
		Total 16-class beats	110,157	23,999	86,158
		Extracted from total 112,647 of labeled ECG beats			

The proposed algorithm was implemented in Python with Tensorflow with GPU [43] and Keras libraries [44]. The experiments were performed on a computer with Intel Core i7-7700 CPU with a total of eight logical processors, memory of 8 GBytes, graphic card Nvidia GeForce GTX 1060 6 GB DDR5, using Microsoft Windows 10 64 bits operating system. The experiments on the training and testing time using this computing platform is elaborated further in Section 5.

4.2. Depthwise Separable and Ensemble of Depthwise Separable CNN Models

Using a heuristic approach and optimization, the proposed CNN model contains 21 layers. Figure 4 shows the comparison between the CNN model with and without depthwise separable CNN in terms of structure and number of train parameters. As shown in the figure, the total train parameters are less with the implementation of depthwise separable. Hence, the training time will be faster for depthwise separable CNN. Table 2 shows the depthwise separable CNN model summary along with its total parameters' calculation.

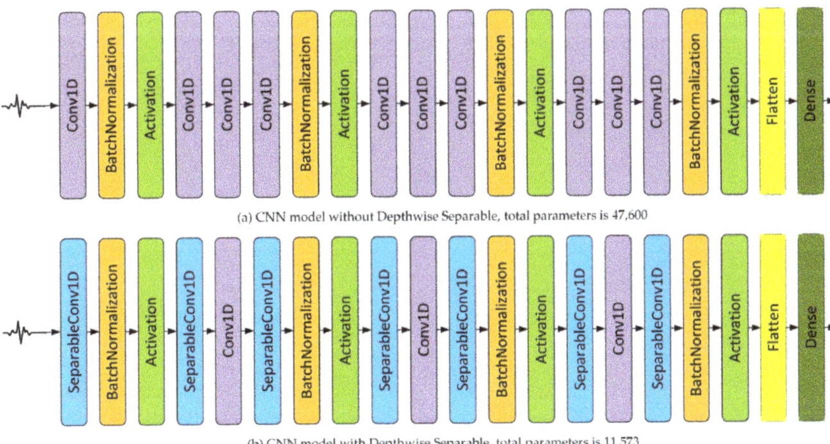

(a) CNN model without Depthwise Separable, total parameters is 47,600

(b) CNN model with Depthwise Separable, total parameters is 11,573

Figure 4. CNN Models with and without the depthwise separable algorithm.

Table 2. Depthwise separable CNN model summary.

#	Layer	Number of Filters, Size, Stride	Output Shape	Number of Parameters
1	input_1		(256, 1)	0
2	separable_conv1d_1	32, 5, 4	(64, 32)	69
3	batch_normalization_1		(64, 32)	128
4	activation_1		(64, 32)	0
5	separable_conv1d_2	32, 5, 1	(64, 32)	1216
6	conv1d_1	32, 1, 1	(64, 32)	1056
7	separable_conv1d_3	32, 5, 4	(16, 32)	1216
8	batch_normalization_2		(16, 32)	128
9	activation_2		(16, 32)	0
10	separable_conv1d_4	32, 5, 1	(16, 32)	1216
11	conv1d_2	32, 1, 1	(16, 32)	1056
12	separable_conv1d_5	32, 5, 4	(4, 32)	1216
13	batch_normalization_3		(4, 32)	128
14	activation_3		(4, 32)	0
15	separable_conv1d_6	32, 5, 1	(4, 32)	1216
16	conv1d_3	32, 1, 1	(4, 32)	1056
17	separable_conv1d_7	32, 5, 4	(1, 32)	1216
18	batch_normalization_4		(1, 32)	128
19	activation_4		(1, 32)	0
20	flatten_1		(32)	0
21	dense_1 (Dense)		(16)	528

Total parameters: 11,573
Trainable parameters: 11,317
Non-trainable parameters: 256

The depthwise separable CNN as described in Table 2 was used to improve the training time, while the ensemble of depthwise separable CNN, as depicted in Figure 5, was used to improve further its classification accuracy. The input layer size is 256 representing the raw ECG beat waveform, while the output layer size is 16 representing the number of classes as described in the MIT-BIH database (see Table 1). There is a group of layers repeated three times, i.e., layer 5 to 9, layer 10 to 14, and layer 15 to 19. These five layers contain three convolution layers with filter size of 5, 1, and 5, respectively. Layer 5 to 7 is configured according to the DSC algorithm. Moreover, layer 2, 7, 12, and 17 were using stride 4 to replace the function of pooling layer according to the ACN algorithm. Suppose that we use an ensemble of three DSC, the total number of parameters will be three times larger, i.e., 34,719 train parameters, which is still less than the train parameters without DSC, i.e., 47,600. As is discussed in Section 5, the use of ensemble improves its accuracy.

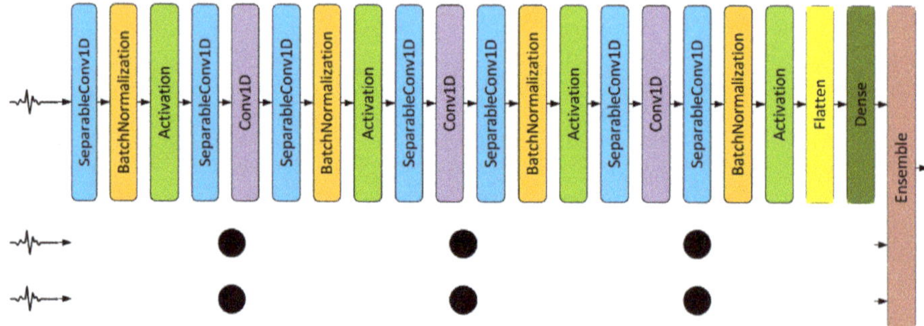

Figure 5. Proposed ensemble of depthwise separable CNNs algorithm.

5. Results and Discussion

This section elaborates beat segmentation and detection, the training process, experiments on the CNN model with and without DSC, the experiments on zero padding and various ensemble configurations, and benchmarking with other algorithms.

5.1. Experiment on ECG Beat Segmentation and Detection

The ECG beat segmentation and detection algorithm described in [39] were implemented in the computing platform described in Section 4. We evaluated the algorithm on the 48 records of the MIT- BIH database. The experiment was repeated 10 times for each record and then the average time was calculated. The experimental result showed that, on average, it requires around 26 µs to detect and segment one ECG beat with zero padding. This result justified the use of two stage ECG classification as proposed in this paper, as the segmentation time is minimal compared to the improvement achieved in the classification stage.

5.2. Training Process

Figure 6 shows the training history for 1000 epochs using the CNN model as described in Table 2 and for the training data described in Table 1. There is a gap of accuracy around 1.5% between training and testing. The accuracy curve for validation shows that the best accuracy can be achieved at epoch of 50. The training accuracy has convergent tendency reaching almost 100% with fluctuation around 99.7%. This result is achieved for a zero-padding input size of 256.

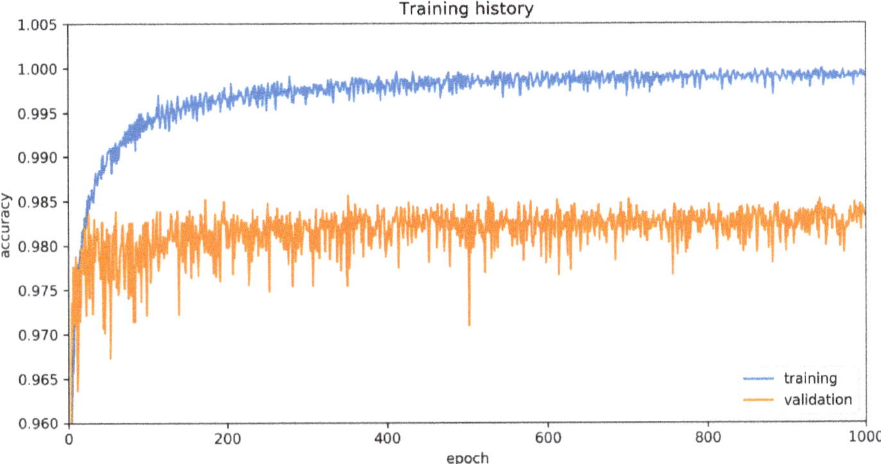

Figure 6. Training history of 1000 epochs iteration of CNN model as described in Table 2. The blue curve is the classification accuracy of the training data. The orange curve is the classification accuracy of the testing data. There is a visible gap around 1.5% between the two curves.

5.3. Performance Measures

We used the performance measure of sensitivity, specificity, positive predictivity, and accuracy as shown in Figure 7 and described in [45]. Sensitivity is the rate of correctly classified classes among all classes. Specificity is the rate of correctly classified nonevents among all events. Positive predictivity is the fraction of real events in all detected events. Finally, the accuracy is the percentage of correctly predicted classes.

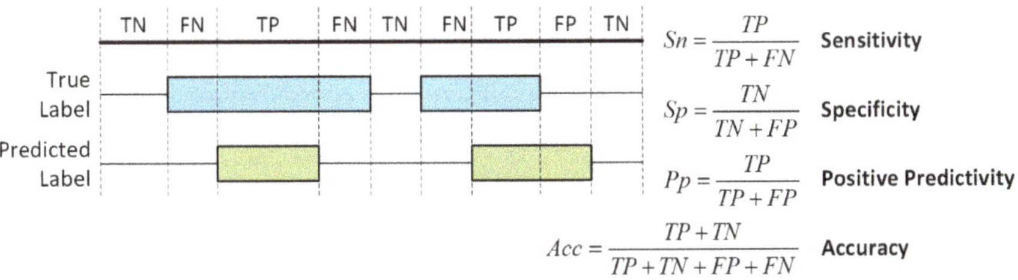

$$Sn = \frac{TP}{TP + FN} \quad \text{Sensitivity}$$

$$Sp = \frac{TN}{TN + FP} \quad \text{Specificity}$$

$$Pp = \frac{TP}{TP + FP} \quad \text{Positive Predictivity}$$

$$Acc = \frac{TP + TN}{TP + TN + FP + FN} \quad \text{Accuracy}$$

Figure 7. Performance measures using sensitivity, specificity, positive predictivity, and accuracy.

5.4. On the Effect of Depthwise Separable CNN

Table 3 shows the experimental result for CNN models with and without depthwise separable algorithms as described in Figure 4 for the input size of 256 samples. The total parameters were 47,600 and 11,573 for without and with DSC algorithm, respectively. The training time for 1000 epochs and the accuracy of both CNN models were comparable. This could be due to the use of GPU during training, in which a faster training time was not apparent in the result. When CPU only was used during training, a faster training time was evident for the CNN model with DSC. Nevertheless, the smaller parameters in the CNN model with DSC provided a faster computation time during the classification stage compared to the model without DSC. Therefore, the CNN model with DSC is used in our next experiments.

Table 3. Effect of DSC implementation on CNN layers

#	DSC	Total Parameters	TN	FN	TP	FP	*Sn*	*Sp*	*Pp*	*Acc*	Training Time	Testing Time
1	Without DSC	47,600	1,291,271	1099	85,059	1099	98.72	99.91	98.72	99.84	711.9 s	81 μs
2	With DSC	11,573	1,290,966	1404	84,754	1404	98.37	99.89	98.37	99.80	745.8 s	62 μs

5.5. On the Effect of Zero Padding

Zero padding in ECG beats was required due to there being a possibility that the distance between RR intervals is less than 256, as well as to avoid more than one R-peak being detected in one segment (see Section 3.1). Furthermore, zero padding has a positive impact on the performance, in terms of sensitivity, specificity, positive predictivity, and accuracy, as evidenced in Table 4. The best performance is highlighted in bold. As shown in Table 4, our proposed ECG segmentation of 256 samples with zero padding has the highest performance. Hence, zero padding is used in our next experiments.

Table 4. Effect of zero padding on various input sizes.

#	Zero Padding	Input Size	TN	FN	TP	FP	*Sn*	*Sp*	*Pp*	*Acc*
1	No	64	1,291,190	1180	84,978	1180	98.63	99.91	98.63	99.83
2	No	128	1,291,413	957	85,201	957	98.89	99.93	98.89	99.86
3	No	256	1,291,411	959	85,199	959	98.89	99.93	98.89	99.86
4	Yes	64	1,291,207	1163	84,995	1163	98.65	99.91	98.65	99.83
5	Yes	128	1,291,389	981	85,177	981	98.86	99.92	98.86	99.86
6	Yes	256	1,291,531	839	85,319	839	**99.03**	**99.94**	**99.03**	**99.88**

5.6. On the Effect of Various Ensemble Configurations

As described in Section 4.2, ensemble of depthwise separable CNNs has been experimented with various input sizes and various numbers of ensembles. The result is shown in Table 5, in terms of sensitivity, specificity, positive predictivity, and accuracy, in which the [256 256 256] ensemble configuration shows the best performance. Hence, the [256 256 256] ensemble configuration is used for benchmarking purposes with other algorithms.

Table 5. Performance of various ensemble of depthwise separable CNN configurations.

#	Ensemble Configuration	TN	FN	TP	FP	*Sn*	*Sp*	*Pp*	*Acc*	Testing Time
1	[128 64]	1,291,194	1176	84,982	1176	98.64	99.91	98.64	99.83	111 μs
2	[256 128]	1,291,369	1001	85,157	1001	98.84	99.92	98.84	99.85	113 μs
3	[256 256]	1,291,460	910	85,248	910	98.94	99.93	98.94	99.87	111 μs
4	[64 64 64]	1,291,207	1163	84,995	1163	98.65	99.91	98.65	99.83	143 μs
5	[128 128 128]	1,291,389	981	85,177	981	98.86	99.92	98.86	99.86	146 μs
6	[256 256 256]	1,291,531	839	85,319	839	**99.03**	**99.94**	**99.03**	**99.88**	150 μs
7	[256 128 64]	1,291,408	962	85,196	962	98.88	99.93	98.88	99.86	150 μs

In the ensemble configuration, the CNN model was trained separately. For example, the [256 256 256] ensemble is trained three times. Therefore, the training time is increased by three times with increased accuracy. As expected, the classification stage (testing time) will be increased around three times as well but within the range of 150 μs. Hence, we can conclude that the proposed two stage ECG classification requires around 180 μs for QRS detection and classification using the computing platform described in Section 4.1.

For more detailed evaluation, the [256 256 256] ensemble confusion matrix is shown in Table 6. The performance assessment of each class using the [256 256 256] ensemble configuration can be seen in Table 7, in which the ECG class is sorted from the largest to the smallest available samples in the MIT-BIH database. As can be seen from the table, fewer samples of data cause inaccurate classification.

In conclusion, the [256 256 256] ensemble achieves the best performance in terms of sensitivity, specificity, positive predictivity, and accuracy.

Table 6. Confusion matrix for [256 256 256] ensemble of CNNs.

							Prediction									
		N	L	R	V	/	A	f	F	!	j	x	a	E	J	e	Q
Ground Truth		0	1	2	3	4	5	6	7	8	9	10	11	12	13	14	15
N	0	63,521	4	13	70	2	79	11	36	3	44	6	2	0	1	0	3
L	1	22	5211	0	13	1	0	0	0	0	0	0	0	1	0	0	1
R	2	11	0	4699	0	0	9	0	0	0	0	0	0	0	0	0	0
V	3	46	10	0	4528	1	11	0	22	10	0	0	7	0	0	0	0
/	4	3	0	0	3	4553	0	9	0	0	0	0	0	1	0	0	0
A	5	143	0	5	16	0	1484	0	2	0	3	0	0	0	2	0	0
f	6	13	0	0	1	6	0	470	0	0	0	0	0	0	1	0	0
F	7	52	0	0	36	0	0	0	313	0	0	1	0	0	0	0	0
!	8	2	0	0	8	0	0	0	0	225	0	0	1	0	0	0	0
j	9	15	0	1	0	0	0	0	0	0	97	0	0	0	2	0	0
x	10	1	0	0	1	0	0	0	0	5	0	89	1	0	0	0	0
a	11	6	0	0	15	0	9	0	0	1	0	0	44	0	0	0	0
E	12	2	0	0	1	0	0	0	0	0	0	0	0	50	0	0	0
J	13	7	0	0	0	0	3	0	0	0	0	0	0	0	32	0	0
e	14	3	0	0	0	0	1	1	0	0	0	0	0	0	0	3	0
Q	15	11	0	2	1	0	0	3	0	0	0	0	0	0	0	0	0

Table 7. Performance evaluation for each class using [256 256 256] ensemble of depthwise separable CNNs.

ECG Class	Total Beats	Train Beats	Test Beats	TN	FN	TP	FP	Sn	Sp	Pp	Acc
N	75,052	11,257	63,795	22,026	274	63,521	337	99.57	98.49	99.47	99.29
L	8075	2826	5249	80,895	38	5211	14	99.28	99.98	99.73	99.94
R	7259	2540	4719	81,418	20	4699	21	99.58	99.97	99.56	99.95
V	7130	2495	4635	81,358	107	4528	165	97.69	99.80	96.48	99.68
/	7028	2459	4569	81,579	16	4553	10	99.65	99.99	99.78	99.97
A	2546	891	1655	84,391	171	1484	112	89.67	99.87	92.98	99.67
f	982	491	491	85,643	21	470	24	95.72	99.97	95.14	99.95
F	803	401	402	85,696	89	313	60	77.86	99.93	83.91	99.83
!	472	236	236	85,903	11	225	19	95.34	99.98	92.21	99.97
j	229	114	115	85,996	18	97	47	84.35	99.95	67.36	99.92
x	193	96	97	86,054	8	89	7	91.75	99.99	92.71	99.98
a	150	75	75	86,072	31	44	11	58.67	99.99	80.00	99.95
E	106	53	53	86,103	3	50	2	94.34	100.00	96.15	99.99
J	83	41	42	86,110	10	32	6	76.19	99.99	84.21	99.98
e	16	8	8	86,150	5	3	0	37.5	100.00	100.00	99.99
Q	33	16	17	86,137	17	0	4	0	100.00	0.00	99.98
Σ	110,157	23,999	86,158	1,291,531	839	85,319	839	99.03	99.94	99.03	99.88

5.7. On Comparison with Other Algorithms

Table 8 shows the benchmarking results with another 10 algorithms, in terms of number of classes (some of the algorithms did not classify all 16 classes), methods, number of prediction stages, and accuracy. In terms of accuracy, refs. [46] and [47] have the closest but still a lower accuracy than our proposed algorithm, i.e., 99.61% and 99.80%, respectively. Note that, ref. [46] only performs classification for eight classes, while [47] only performs classification for four classes, while we perform classification for sixteen classes. Hence, it is evident that our proposed algorithm with the [256 256 256] ensemble of CNNs outperforms the other algorithms. Particularly, we managed to classify all 16 classes from the MIT-BIH database, while reducing the number of stages to two (lower computational cost), and achieving highest accuracy.

Table 8. Benchmarking of the proposed algorithm with other algorithms using MIT-BIH database.

No	Author, Year	#Class	Methods	Prediction Stage	Accuracy
1	Melgani and Bazi, 2008 [10]	6	SVM and PSO	3	89.72%
2	Ince et al., 2009 [36]	5	DWT, PCA, and ANN	4	98.30%
3	Wen et al., 2009 [37]	16	Self Organizing CMAC Neural Network	2	98.21%
4	Sarfraz et al., 2014 [46]	8	ICA and BPNN	3	99.61%
5	Kiranyaz et al., 2015 [27]	5	1D-CNN	2	95.14%
6	Raj and Ray, 2017 [13]	16	DCT_DOST, PCA, SVM_PSO	4	98.82%
7	Nanjun and Meshram, 2018 [48]	2	DWT and DNN	3	98.33%
8	Zhai and Tin, 2018 [29]	5	2D-CNN	2	96.05%
9	Rangappa and Agarwal, 2018 [49]	2	k-NN	3	98.40%
10	Xia et al., 2018 [47]	4	SDAE, DNN	4	99.80%
11	Proposed Algorithm	16	Ensemble CNNs	2	**99.88%**

6. Conclusions

In this paper, we presented an efficient algorithm for cardiac arrhythmia classification using the ensemble of depthwise separable convolutional neural networks. First, we optimized the beat segmentation by taking ECG samples centered around the R-peak. Second, we used all convolutional network, batch normalization, and depthwise separable convolution, to achieve the best accuracy while reducing the computational cost. Finally, we ensembled three depthwise separable CNNs by averaging three CNNs of 256 sample input size. Performance evaluation showed that our proposed algorithms achieved around 99.88% accuracy in 16 classes classification. The proposed two-stage ECG classification required around 180 μs, which can be implemented in a real time application. Future work will include the implementation of the current CNN on GPU to speed up its training, as well as to vary the input segment for various patients, the use of different databases, the use of other optimization methods, and the implementation in clinical application validated by a cardiologist.

Author Contributions: Formal analysis, E.I.; Funding acquisition, K.R.; Methodology, E.I.; Supervision, K.R.; Writing—original draft, E.I.; Writing—review and editing, K.R., D.S. and T.S.G. All authors have read and agreed to the published version of the manuscript.

Funding: This article publication is supported by Universitas Indonesia through Q1Q2 International Journal Publication Grant Scheme under the contract number NKB-0306/UN2.R3.1/HKP.05.00/2019.

Conflicts of Interest: The authors declare no conflict of interest.

References

1. Evans, G.F.; Shirk, A.; Muturi, P.; Soliman, E.Z. Feasibility of using mobile ECG recording technology to detect atrial fibrillation in low-resource settings. *Glob. Heart* **2017**, *12*, 285–289. [CrossRef] [PubMed]
2. Sun, W.; Zeng, N.; He, Y. Morphological Arrhythmia Automated Diagnosis Method Using Gray-Level Co-occurrence Matrix Enhanced Convolutional Neural Network. *IEEE Access* **2019**, *7*, 67123–67129. [CrossRef]
3. Laguna, P.; Cortés, J.P.M.; Pueyo, E. Techniques for ventricular repolarization instability assessment from the ECG. *Proc. IEEE* **2016**, *104*, 392–415. [CrossRef]
4. Satija, U.; Ramkumar, B.; Manikandan, M.S. A New Automated Signal Quality-Aware ECG Beat Classification Method for Unsupervised ECG Diagnosis Environments. *IEEE Sens. J.* **2018**, *19*, 277–286. [CrossRef]

5.   Lynn, H.M.; Pan, S.B.; Kim, P. A Deep Bidirectional GRU Network Model for Biometric Electrocardiogram Classification Based on Recurrent Neural Networks. *IEEE Access* **2019**, *7*, 145395–145405. [CrossRef]

6.   Chu, Y.; Shen, H.; Huang, K. ECG Authentication Method Based on Parallel Multi-Scale One-Dimensional Residual Network With Center and Margin Loss. *IEEE Access* **2019**, *7*, 51598–51607. [CrossRef]

7.   Kim, H.; Chun, S.Y. Cancelable ECG Biometrics Using Compressive Sensing-Generalized Likelihood Ratio Test. *IEEE Access* **2019**, *7*, 9232–9242. [CrossRef]

8.   Dokur, Z.; Olmez, T.; Yazgan, E. Comparison of discrete wavelet and Fourier transforms for ECG beat classification. *Electron. Lett.* **1999**, *35*, 1502–1504. [CrossRef]

9.   Banerjee, S.; Mitra, M. Application of cross wavelet transform for ECG pattern analysis and classification. *IEEE Trans. Instrum. Meas.* **2013**, *63*, 326–333. [CrossRef]

10.   Melgani, F.; Bazi, Y. Classification of electrocardiogram signals with support vector machines and particle swarm optimization. *IEEE Trans. Inf. Technol. Biomed.* **2008**, *12*, 667–677. [CrossRef]

11.   Venkatesan, C.; Karthigaikumar, P.; Paul, A.; Satheeskumaran, S.; Kumar, R. ECG signal preprocessing and SVM classifier-based abnormality detection in remote healthcare applications. *IEEE Access* **2018**, *6*, 9767–9773. [CrossRef]

12.   Chen, X.; Wang, Y.; Wang, L. Arrhythmia Recognition and Classification Using ECG Morphology and Segment Feature Analysis. *IEEE/ACM Trans. Comput. Biol. Bioinform.* **2018**, *16*, 131–138.

13.   Raj, S.; Ray, K.C. ECG signal analysis using DCT-based DOST and PSO optimized SVM. *IEEE Trans. Instrum. Meas.* **2017**, *66*, 470–478. [CrossRef]

14.   Pasolli, E.; Melgani, F. Active learning methods for electrocardiographic signal classification. *IEEE Trans. Inf. Technol. Biomed.* **2010**, *14*, 1405–1416. [CrossRef]

15.   Li, Z.; Feng, X.; Wu, Z.; Yang, C.; Bai, B.; Yang, Q. Classification of Atrial Fibrillation Recurrence Based on a Convolution Neural Network With SVM Architecture. *IEEE Access* **2019**, *7*, 77849–77856. [CrossRef]

16.   Zigel, Y.; Cohen, A.; Katz, A. The weighted diagnostic distortion (WDD) measure for ECG signal compression. *IEEE Trans. Biomed. Eng.* **2000**, *47*, 1422–1430.

17.   Gerencsér, L.; Kozmann, G.; Vago, Z.; Haraszti, K. The use of the SPSA method in ECG analysis. *IEEE Trans. Biomed. Eng.* **2002**, *49*, 1094–1101. [CrossRef]

18.   Rahman, Q.A.; Tereshchenko, L.G.; Kongkatong, M.; Abraham, T.; Abraham, M.R.; Shatkay, H. Utilizing ECG-based heartbeat classification for hypertrophic cardiomyopathy identification. *IEEE Trans. Nanobiosci.* **2015**, *14*, 505–512. [CrossRef]

19.   Sopic, D.; Aminifar, A.; Aminifar, A.; Atienza, D. Real-time event-driven classification technique for early detection and prevention of myocardial infarction on wearable systems. *IEEE Trans. Biomed. Circuits Syst.* **2018**, *12*, 982–992. [CrossRef]

20.   Lai, D.; Zhang, Y.; Zhang, X. An automated strategy for early risk identification of sudden cardiac death by using machine learning approach on measurable arrhythmic risk markers. *IEEE Access* **2019**, *7*, 94701–94716. [CrossRef]

21.   Rad, A.B.; Eftestol, T.; Engan, K.; Irusta, U.; Kvaloy, J.T.; Kramer-Johansen, J.; Wik, L.; Katsaggelos, A.K. ECG-based classification of resuscitation cardiac rhythms for retrospective data analysis. *IEEE Trans. Biomed. Eng.* **2017**, *64*, 2411–2418. [CrossRef] [PubMed]

22.   Fira, C.M.; Goras, L. An ECG signals compression method and its validation using NNs. *IEEE Trans. Biomed. Eng.* **2008**, *55*, 1319–1326. [CrossRef] [PubMed]

23.   Mar, T.; Zaunseder, S.; Martínez, J.P.; Llamedo, M.; Poll, R. Optimization of ECG classification by means of feature selection. *IEEE Trans. Biomed. Eng.* **2011**, *58*, 2168–2177. [CrossRef]

24.   Bouaziz, F.; Oulhadj, H.; Boutana, D.; Siarry, P. Automatic ECG arrhythmias classification scheme based on the conjoint use of the multi-layer perceptron neural network and a new improved metaheuristic approach. *IET Signal Process.* **2019**, *13*, 726–735. [CrossRef]

25.   Huang, J.; Chen, B.; Yao, B.; He, B. ECG arrhythmia classification using STFT-based spectrogram and convolutional neural network. *IEEE Access* **2019**, *7*, 92871–92880. [CrossRef]

26.   Mai, V.; Khalil, I.; Meli, C. ECG biometric using multilayer perceptron and radial basis function neural networks. In Proceedings of the 2011 Annual International Conference of the IEEE Engineering in Medicine and Biology Society, Boston, MA, USA, 30 August–3 September 2011.

27.   Kiranyaz, S.; Ince, T.; Gabbouj, M. Real-time patient-specific ECG classification by 1-D convolutional neural networks. *IEEE Trans. Biomed. Eng.* **2015**, *63*, 664–675. [CrossRef]

28. Nurmaini, S.; Partan, R.U.; Caesarendra, W.; Dewi, T.; Rahmatullah, M.N.; Darmawahyuni, A.; Bhayyu, V.; Firdaus, F. An Automated ECG Beat Classification System Using Deep Neural Networks with an Unsupervised Feature Extraction Technique. *Appl. Sci.* **2019**, *9*, 2921. [CrossRef]

29. Zhai, X.; Tin, C. Automated ECG classification using dual heartbeat coupling based on convolutional neural network. *IEEE Access* **2018**, *6*, 27465–27472. [CrossRef]

30. Springenberg, J.T.; Dosovitskiy, A.; Brox, T.; Riedmiller, M. Striving for simplicity: The all convolutional net. *arXiv* **2014**, arXiv:1412.6806.

31. Ioffe, S.; Szegedy, C. Batch normalization: Accelerating deep network training by reducing internal covariate shift. *arXiv* **2015**, arXiv:1502.03167.

32. Kaiser, L.; Gomez, A.N.; Chollet, F. Depthwise separable convolutions for neural machine translation. *arXiv* **2017**, arXiv:1706.03059,.

33. Polikar, R. Ensemble learning. In *Ensemble Machine Learning*; Springer: Berlin, Germany, 2012; pp. 1–34.

34. Chollet, F. Xception: Deep learning with depthwise separable convolutions. In Proceedings of the IEEE Conference on Computer Vision and Pattern Recognition, Honolulu, HI, USA, 21–26 July 2017.

35. Zhang, T.; Zhang, X.; Shi, J.; Wei, S. Depthwise Separable Convolution Neural Network for High-Speed SAR Ship Detection. *Remote Sens.* **2019**, *11*, 2483. [CrossRef]

36. Ince, T.; Kiranyaz, S.; Gabbouj, M. A generic and robust system for automated patient-specific classification of ECG signals. *IEEE Trans. Biomed. Eng.* **2009**, *56*, 1415–1426. [CrossRef] [PubMed]

37. Wen, C.; Lib, T.-C.; Chang, K.-C.; Huang, C.-H. Classification of ECG complexes using self-organizing CMAC. *Measurement* **2009**, *42*, 399–407. [CrossRef]

38. Rajpurkar, P.; Hannun, A.Y.; Rajpurkar, P.; Haghpanahi, M.; Tison, G.H.; Bourn, C.; Turakhia, M.P.; Ng, A.Y. Cardiologist-level arrhythmia detection with convolutional neural networks. *arXiv* **2017**, arXiv:1707.01836.

39. Pan, J.; Tompkins, W.J. A real-time QRS detection algorithm. *IEEE Trans. Biomed. Eng.* **1985**, *32*, 230–236. [CrossRef]

40. Li, C.; Zheng, C.; Tai, C. Detection of ECG characteristic points using wavelet transforms. *IEEE Trans. Biomed. Eng.* **1995**, *42*, 21–28.

41. Moody, G.B.; Mark, A.R.G. The impact of the MIT-BIH arrhythmia database. *IEEE Eng. Med. Biol. Mag.* **2001**, *20*, 45–50. [CrossRef]

42. Dupre, A.; Vincent, S.; Iaizzo, P.A. Basic ECG Theory, Recordings, and Interpretation. In *Handbook of Cardiac Anatomy, Physiology, and Devices*; Springer: Berlin, Germany, 2005; pp. 191–201.

43. Abadi, M.; Barham, P.; Chen, J.; Chen, Z.; Davis, A.; Dean, J.; Devin, M.; Ghemawat, S.; Irving, G.; Isard, M.; et al. Tensorflow: A system for large-scale machine learning. In Proceedings of the 12th USENIX Symposium on Operating Systems Design and Implementation (OSDI16), Savannah, GA, USA, 2–4 November 2016.

44. Chollet, F. Keras: Deep Learning Library for Theano and Tensorflow. 2015. Available online: https://keras.io/ (accessed on 20 May 2019).

45. Hu, Y.H.; Palreddy, S.; Tompkins, W.J. A patient-adaptable ECG beat classifier using a mixture of experts approach. *IEEE Trans. Biomed. Eng.* **1997**, *44*, 891–900.

46. Sarfraz, M.; Khan, A.A.; Li, F.F. Using independent component analysis to obtain feature space for reliable ECG Arrhythmia classification. In Proceedings of the 2014 IEEE International Conference on Bioinformatics and Biomedicine (BIBM), Belfast, UK, 2–5 November 2014.

47. Xia, Y.; Zhang, H.; Xu, L.; Gao, Z.; Zhang, H.; Liu, H.; Li, S. An automatic cardiac arrhythmia classification system with wearable electrocardiogram. *IEEE Access* **2018**, *6*, 16529–16538. [CrossRef]

48. Nanjundegowda, R.; Meshram, V.A. Arrhythmia Detection Based on Hybrid Features of T-wave in Electrocardiogram. *Int. J. Intell. Eng. Syst.* **2018**, *11*, 153–162. [CrossRef]

49. Rangappa, V.G.; Prasad, S.V.A.V.; Agarwal, A. Classification of Cardiac Arrhythmia stages using Hybrid Features Extraction with K-Nearest Neighbour classifier of ECG Signals. *Learning* **2018**, *11*, 21–32.

Article

Automatic Cephalometric Landmark Detection on X-ray Images Using a Deep-Learning Method

Yu Song [1,†], Xu Qiao [2,†], Yutaro Iwamoto [1] and Yen-wei Chen [1,*]

[1]   Graduate School of Information Science and Eng., Ritsumeikan University, Shiga 603-8577, Japan;
       gr0398ep@ed.ritsumei.ac.jp (Y.S.); yiwamoto@fc.ritsumei.ac.jp (Y.I.)
[2]   Department of Biomedical Engineering, Shandong University, Shandong 250100, China; qiaoxu@sdu.edu.cn
*   Correspondence: chen@is.ritsumei.ac.jp
†   These authors contributed equally to this work.

Received: 30 November 2019; Accepted: 6 February 2020; Published: 7 April 2020

Abstract: Accurate automatic quantitative cephalometry are essential for orthodontics. However, manual labeling of cephalometric landmarks is tedious and subjective, which also must be performed by professional doctors. In recent years, deep learning has gained attention for its success in computer vision field. It has achieved large progress in resolving problems like image classification or image segmentation. In this paper, we propose a two-step method which can automatically detect cephalometric landmarks on skeletal X-ray images. First, we roughly extract a region of interest (ROI) patch for each landmark by registering the testing image to training images, which have annotated landmarks. Then, we utilize pre-trained networks with a backbone of ResNet50, which is a state-of-the-art convolutional neural network, to detect each landmark in each ROI patch. The network directly outputs the coordinates of the landmarks. We evaluate our method on two datasets: ISBI 2015 Grand Challenge in Dental X-ray Image Analysis and our own dataset provided by Shandong University. The experiments demonstrate that the proposed method can achieve satisfying results on both SDR (Successful Detection Rate) and SCR (Successful Classification Rate). However, the computational time issue remains to be improved in the future.

Keywords: cephalometric landmark; X-ray; deep learning; ResNet; registration

---

1. Introduction

Cephalometric analysis is performed on skeletal X-ray images. This is necessary for doctors to make orthodontic diagnoses [1–3]. In cephalometric analysis, the first step is to detect landmarks in X-ray images. Experienced doctors are needed to identify the locations of the landmarks. Measurements of the angles and distances between these landmarks greatly assist diagnosis and treatment plans. The work is time-consuming and tedious, and the problem of intra-observer variability arises since different doctors may differ considerably in their identification of landmarks [4].

Under these circumstances, computer-aided detection, which can automatically identify landmarks objectively, is highly desirable. Studies on anatomical landmark detection have a long history. In 2014 and 2015, International Symposium on Biomedical Imaging (ISBI) launched challenges on cephalometry landmark detection [5,6], and several state-of-the-art methods were proposed. In 2015's challenge, Lindner and Cootes's method based on random-forest achieved the highest results, with a successful detection rate of 74.84% for a 2-mm precision range [7] (since landmark location cannot be exactly same with manual annotation, Errors within a certain range are acceptable. Usually, medically acceptable range is 2 mm). Ibragimov et al. achieved the second highest result in 2015's challenge, with a successful detection accuracy of 71.7% within a 2-mm range [8]. They applied Haar-like feature extraction with a random-forest regression, then making refinements with a global-context shape model. In 2017, Ibragimov et al. added a convolutional neural network for binary classification to their conventional method. They surpass the result of Lindner's a little bit, with around 75.3% prediction accuracy within a 2-mm range [9]. In 2017, Hansang Lee et al. proposed a deep learning based method which achieved not bad results but in a small resized image. He trained two networks to regress the landmark's x and y coordinate directly [10]. In 2019, Jiahong Qian et al. proposed a new architecture called CephaNet which improves the architecture of Faster R-CNN [11,12]. Its accuracy is nearly 6% higher than other conventional methods.

Despite the variety of techniques available, automatic cephalometric landmark detection remains insufficient due to its limited accuracy. In recent years, deep learning has gained attention for its success in computer vision field. For example, convolutional neural network models are widely used in problems like landmark detection [13,14], image classification [15–17] and image segmentation [18,19]. Trained models' performances surpass that of human beings in many applications. Since direct regression of several landmarks is a highly non-linear mapping, which is difficult to learn [20–23]. In our method, we only try to detect one key point in one patch image. We learn a non-linear mapping function for only one key point. Each key point has its corresponding non-linear mapping function. So we can achieve more accurate detection of key point than other methods.

In this paper, we propose a two-step method for the automatic detection of cephalometric landmarks. First, we get the coarse landmark location by registering the test image to a most similar image in the training set. Based on the registration result, we extract an ROI patch centered at the coarse landmark location. Then, by using our pre-trained network with the backbone of ResNet50, which is a state-of-the-art convolutional neural network, we detect the landmarks in the extracted ROI patches to make refinements. The experiments demonstrate that our two-step method can achieve satisfying results compared with other state-of-the-art methods.

The remainder of this paper is organized as follows: We describe the study materials in Section 2. In Section 3, we explain the proposed method in detail. Then, in Section 4, we describe the experiments and discuss the results. Lastly, we present the conclusion in Section 5.

2. Materials

2.1. Training Dataset

The training dataset is the 2015 ISBI Grand Challenge training dataset [6]. It contains 150 images, each of which is 1935 × 2400 pixels in TIFF format. Each pixel's size is 0.1 × 0.1 mm. The images were labeled by two experienced doctors. We use the average coordinate value of the two doctors as ground-truth for training.

We aim to detect 19 landmarks for each X-ray image, as presented in Figure 1, which are as follows: sella turcica, nasion, orbitale, porion, subspinale, supramentale, pogonion, menton, gnathion, gonion, lower incisal incision, upper incisal incision, upper lip, lower lip, subnasale, soft tissue pogonion, posterior nasal spine, anterior nasal spine, and articulate.

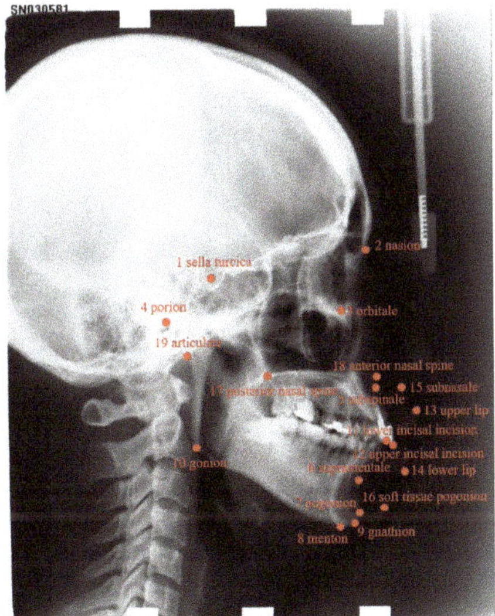

Figure 1. All 19 landmarks on x-ray images which we aim to detect.

Data Augmentation

The size of an X-ray image is 1935 × 2400 pixels. The scale is too huge to input the whole image into a CNN to detect the locations of all 19 landmarks simultaneously. Therefore, we decide to detect landmarks in small patches. In each patch, only one landmark will be detected. We train one model for one landmark. Overall, we have 19 landmarks since there are 19 landmarks to be detected. These models have the same architecture but different weights, as presented in Figure 2.

In order to do data augmentation, we select 400 patches randomly for each landmark in one training image. Each patch is 512 × 512 pixels. The landmark could be everywhere in the patch. Its coordinate (x,y) in the ROI patch is used as teaching signal (ground truth) for training. Then, we resize them into 256 × 256 patches, as presented in Figure 3. This means we will have 60,000 (400 × 150) images to train one model for landmark i (i = 1, 2, ... 19).

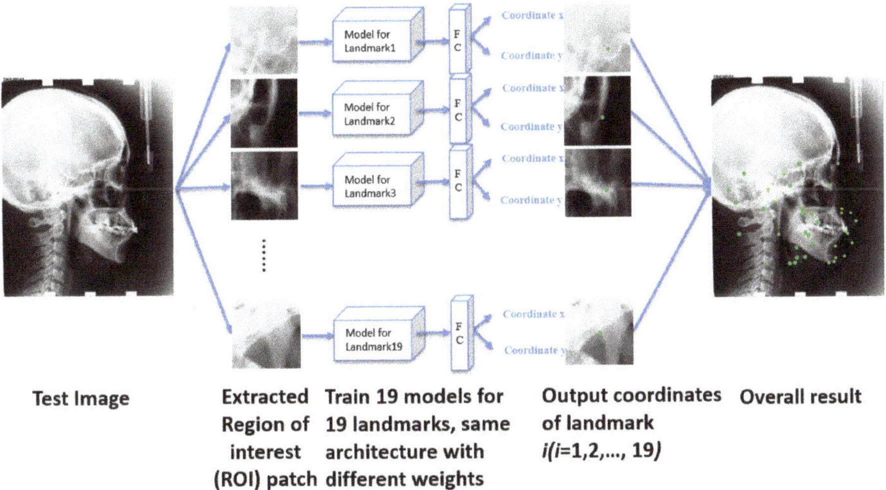

Figure 2. We train 19 networks with same architecture but with different weights.

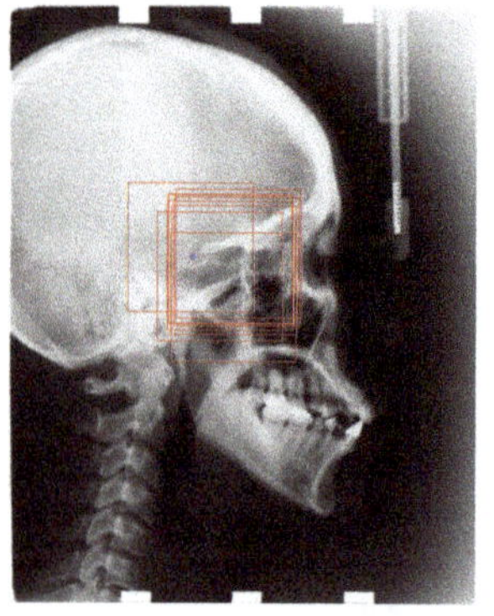 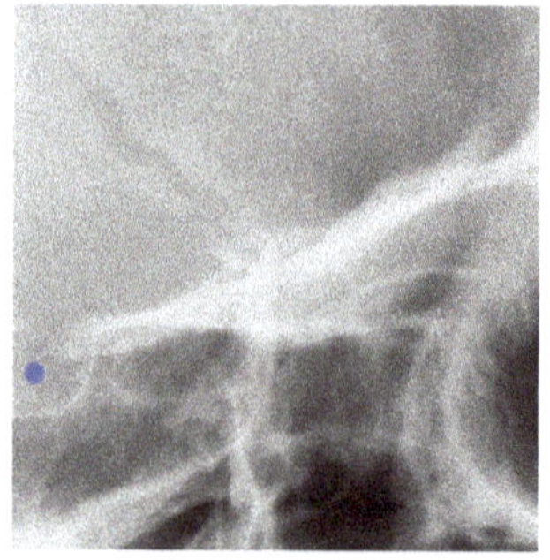 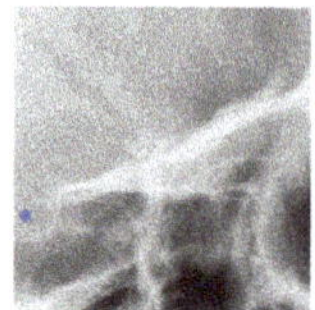

(a) Example of our training image, the red boxes are cropped training images.

(b) The cropped 512 × 512 patch

(c) The resized 256 × 256 patch

Figure 3. The way we crop ROI patches for each landmark: the blue dot is a target landmark, the red boxes are extracted 512 × 512 ROI patches. We can see the landmark is included in all the ROI patches. We randomly extract 400 patches for one landmark in one x-ray image for training. Then we resize the ROI patches into 256 × 256 to make predictions.

2.2. Test Dataset

We have two test datasets. Both datasets are tested through the models, which are trained with the ISBI training dataset.

2.2.1. International Symposium on Biomedical Imaging (ISBI) Test Dataset

The first test dataset from the ISBI Grand Challenge includes two parts: Test Dataset 1 with 150 test images and test dataset 2 with 100 images. The ISBI test images are collected with the same machine as the training data, and each image is 1935 × 2400 in TIFF format.

2.2.2. Our Dataset

Our own dataset is provided by Shandong University, China. In contrast to the ISBI dataset, the data are taken with different machines. There are 100 images, each of which is 1804 × 2136 in JPEG format.

Also, the landmarks labeled by the doctors are not exactly the same as those in the ISBI dataset (13 of them are same). We used labels for sella turcica, nasion, orbitale, porion, subspinale, supramentale, pogonion, menton, gnathion, gonion, lower incisal incision, upper incisal incision, and anterior nasal spine. Other landmarks were not labeled by the doctors.

2.2.3. Region Of Interest (ROI)

Extracted ROI patches are used as input images in our corresponding models. The method of extraction is described in detail in Section 3.

3. Proposed Method

3.1. Overview

The method is a two-step method: ROI extraction and Landmark detection. For ROI extraction, we crop patches by registering the test image to training images, which have annotated landmarks. Then, we use pre-trained CNN models with the backbone of ResNet50, which is a state-of-the-art CNN, to detect the landmarks in the extracted ROI patches. The procedure is presented in Figure 4.

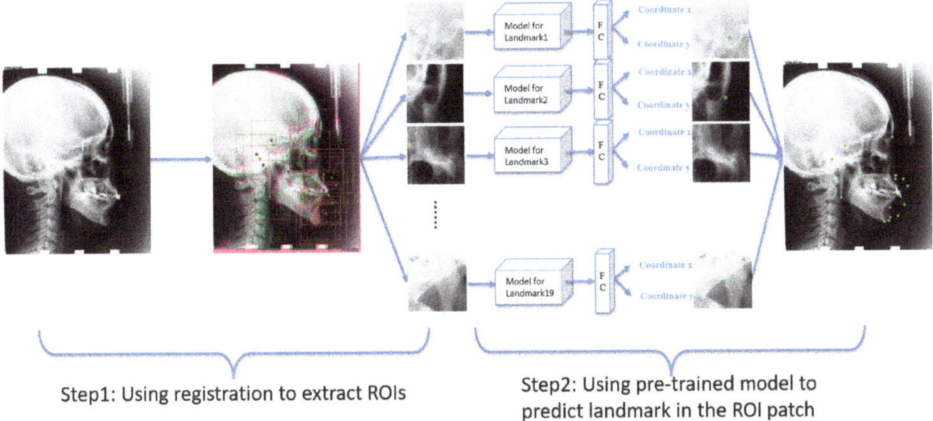

Figure 4. The overview of our two-step method.

3.2. Automatic ROI Extraction

Because of the limitation of the input image size, we aim to extract small ROI patches automatically. That is, we extract ROI patches that include landmark $i(i = 1,2,3\ldots19)$. To extract the ROI automatically, we use registration to locate a coarse location of the landmark. Next, we extract the ROI patch centered on the coarse location. Thus, the landmark location we aim to detect will be included in the ROI patch.

In this process, the test image is registered to training images, which have annotated landmarks. We consider the test images as moving images and the training images as reference images. The type of transform that we used is translation transform. The optimizer is gradient descent, which we used to update the locations of the moving images. We apply the intensity-based registration method and calculate the mean squared error (MSE) between the reference image and moving image.

After registration, we copy the landmark locations of the reference images (training images) to moving images (test images). We consider the landmark location of a reference image as the center of the ROI patch, extract a 512×512 resolution patch image on the moving image, and then resize it to a 256×256 resolution patch for the test. The extracted images will be treated as input to our trained convolutional neural network (CNN); thus, we can detect the corresponding landmark in the patch image. The registration results are presented in Figure 5.

Since people's heads vary in shape, randomly choosing one image among the training images as a reference image is not sufficient, and the landmark to be detected will not be included in the ROI. To avoid this situation, we iterate all the training images, which means we perform registration 150 times for one test image (the total number of training images is 150). Then, as the reference image, we choose the training image that has the smallest square error with test images. This enables us to find the closest training sample to the test images. For computational time, different references and moving images vary a lot. The shortest only took a few seconds for one registration, while the longest could take more than one minute. In all, the average time for registering one image to all training samples is around twenty minutes. The procedure is presented in Figure 6.

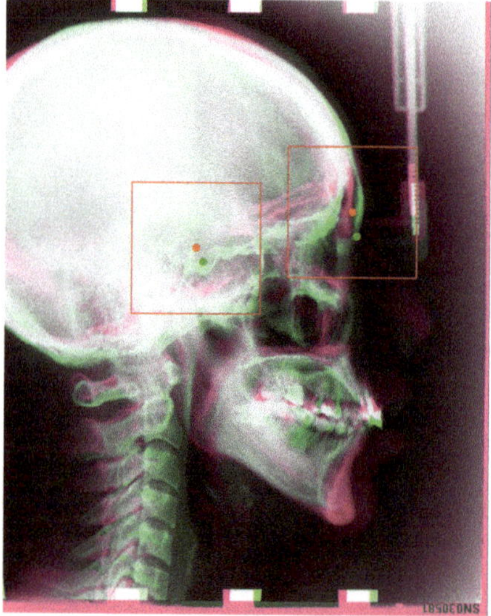

Figure 5. In the image, we take the reference image's landmarks(Training images) as center of the patch image, then we draw a bounding box (ROI) in the test image, as the input to our corresponding trained models. Red dots are the reference images' landmarks, green dots are test image's landmarks, which are the landmarks we aim to detect.

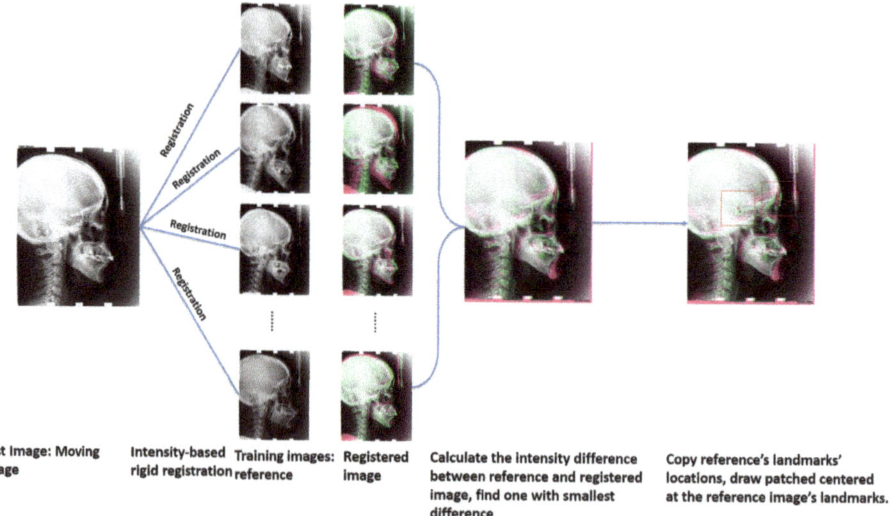

Figure 6. The architecture of Automatic registration.

3.3. Landmark Detection

3.3.1. Network Architecture

We use the ResNet50 architecture [24] as our backbone to extract features in ROI patches. CNN can extract useful features automatically for different computer vision tasks. We use regression method to estimate the landmark location after feature extraction. In our experiment, we choose to use fully connected layer for the estimation of landmark location as a regression problem. We first flatten all the features. Then, we add one fully connected layer, which directly outputs the coordinate of the landmark in the patch.

The state-of-the-art architecture ResNet is widely used for its unique architecture. It has convolutional blocks and identity blocks. By adding former information to the later layers, ResNet solves the gradient vanishing problems effectively. The overall architecture of ResNet50 is presented in Figure 7.

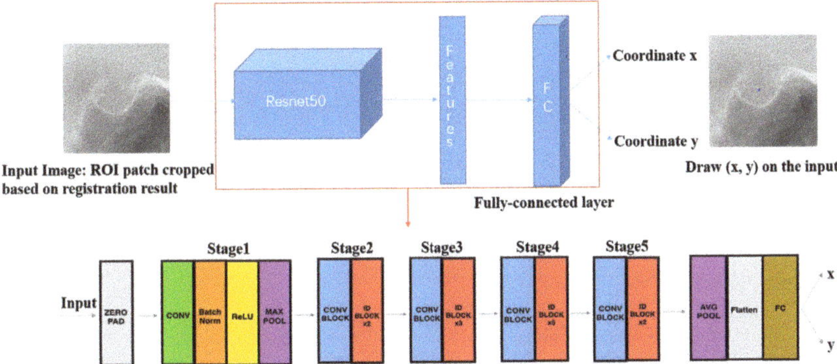

Figure 7. The architecture of ResNet50, the identity block and convolutional block add the former input to the later output. For identity block, it directly adds the former input to the later output. While for convolutional block, before adding, the former input first goes through a convolutional layer to make it the same size with later output.

3.3.2. Cost function

The cost function we used during training is mean squared error (MSE), since we want to make the predicted landmark become as close as possible to the ground-truth landmark location during training. In this case, MSE can be written as in Equation (1):

$$MSE = \frac{1}{n} \sum_{i=1}^{n} ((x_i - \hat{x}_i)^2 + (y_i - \hat{y}_i)^2) \tag{1}$$

where x_i and y_i indicate the ground-truth coordinate, and $\hat{x}_i$ and $\hat{y}_i$ indicate the estimated coordinate.

4. Experiments and Discussion

4.1. Implementation

For the implementation of our experiments, we use Titan X GPU (12 GB) to help accelerating training. We use Python programming language along with deep learning tools Tensorflow and Keras in our experiments. We use Adam optimizer and set learning rate to 0.001. Ten percent of our training data (60,000 × 10% = 6000) are separated for validation during training procedure and early stopping is applied to find the best weights.

4.2. Evaluation measures

According to the 2015 ISBI challenges on cephalometry landmark detection [6], we use the mean radial error (MRE) to make evaluation. The radial error is defined in Equation (2). Δx and Δy are the absolute differences of the estimated coordinate and ground-truth coordinate of x-axis and y-axis.

$$R = \sqrt{\Delta x^2 + \Delta y^2} \tag{2}$$

The mean radial error (MRE) is defined in Equation (3):

$$MRE = \frac{\sum_{i=1}^{n} R_i}{n} \tag{3}$$

Because the predicted results have some difference with the ground-truth. If the difference is within some range, we consider it correct in that range. In our experiment, we evaluate the range of z mm (where $z = 2, 2.5, 3, 4$). For example, if the radial error is 1.5 mm, we consider it as a 2 mm success. Equation (4) explains the meaning of successful detection rate (SDR), where N_a indicates the number of accurate detections, and N indicates total number of detections.

$$SDR = \frac{N_a}{N} * 100\% \tag{4}$$

4.3. Landmark Detection Results

4.3.1. ISBI Test Dataset

The results for ISBI public data are presented in Tables 1 and 2. Table 1 presents the SDR, and Table 2 presents the MRE results.

Table 1. Successful Detection Rate (SDR) Results on Testset1 and Testset2 of range 2 mm, 2.5 mm, 3 mm, 4 mm.

	SDR (Successful Detection Rate)							
Anatomical Landmarks	**2 mm (%)**		**2.5 mm (%)**		**3 mm (%)**		**4 mm (%)**	
	Testset1	**Testset2**	**Testset1**	**Testset2**	**Testset1**	**Testset2**	**Testset1**	**Testset2**
1. sella turcica	98.7	99.0	98.7	99.0	99.3	99.0	99.3	99.0
2. nasion	86.7	89.0	89.3	90.0	93.3	92.0	96.0	99.0
3. orbitale	84.7	36.0	92.0	61.0	96.0	80.0	99.3	96.0
4. porion	68.0	78.0	78.0	81.0	84.7	86.0	93.3	93.0
5. subspinale	66.0	84.0	76.0	94.0	87.3	99.0	95.3	100.0
6. supramentale	87.3	27.0	93.3	43.0	97.3	59.0	98.7	82.0
7. pogonion	93.3	98.0	97.3	100.0	98.7	100.0	99.3	100.0
8. menton	94.0	96.0	98.0	98.0	98.0	99.0	99.3	100.0
9. gnathion	94.0	100.0	98.0	100.0	98.0	100.0	99.3	100.0
10. gonion	62.7	75.0	75.3	87.0	84.7	94.0	92.0	98.0
11. lower incisal incision	92.7	93.0	97.3	96.0	97.3	97.0	99.3	99.0
12. upper incisal incision	97.3	95.0	98.0	98.0	98.7	98.0	99.3	98.0
13. upper lip	89.3	13.0	98.0	35.0	99.3	69.0	99.3	93.0
14. lower lip	99.3	63.0	99.3	80.0	100.0	93.0	100.0	98.0
15. subnasale	92.7	92.0	95.3	94.0	96.7	95.0	99.3	99.0
16. soft tissue pogonion	86.7	4.0	92.7	7.0	95.3	13.0	99.3	41.0
17. posterior nasal spine	94.7	91.0	97.3	98.0	98.7	100.0	99.3	100.0
18. anterior nasal spine	87.3	93.0	92.0	96.0	95.3	97.0	98.7	100.0
19. articulate	66.0	80.0	76.0	87.0	83.0	93.0	91.3	97.0
Average:	86.4	74.0	91.7	81.3	94.8	87.5	97.8	94.3

Table 2. Results on Testset1 and Testset2 of mean radial error.

Anatomical Landmarks	Mean Radial Error (mm)	
	Testset1	**Testset2**
1. sella turcica	0.577	0.613
2. nasion	1.133	0.981
3. orbitale	1.243	2.356
4. porion	1.828	1.702
5. subspinale	1.752	1.253
6. supramentale	1.136	2.738
7. pogonion	0.836	0.594
8. menton	0.952	0.705
9. gnathion	0.952	0.594
10. gonion	1.817	1.431
11. lower incisal incision	0.697	0.771
12. upper incisal incision	0.532	0.533
13. upper lip	1.332	2.841
14. lower lip	0.839	1.931
15. subnasale	0.853	0.999
16. soft tissue pogonion	1.173	4.265
17. posterior nasal spine	0.819	1.028
18. anterior nasal spine	1.157	0.995
19. articulate	1.780	1.310
Average:	1.077	1.542

We compare our detection results with other benchmarks' results. Table 3 shows the comparison.

Table 3. SDR of the proposed method compared with other benchmarks for ISBI 2015 challenges on cephalometry landmark detection Testset1 and Testset2.

	Comparisons of SDR			
Method	**2 mm (%)**	**2.5 mm (%)**	**3 mm (%)**	**4 mm (%)**
Testset1				
Ibrgimov(2015) [8]	71.7	77.4	81.9	88.0
Lindner [7]	73.7	80.2	85.2	91.5
Ibragimov [9]	75.4	80.9	84.3	88.3
Jiahong Qian [11]	82.5	86.2	89.3	90.6
Proposed:	**86.4**	**91.7**	**94.8**	**97.8**
Testset2				
Ibrgimov [8]	62.7	70.5	76.5	85.1
Lindner [7]	66.1	72.0	77.6	87.4
Ibragimov [9]	67.7	74.2	79.1	84.6
Jiahong Qian [11]	72.4	76.2	80.0	85.9
Proposed:	**74.0**	**81.3**	**87.5**	**94.3**

Figure 8 shows one of the examples of our results.

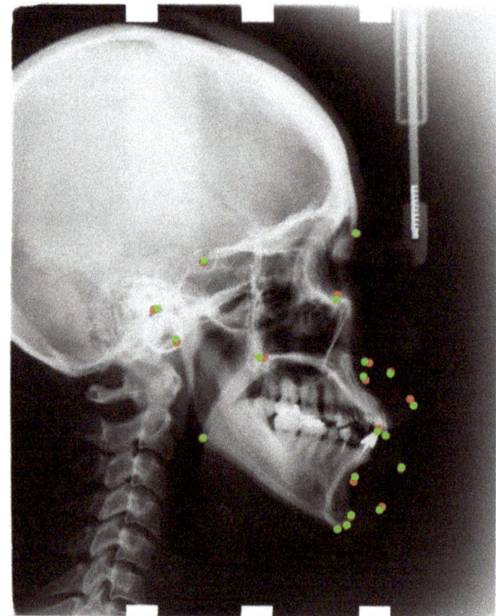

Figure 8. Green points: predicted landmarks' locations. Red points: ground-truth of landmarks' locations.

4.3.2. Our Dataset

We use our trained model(use ISBI training dataset) to test our own dataset directly. We also register all images first. Since there are some different landmarks labeled in our own dataset, in the experiment, we only test same parts' landmarks with ISBI dataset. The results are presented in Tables 4 and 5.

Table 4. Results of 2 mm, 2.5 mm, 3 mm and 4 mm accuracy on our own dataset, since we do not have label for landmark 13, 14, 15, 16, 17, 19, we test other landmarks exclude these ones.

Our own data from Shandong University			
Anatomical Landmarks	**2 mm (%)**	**2.5 mm (%)**	**3 mm (%)**	**4 mm (%)**
---	---	---	---	---
1. sella turcica	68.0	71.0	76.0	82.0
2. nasion	83.0	90.0	95.0	99.0
3. orbitale	40.0	58.0	74.0	88.0
4. porion	48.0	56.0	66.0	72.0
5. subspinale	32.0	39.0	50.0	67.0
6. supramentale	33.0	43.0	55.0	71.0
7. pogonion	68.0	77.0	81.0	90.0
8. menton	82.0	86.0	95.0	99.0
9. gnathion	91.0	97.0	99.0	100.0
10.gonion	51.0	63.0	75.0	86.0
11.lower incisal incision	68.0	76.0	78.0	90.0
12.upper incisal incision	75.0	86.0	90.0	97.0
18.anterior nasal spine	67.0	74.0	81.0	85.0
Average:	62.0	70.5	78.1	86.6

Table 5. Results of Mean Radial Error on our own dataset.

Our own data from Shandong University	
Anatomical Landmarks	**Mean Radial Error (mm)**
1. sella turcica	2.4
2. nasion	1.4
3. orbitale	2.5
4. porion	3.0
5. subspinale	3.6
6. supramentale	3.3
7. pogonion	1.7
8. menton	1.2
9. gnathion	1.2
10.gonion	2.4
11.lower incisal incision	1.8
12.upper incisal incision	1.3
18.anterior nasal spine	2.0
Average:	2.1

4.4. Skeletal Malformations Classification

For the ISBI Grand Challenge, the class labels of skeletal malformations based on ANB, SNB, SNA, ODI, APDI, FHI, FMA, and MW were also annotated for each subject. The skeletal malformation classifications for ANB, SNB, SNA, ODI, APDI, FHI, FMA, and MW are defined in Table 6 and the definitions (calculations) of ANB, SNB, SNA, ODI, APDI, FHI, FMA, and MW are summarized as follows:

1. ANB: The angle between Landmark 5, Landmark 2 and Landmark 6
2. SNB: The angle between Landmark 1, Landmark 2 and Landmark 6
3. SNA: The angle between Landmark 1, Landmark 2 and Landmark 5
4. ODI (Overbite Depth Indicator): Sum of the angle between the lines from Landmark 5 to Landmark 6 and from Landmark 8 to Landmark 10 and the angle between the lines from Landmark 3 to Landmark 4 and from Landmark 17 to Landmark 18
5. APDI (AnteroPosterior Dysplasia Indicator): Sum of the angle between the lines from Landmark 3 to Landmark 4 and from Landmark 2 to Landmark 7, the angle between the lines from Landmark 2 to Landmark 7 and from Landmark 5 to Landmark 6 and the angle between the lines from Landmark 3 to Landmark 4 and from Landmark 17 to Landmark 18
6. FHI: Obtained by the ratio of the Posterior Face Height (PFH = the distance from Landmark 1 to Landmark 10) to the Anterior Face Height (AFH = the distance from Landmark 2 to Landmark 8). FHI = PFH/AFH.
7. FMA: The angle between the line from Landmark 1 to Landmark 2 and the line from Landmark 10 to Landmark 9.
8. Modify-Wits (MW): The distance between Landmark 12 and Landmark 11.

$$MW = \sqrt{(x_{L12} - x_{L11})^2 + (y_{L12} - y_{L11})^2}.$$

Table 6. Classifications for 8 methods, ANB, SNB, SNA, ODI, APDI, FHI, FMA and MW.

	ANB	SNB	SNA	ODI	APDI	FHI	FMA	MW
Class1 (C1)	3.2°–5.7°	74.6°–78.7°	79.4°–83.2°	68.4°–80.5°	77.6°–85.2°	0.65–0.75	26.8°–31.4°	2–4.5 mm
Class2 (C2)	>5.7°	<74.6°	>83.2°	>80.5°	<77.6°	>0.75	>31.4°	=0 mm
Class3 (C3)	<3.2°	>78.7°	<79.4°	<68.4°	>85.2°	<0.65	<26.8°	<0 mm
Class4 (C4)	–	–	–	–	–	–	–	>4.5 mm

The Successful Classification Rate (SCR) for ANB, SNB, SNA, ODI, APDI, FHI, FMA, and MW with the detected landmarks are presented in Table 7 (We run the official python program offered by ISBI Grand Challenge [6] to automatically get the SCR results). To make a comparison, we also show the SCR by the state-of-the-art methods in Table 7. As presented in Table 7, some of the SCRs of the proposed method are much higher than those of the existing methods (eg: ANB, SNA, ODI, FHI). For SNB measure, Lindner [7] is 0.1% percent higher than our result.

Table 7. Classification results for 8 measures, ANB, SNB, SNA, ODI, APDI, FHI, FMA and MW. We used junior doctors' annotations as ground-truth value.

	SCR (%)							
	ANB	SNB	SNA	ODI	APDI	FHI	FMA	MW
Ibrgimov(2015) [8]	68.0	78.7	56.7	77.3	83.3	76.0	81.3	84.0
Lindner [7]	71.3	**83.3**	60.0	80.0	83.3	77.3	81.3	85.3
Ibragimov [9]	72.9	73.7	64.1	75.9	84.8	79.7	**82.0**	80.2
Proposed:	**82.0**	83.2	**75.2**	**85.2**	**86.4**	**86.0**	82.0	**86.8**

4.5. Discussion

For registration, since people's heads vary in shape, even though we selected the closest image to the training data as the reference image for each test image, there are still missed situations. This means that after the registration, the patch we created for the test does not include the ground-truth landmark. For the ISBI dataset, there is only one missed patch, and the rate is about 0.0002 (1/(19 × 250)), as presented in Figure 9. Overall, it has little impact on the results. For Testset2 of the ISBI Grand Challenge, we can see that Landmark 3, Landmark 6, Landmark 13, and Landmark 16's have relatively low accuracy. However, the process works fine on Testset1. After we visualize the testing result, we find out that the anatomy of those failed cases are very different from the successfully detected ones. Since our training dataset has only 150 images, even though we did data augmentation, we still cannot deny the fact that people's skeletal anatomies are so different that many types are not included in the training dataset. When we use our trained models to make predictions on those with huge anatomy differences, we will achieve unsatisfying results. We think this explains the poor performance on those landmarks in Testset2. We show an example in Figure 10.

At first, we try to input all the X-ray images to predict all 19 landmarks simultaneously. We performed conventional data augmentation on training data, such as rotation, cropping, and so on. However, the result is not satisfying; although the location of all the landmarks is very similar to the ground truth, the predicted results are very far from the ground-truth result, as presented in Figure 11.

When we performed skeletal malformation classification based on the detected result, we found out that the result varies a lot when we use the annotation data of a different doctor as the ground truth (There are two annotation data made by two doctors (junior doctor and senior doctor) in ISBI Grand Challenge 2015; because their landmark locations are different, this results in the variability of the skeletal malformation classification. Table 8 presents the proportion that two doctors have the same classification label.). The result we showed previously was based on the annotation data of the junior doctor. If we used the annotation data of the senior doctor as the ground truth when performing classification, the SCR would be lower. It proves that annotation data have great influence on the results. During training, we use the average of the annotated landmark locations of the two doctors as the ground truth. When we use only the senior doctor's annotated landmark locations as the ground truth during the training, the SDR would also drop, which also proves that the annotation data greatly affects the final results.

Table 8. The proportion that two doctors made the same classification label over 400 images (training and test images).

Proportion (%)							
ANB	**SNB**	**SNA**	**ODI**	**APDI**	**FHI**	**FMA**	**MW**
78.0	82.0	67.75	77.0	82.25	84.25	82.0	87.5

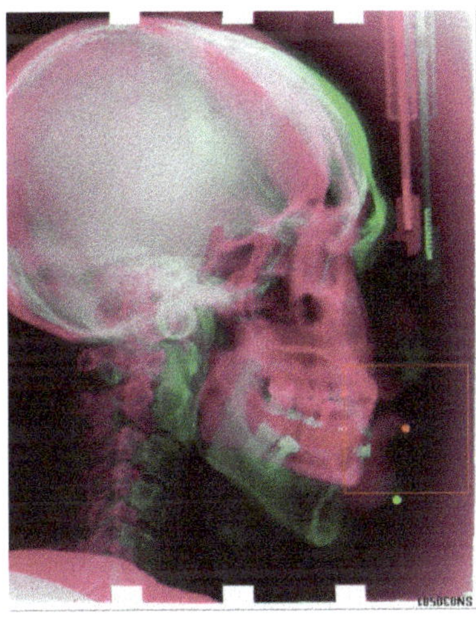

Figure 9. Green point: test image's landmark location. Red point: reference image's landmark location. We can see that the green dot is not included in the bounding box we cropped.

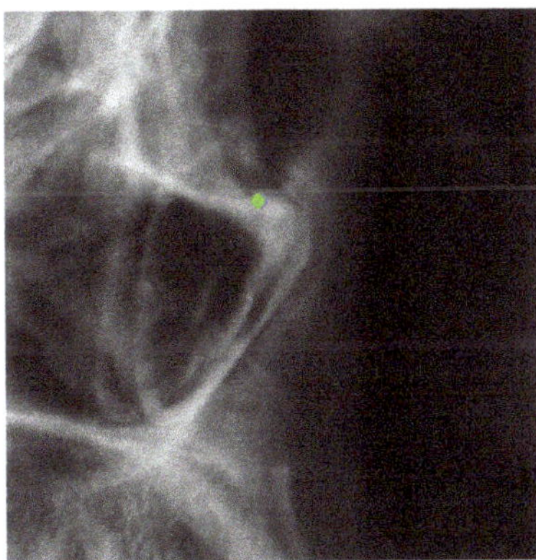
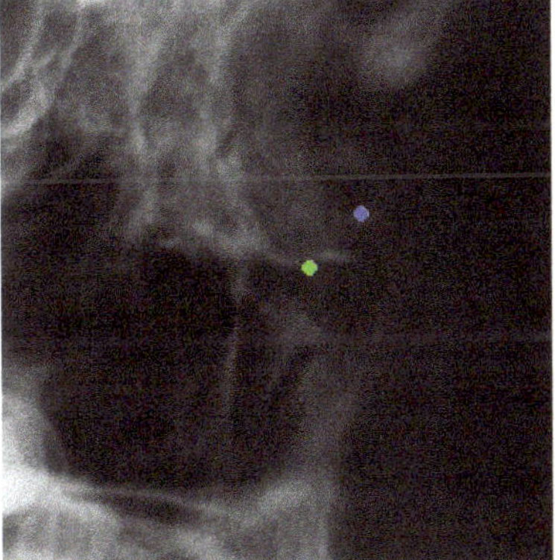

Figure 10. Successful detection example from Testset1 and failed detection example from Testset2 on Landmark 3. Green points: predicted result. Blue points: ground-truth location.

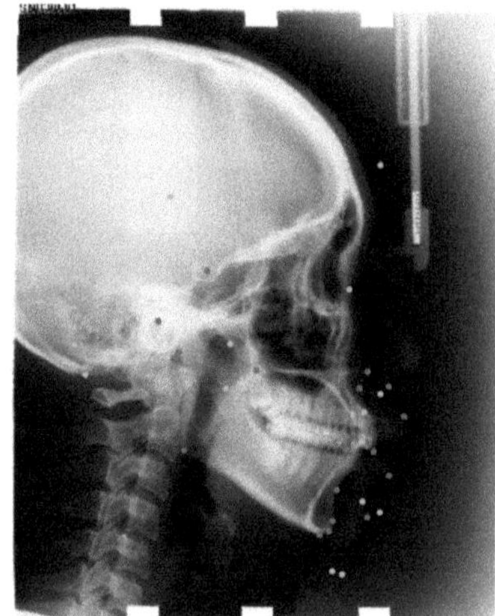

Figure 11. Green points: predicted landmarks' locations. Red points: ground-truth locations of landmarks. We can see the result is not satisfying at all.

5. Conclusions

We proposed an approach to automatically predict landmark location and used a deep learning method with very small training data, with only 150 X-ray training images. The results outperformed other benchmarks' results in all SDR and nearly all SCR, which proves that the proposed method are effective for cephalometric landmark detection. The proposed method could be used for landmark detection in clinical practice under the supervision of doctors.

For future work, we are developing a GUI for doctors to use. The system offers either automatic or semi-automatic landmark detection. In the automatic mode, the system will automatically extract an ROI (based on registration) and select a proper model for each landmark. In the semi-automatic mode, the doctor needs to give a bounding box extract ROI manually and select a corresponding model for each landmark detection, which can reduce computational time. Since we used simple registration in this study and it took a long time to register one test image with all the training images, around twenty minutes as we mentioned before. Under such case, we aim to design a better method that reduces the registration time in testing phase, to make the automatic detection more efficient. For example, maybe we will also utilize deep learning method, either to do registration or just simply to regress a coarse landmark region, to replace the rigid registration we used in this paper. Moreover, because we detected all the landmarks in patch images, we did not take global-context information (i.e., the relationship among all the landmarks) into consideration. In future work, we want to utilize this global-context information to check whether we can achieve better performance. For example, maybe we will try to change the network architecture to allow the network to utilize whole image's features to better and faster locate the coarse locations of landmarks.

Author Contributions: Conceptualization, Y.S.; Data curation, Y.S. and X.Q.; Formal analysis, Y.S.; Funding acquisition, X.Q. and Y.-w.C.; Investigation, Y.S.; Methodology, Y.S., X.Q., Y.I. and Y.-w.C.; Project administration, X.Q.; Supervision, Y.I. and Y.-w.C.; Validation, Y.S.; Writing—original draft, Y.S.; Writing—review and editing, Y.S., X.Q., Y.I. and Y.-w.C. All authors have read and agreed to the published version of the manuscript.

Funding: This work was supported in part by the National Natural Science Foundation of China under the Grant No. 61603218, and in part by the Grant-in Aid for Scientific Research from the Japanese Ministry for Education, Science, Culture and Sports (MEXT) under the Grant No. 18H03267 and No. 17K00420.

Conflicts of Interest: The authors declare no conflict of interest.

Appl. Sci. **2020**, *10*, 2547

Abbreviations

The following abbreviations are used in this manuscript:

GUI   The graphical user interface
SCR   Successful Classification Rate
SDR   Successful Detection Rate
MRE   Mean squared error
ROI   Region of interest

References

1. Kaur, A.; Chandan Singh, T. Automatic cephalometric landmark detection using Zernike moments and template matching. *Signal Image Video Process.* **2015**, *29*, 117–132. [CrossRef]
2. Grau, V.; Alcaniz, M.; Juan, M.C.; Monserrat, C.; Knol, T. Automatic localization of cephalometric landmarks. *J. Biomed. Inform.* **2001**, *34*, 146–156. [CrossRef]
3. Ferreira, J.T.L.; Carlos de Souza Telles, T. Evaluation of the reliability of computerized profile cephalometric analysis. *Braz. Dent. J.* **2002**, *13*, 201–204. [CrossRef] [PubMed]
4. Weining, Y.; Yin, D.; Li, C.; Wang, G.; Tianmin, X.T. Automated 2-D cephalometric analysis on X-ray images by a model-based approach. *IEEE Trans. Biomed. Eng.* **2006**, *13*, 1615–1623. [CrossRef] [PubMed]
5. Ching-Wei, W.; Huang, C.; Hsieh, M.; Li, C.; Chang, S.; Li, W.; Vandaele, R. Evaluation and comparison of anatomical landmark detection methods for cephalometric X-ray images: A grand challenge. *IEEE Trans. Biomed. Eng.* **2015**, *53*, 1890–1900.
6. Wang, C.W.; Huang, C.T.; Lee, J.H.; Li, C.H.; Chang, S.W.; Siao, M.J.; Lai, T.M.; Ibragimov, B.; Vrtovec, T.; Ronneberger, O.; et al. A benchmark for comparison of dental radiography analysis algorithms. *IEEE Trans. Biomed. Eng.* **2016**, *31*, 63–76. [CrossRef] [PubMed]
7. Lindner, C.; Tim, F.; Cootes, T. Fully automatic cephalometric evaluation using Random Forest regression-voting. In Proceedings of the IEEE International Symposium on Biomedical Imaging (ISBI), Brooklyn Bridge, NY, USA, 16–19 April 2015.
8. Ibragimov, B.; Boštjan, L.; Pernus, F.; Tomaž Vrtovec, T. Computerized cephalometry by game theory with shape-and appearance-based landmark refinement. In Proceedings of the IEEE International Symposium on Biomedical Imaging (ISBI), Brooklyn Bridge, NY, USA, 16–19 April 2015.
9. Arik, S.Ö.; Bulat, I.; Lei, X.T. Fully automated quantitative cephalometry using convolutional neural networks. *J. Med. Imaging* **2017**, *4*, 014501
10. Lee, H.; Park, M.; Kim, J. Cephalometric landmark detection in dental x-ray images using convolutional neural networks. In Proceedings of the Medical Imaging 2017: Computer-Aided Diagnosis, Orlando, FL, USA, 3 March 2017.
11. Qian, J.; Cheng, M.; Tao, Y.; Lin, J.; Lin, H. CephaNet: An Improved Faster R-CNN for Cephalometric Landmark Detection. In Proceedings of the 2019 IEEE 16th International Symposium on Biomedical Imaging (ISBI 2019), Venice, Italy, 8–11 April 2019; pp. 868–871.
12. Ren, S.; He, K.; Girshick, R; Sun, J., Faster r-cnn: Towards real-time object detection with region proposal networks. In Proceedings of the Advances in neural information processing systems(NIPS), Montreal, QC, Canada, 7–12 December 2015; pp. 91–99.
13. Ramanan, D.; Zhu, T. Face detection, pose estimation, and landmark localization in the wild. In Proceedings of the 2012 IEEE Conference on Computer Vision and Pattern Recognition (CVPR), Providence, RI, USA, 16–21 June 2012; pp. 2879–2886.
14. Belhumeur, P.N.; David, W.; David, J.; Kriegman, J.; Neeraj Kumar, T. Localizing parts of faces using a consensus of exemplars. *IEEE Trans. Pattern Anal. Mach. Intell.* **2013**, *35*, 2930–2940. [CrossRef] [PubMed]
15. Simonyan, K.; Andrew, Z.T. Very deep convolutional networks for large-scale image recognition. *arXiv* **2014**, arXiv:1409.1556.
16. Krizhevsky, A.; Ilya, S.; Geoffrey, E.; Hinton, T. Imagenet classification with deep convolutional neural networks. *Adv. Neural Inf. Process. Syst.* **2012**, 1097–1105. [CrossRef]

17. Claudiu, C.D.; Meier, U.; Masci, J.; Gambardella, L.M.; Schmidhuber, J. Flexible, High Performance Convolutional Neural Networks for Image Classification. In Proceedings of the Twenty-Second International Joint Conference on Artificial Intelligence, Barcelona, Spain, 16–22 July 2011.

18. Long, J.; Shelhamer, E.; Darrell, T. Fully Convolutional Networks for Semantic Segmentation. In Proceedings of the IEEE Conference on Computer Vision and Pattern Recognition (CVPR), Boston, MA, USA, 7–12 June 2015; pp. 3431–3440.

19. Olaf, R.; Fischer, P.; Brox, T. U-net: Convolutional networks for biomedical image segmentation. In Proceedings of the International Conference on Medical image computing and computer-assisted intervention, Munich, Germany, 5–9 October 2015; pp. 234–241.

20. Pfister, T.; Charles, J.; Zisserman, A. Flowing convnets for human pose estimation in videos. In Proceedings of the IEEE International Conference on Computer Vision (ICCV), Santiago, Chile, 13–16 December 2015.

21. Payer, C.; Štern, D.; Bischof, H.; Urschler, M. Regressing heatmaps for multiple landmark localization using CNNs. In Proceedings of the International Conference on Medical Image Computing and Computer-Assisted Intervention (MICCAI). Athens, Greece, 17–21 October 2016.

22. Davison, A. K.; Lindner, C.; Perry, D. C.; Luo, W.; Cootes, T. F. Landmark localisation in radiographs using weighted heatmap displacement voting. In Proceedings of the International Workshop on Computational Methods and Clinical Applications in Musculoskeletal Imaging (MSKI), Granada, Spain, 16 September 2018; pp. 73–85.

23. Tompson, J. J.; Jain, A.; LeCun, Y.; Bregler, C. Joint training of a convolutional network and a graphical model for human pose estimation. In Proceedings of the Advances in neural information processing systems (NIPS), Montreal, QC, Canada, 8–13 December 2014; pp. 1799–1807.

24. Kaiming, H.; Zhang, X.; Ren, S.; Sun, J. Deep residual learning for image recognition. In Proceedings of the IEEE Conference on Computer Vision and Pattern Recognition (CVPR), Las Vegas, NV, USA, 27–30 June 2016; pp. 770–778.

Article

Person Independent Recognition of Head Gestures from Parametrised and Raw Signals Recorded from Inertial Measurement Unit

Anna Borowska-Terka * and Pawel Strumillo

Faculty of Electrical, Electronic, Computer and Control Engineering, Institute of Electronics, Lodz University of Technology, 211/215 Wolczanska Str., 90-924 Lodz, Poland; pawel.strumillo@p.lodz.pl
* Correspondence: anna.borowska-terka@p.lodz.pl

Received: 10 May 2020; Accepted: 17 June 2020; Published: 19 June 2020

Abstract: Numerous applications of human–machine interfaces, e.g., dedicated to persons with disabilities, require contactless handling of devices or systems. The purpose of this research is to develop a hands-free head-gesture-controlled interface that can support persons with disabilities to communicate with other people and devices, e.g., the paralyzed to signal messages or the visually impaired to handle travel aids. The hardware of the interface consists of a small stereovision rig with a built-in inertial measurement unit (IMU). The device is to be positioned on a user's forehead. Two approaches to recognize head movements were considered. In the first approach, for various time window sizes of the signals recorded from a three-axis accelerometer and a three-axis gyroscope, statistical parameters were calculated such as: average, minimum and maximum amplitude, standard deviation, kurtosis, correlation coefficient, and signal energy. For the second approach, the focus was put onto direct analysis of signal samples recorded from the IMU. In both approaches, the accuracies of 16 different data classifiers for distinguishing the head movements: pitch, roll, yaw, and immobility were evaluated. The recordings of head gestures were collected from 65 individuals. The best results for the testing data were obtained for the non-parametric approach, i.e., direct classification of unprocessed samples of IMU signals for Support Vector Machine (SVM) classifier (95% correct recognitions). Slightly worse results, in this approach, were obtained for the random forests classifier (93%). The achieved high recognition rates of the head gestures suggest that a person with physical or sensory disability can efficiently communicate with other people or manage applications using simple head gesture sequences.

Keywords: electronic human-machine interface; blindness; gesture recognition; inertial sensors; IMU

---

1. Introduction

Human–System Interaction (HSI) is currently actively pursued as a separate research field dedicated to the development of new technologies facilitating human communication with systems [1]. Depending on the application, such interaction systems are also referred to as Human–Computer Interfaces (HCI) or more generally human–machine interfaces. The design and construction of such systems requires an interdisciplinary research approach and involves knowledge of sensory perception mechanisms in humans, cognitive processes, and information processing, as well as basics of ergonomics. A well-designed user interface often determines the usefulness of an entire system [2].

Designing interfaces accessible for persons with sensory and motor disabilities is a particularly challenging research issue [3,4]. It is necessary to develop innovative solutions that enable handicapped users to communicate with such devices or systems. The interfaces use alternative, often multimodal communication methods that compensate for the diminished or lost sensorial or physical functions [5]. Individuals with hearing loss use sign language (one of the so-called visual-spatial languages) and are aided by sign language computer applications. For people with physical disabilities, special input-output devices utilizing inertial sensors and innovative interfaces are built so that "contactless" communication with a computer via, among others, Brain–Computer Interfaces (BCI) [6] and video interfaces, which are capable of tracking eye movements [7] or detecting intentional blinks [8], have been developed. The work reported in [9] shows that people with cerebral palsy can communicate with the use of personalized gesture datasets while aided by an intelligent user interface. The visually impaired, on the other hand, interact with computers using speech synthesizers and the so-called Braille displays that allow them to type in and display the Braille alphabet. A more complex problem for a blind person is handling devices while in motion. The basic mobility aid that the blind use is a white cane, which engages one hand and hence, complicates the operation of an additional device, e.g., a navigation tool during a walk.

The control of computer applications is traditionally performed using a keyboard and/or a computer mouse or a touch screen. However, solutions that enable people with serious physical disabilities to work with computers are increasingly more common. In [10], an interface based on the recognition of head movements and simultaneous lip movements designed for users with limb disability was reported. It is, however, not only computer applications that can be controlled in a non-contact manner. It was shown that the movement of an electric wheelchair of a disabled person can be controlled by means of head movements [11–13] or eye gaze [14] and even EEG signals [15].

Recent advances in electromechanical MEMS (Micro Electro Mechanical Systems) technologies have made it possible to develop miniature and cheap inertial sensors, and consequently use them in human–computer interfaces [16]. In [17], in order to identify eight different types of physical activity, a three-axis accelerometer placed on the wrist of the dominant hand was used. In [18], five acceleration sensors were used to recognize the movement of hands and to distinguish gait from immobility. The sensors were placed on the chest and on all the limbs. Similarly to our work, 3-axis compass, 3-axis accelerometer and 3-axis gyroscope were mounted on the user's head and used to identify six different head gestures [19]. In [20], a system that can be utilized to monitor the body movements of people with neurodegenerative diseases was reported. In this system, signals recorded from three sensors were analyzed, one of them was mounted on the head and the other two on the shins. The fusion of signals acquired from an accelerometer and surface EMG electrodes was used in [21] to assess Parkinson's disease-related symptoms. In [22], simultaneous processing of images and signals from inertial sensors was employed to estimate six degree-of-freedom head movements. In the above-mentioned studies, the first processing step in analyzing IMU signals is to parametrize the signals for predefined analysis time windows [17,18]. The extracted parameters are then used to build appropriate classifiers that recognize given types of movements.

In a recent work [23], novel methods for recognizing human physical activity that are based on the so called symbolic representation algorithms were presented. Using three database sets, the authors have shown that their approach performs best in terms of accuracy, processing time, and memory consumption in comparison to other classic approaches that were based on supervised classification techniques. In another very recent work [23], IMU were applied to measure the 3D range of motion of the trunk and lower limb joints. The results of this study show that inertial sensors can be successfully applied to investigate maladaptive movement strategies.

In our study, we propose a hands-free interface enabling blind persons to control the menu of a navigation device by means of head movements. We also envision that such an interface can serve as a communication aid for people with serious motor disabilities, e.g., for the paraplegic individuals.

The main contribution of our work, which has not been addressed in earlier studies [17–19,24], is the comparison of the parametric and the time domain representations of IMU signals on which different classifiers are trained to recognize four head movement patterns. Similarly to the work reported in [25], in which IMU were used to recognize types of physical activities, we studied the dependence of different lengths of time windows on the parametric representation of inertial signals to recognize different types of head gestures. In our work, we show that time domain approaches to analysis of inertial signals can outperform parameter-based ones and are less computationally demanding. The latter feature is particularly important for mobile implementations of the interface. The ultimate motivation for our research is to develop user-friendly and efficient electronic travel aids that will help the visually impaired retain orientation and mobility in unfamiliar environments [26,27].

The remainder of this paper is organized as follows. In Section 2, we describe the experimental methods that we used for data recording and data pre-processing. We also shortly review the applied data classifiers. In Section 3, we present the head gesture recognition results for the parametric and time domain representations of IMU signals. Finally, in Section 4 we apprise the achieved results, show the envisioned applications of our study, and point out limitations of the presented work.

2. Materials and Methods

In our proof of concept approach, we have applied a DUO MLX device (see Figure 1b) equipped with a stereovision camera, a three-axis electronic gyroscope, three-axis acceleration sensor, a thermometer, and a magnetometer [28]. The device is of small form factor: $52 \times 25 \times 13$ mm and weighs 12.5 g. The device is mounted on the user's forehead (see Figure 1a). The study was conducted with 65 persons. The participants were mainly second year university students, 44 of whom were women. The trials were approved by the bioethics commission of the Medical University of Lodz (No. RNN/261/16/KE). All the trial participants were informed about the purpose of the study, the materials used and their role in the trial sessions. After fixing the DUO MLX device on their forehead, the users were asked to perform rehearsal head movements and were instructed to proceed with the movements just within a comfort zone of their head positions. During collection of the data, the participants remained in a sitting position. Each participant was sitting straight in a chair and did not change position during the experiment and performed only the given motions (yaw, pitch, roll, and immobility). Each user had the DUO MLX device mounted rigidly on their forehead. The participant did not touch the DUO MLX device or change its position on the forehead during the experiment.

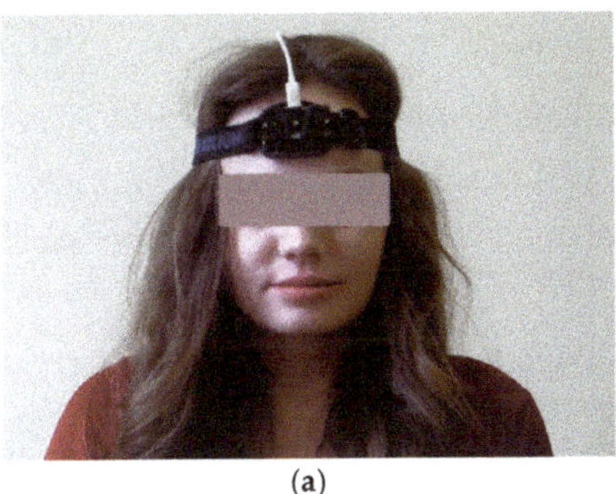
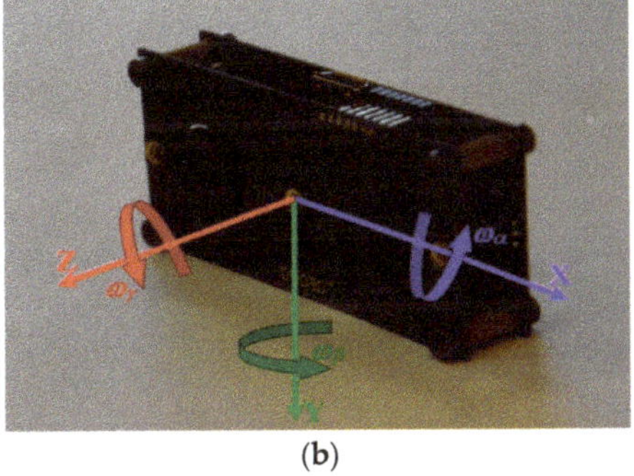

Figure 1. (**a**) Placement of the inertial measurement unit on the user's head; (**b**) DUO MLX Stereo System equipped with inertial sensors with marked coordinate axes Ox, Oy, and Oz, and angular velocities ω_α, ω_β, and ω_γ.

The aim of the study was to automatically classify three basic head gestures: *roll* (head movement "sideways"), *pitch* (head movement "up-down") and *yaw* (head movement "left-right"), and *immobility* (head at rest). A single movement cycle lasted 2.5 s on average with 1.2 s standard deviation of the movement period. The recordings of test and learning datasets from one individual took approx. 4 min each.

Sample plots of the signals recorded by the acceleration sensor and the gyroscope are shown in Figure 2. The recorded accelerations a_x, a_y, and a_z along axes Ox, Oy, and Oz respectively are plotted in Figure 2a and angular velocities ω_α, ω_β, and ω_γ (as shown in Figure 1b) recorded from the gyroscope are plotted in Figure 2b. All the signals are recorded at a sampling rate of 100 Hz.

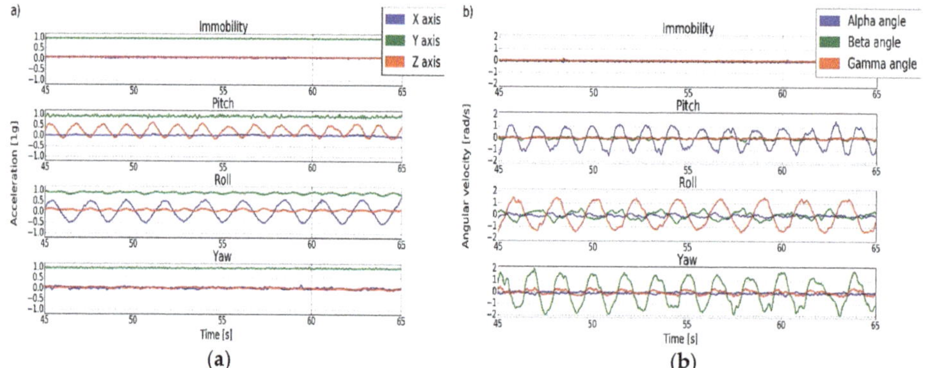

Figure 2. Example signal waveforms recorded by: (**a**) acceleration sensors and (**b**) gyroscope sensors.

2.1. Signal Recordings and Pre-Processing

Signal recordings from 65 participants were grouped into the following datasets:

1. The learning dataset recorded from a randomly selected 53, i.e., approx. 80% of all the 65 trial participants. Signals that were used as the learning set were recorded as follows: *10 s immobility—2.5 min of a specific motion type—10 s immobility*, the type of gestures performed by test participants were: yaw, roll, pitch, and immobility. The collection of the learning signals therefore consisted of $53 \times 4 = 212$ recordings.

2. Two types of testing datasets that have not been used for training the classifiers:

    *   *T_53_set* that was recorded for 53 trial participants for whom the learning data were recorded,
    *   *T_12_set* that was recorded for the rest of the trial participants, i.e., 12 individuals who did not take part in the recordings of the learning datasets,

    further, for each of the two testing data sets there were two different testing scenarios applied:

    *   testing scenario *T1* for which the dataset was recorded following the sequence of: *20 s of immobility—40 s of yaw head movement—20 s of immobility—40 s of roll—20 s of immobility—40 s of pitch—20 s of immobility*; thus, two types of the testing datasets were recorded *T1_53_set* and *T1_12_set*, respectively
    *   testing scenario *T2* for which the datasets were recorded according to the following procedure: head gesture sequences of *roll*, *yaw* and *pitch* gestures in random order for time periods lasting from 5 to 20 s and no immobility time gaps between the gestures; thus, two other types of the testing datasets were recorded: *T2_53_set* and *T2_12_set* respectively.

The signals recorded from the IMU were processed according to the two following procedures:

1. *With signal parameterization (SP):* for each of the six recorded IMU signals, i.e., three signals from the accelerometer (a_x, a_y, a_z) and three signals from the gyroscope (ω_α, ω_β, ω_γ), the following features were extracted: (1) average, (2) minimum, (3) maximum, (4) standard deviation, (5) kurtosis, (6) correlation coefficients for pairs of signals from the accelerometer and pairs of signals from the gyroscope, and (7) signal energy. Thus, the total number of parameters derived from the signals was: 6 signals $\times$ 7 parameters = 42 parameters.

These parameters were calculated for different time window widths, i.e., $T \in \{0.1, 0.2, 0.3, 0.4, 0.5, 0.6, 0.8, 1.0, 2.0\}$ given in seconds. The work focuses on the causal recognition mode of the gestures for which the time windows are determined for current and past samples only. The consecutive windows were shifted by a 0.1 s time step as shown in Figure 3.

2. *Time domain representation (TDR)*: current samples of signals recorded from the accelerometer (a_x, a_y, a_z) and gyroscope (ω_α, ω_β, ω_γ) were used directly as six-element training vectors for the classifiers.

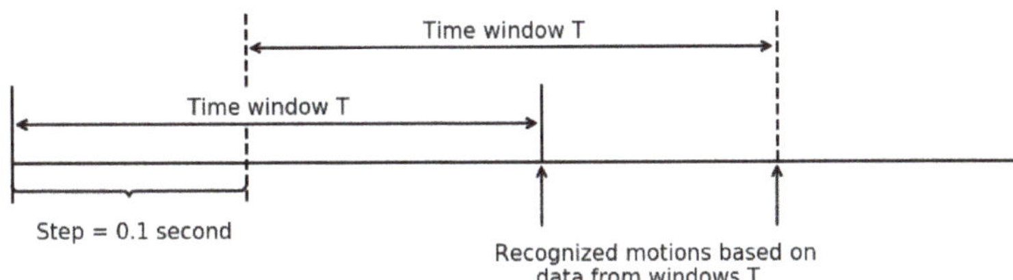

Figure 3. An example of two consecutive time windows for which signal parameters were computed.

In order to eliminate user errors related to an excessively slow start of the selected type of movement, too rapid cessation of motion or incorrect initial movement, the learning datasets comprised 90 s recordings approx. in the middle of the 150 s time recording span.

According to the described recording procedure of the learning datasets (collected from 53 individuals), the following datasets were built:

1. For the procedure with signal parametrization: 900 42-element vectors for each individual. The calculations of these 900 vectors were made for each time window width $T \in \{0.1, 0.2, 0.3, 0.4, 0.5, 0.6, 0.8, 1.0, 2.0\}$, given in seconds, that were shifted with time steps of 0.1 s.
2. For the procedure based on the time domain: 9000 vectors for each individual, with six signal samples representing a "time capture" of the given head motion.

Finally, the test datasets on which the performance of individual classifiers were evaluated were correspondingly as follows:

1. For the *SP* procedure—set with 42-element vectors. Calculations for these testing vectors were made for time windows of widths $T \in \{0.1, 0.2, 0.3, 0.4, 0.5, 0.6, 0.8, 1.0, 2.0\}$ seconds, that were shifted with time steps of 0.1 s.
2. For the *TDR* procedure—set of vectors each with six signal samples.

The datasets are available for download at [29].

2.2. The Classifiers

For the purpose of classification of the head gestures on the basis of the recorded IMU signals, data classifiers generally recognized to be very efficient were applied [30]. In particular, various architectures of decision trees and decision forests were used since they can provide an insight into the data decision process, i.e., these classifiers can decompose the classification task into decision rules of the input data features. The data classifiers that were applied were the following: (1) a decision tree, (2) a decision tree with a minimum number of samples per leaf equal to 5, (3) a random forest consisting of 10 decision trees, and (4) a random forest consisting of 10 decision trees, each of which contained a minimum number of leaf samples equal to 5, (5) k-nearest neighbors method (k-NN) for $k \in \{1, 3, 5, 7, 9, 11, 13, 15, 19\}$, (6) Support Vector Machines (SVM) with a radial basis function kernel, and lastly (7) the SVM with a third degree polynomial as the kernel function.

2.2.1. Decision Trees and Random Forest

Decision tree is a supervised non-parametric classifier that allows building decision rules by multistage divisions of the dataset into disjoint classes [31].

Different algorithms can be used for building decision trees. In this work, we have employed the CART (Classification and Regression Trees) algorithm and used the so-called Gini coefficient as a measure of the diversity of classes in the tree nodes [25]. Classification trees can be combined into groups to form the so-called random forests. The final classification of the vector is performed by voting, i.e., a vector is assigned to the class for which it receives the largest number of votes from individual decision trees. Random forests are efficient classifiers for very large data sets [32,33].

2.2.2. The k-Nearest Neighbor classifier

The k-Nearest Neighbor Classifier (k-NN) is another non-parametric data classification technique. This is an instance-based learning algorithm that works on a simple scheme in which data sample x is assigned to a class for which there is a majority of data prototypes within the k nearest neighbors [34]. The k value should be selected in such a way as to retain the balance between overtraining and insufficient fit. We have evaluated the performance of this classifier for $k \in \{1, 3, 5, 7, 9, 11, 13, 15, 17, 19\}$ neighbors. However, we present just the best classification results obtained for $k = 7$ for training procedure SP, i.e., with IMU signal parametrization and for $k = 19$ obtained for the time domain analysis. In all the tested k-NN classifiers, the Euclidean distance was used as the distance metric.

2.2.3. Support Vector Machines (SVM)

The SVM classifier is particularly suitable for classifying large dimensional datasets [35,36]. The basic optimization goal in this classifier is to maximize the margin, i.e., the between-class separation region. The margin is defined as the distance between the decision boundary (separation hyperspace) and the nearest learning samples, called the support vectors. The aim is to obtain decision boundaries with the widest possible margins. In our implementation of the SVM classifier for the multiclass problem, a scheme in which the one-vs-rest classification approach was adopted, we tested the performance of these classifiers for regularization parameter values $C \in \{0.1, 1.0, 5.0, 10.0\}$ for the two types of kernels, a RBF kernel and a 3rd degree polynomial. The best classification results were obtained for $C = 1.0$ for the parametrized signals and for $C = 10.0$ for time domain analysis. Thus, we present the classification results for the regularization parameter values for which the corresponding SVM classifiers performed best.

3. Classifier Training and Results

The classifiers were trained according to the procedures described in Section 2.1. The main aim was to compare the performance of different classifiers for parametrized signals and raw signal samples recorded from the inertial sensors.

The 10-fold cross validation method was used to evaluate the classifiers' performance in the head gesture classification task. The training sets (for both the *SP* and *TDR* training procedure) were therefore randomly divided into 10 different subsets of equal cardinality. Each subset was sequentially used as the validation set while the other nine subsets served as a training set for the classifiers. Each classifier was cross validated on the same subsets, i.e., the division of the training and validation sets was made first and then each classifier was trained on identical data subsets. Tables 1 and 2 report the results of the 10-fold cross-validation for the *SP* and *TDR* training procedures, respectively.

Table 1. Results of 10-fold cross-validation training—classification accuracy for the classifiers trained on the parametrized data consisting of 42 parameters (training procedure SP).

No.	Decision Tree	Cropped Decision Tree	Random Forest	Cropped Random Forest	k-NN for $k = 7$	SVM with RBF Kernel	SVM with 3rd Degree Polynomial
1.	97.63	97.84	98.57	98.52	95.40	97.82	98.14
2.	97.58	97.79	98.56	98.55	95.57	97.80	98.22
3.	97.50	97.74	98.64	98.58	95.51	97.86	98.22
4.	97.66	97.88	98.62	98.55	95.55	97.76	98.08
5.	97.55	97.87	98.71	98.60	95.52	97.68	98.17
6.	97.45	97.86	98.56	98.52	95.41	97.79	98.10
7.	97.49	97.81	98.57	98.57	95.50	97.77	98.03
8.	97.52	98.01	98.64	98.61	95.64	97.88	98.23
9.	97.62	97.86	98.65	98.55	95.49	97.84	98.17
10.	97.85	98.04	98.79	98.78	95.45	97.87	98.23
Mean ± Std	97.59 ± 0.11	97.87 ± 0.09	98.63 ± 0.07	98.58 ± 0.07	95.50 ± 0.07	97.81 ± 0.06	98.16 ± 0.07

Table 2. Results of 10-fold cross-validation training—classification accuracy for the classifiers trained on the IMU signal samples (training procedure *TDR*).

No.	Decision Tree	Cropped Decision Tree	Random Forest	Cropped Random Forest	k-NN for $k = 19$	SVM with RBF Kernel	SVM with 3rd Degree Polynomial
1.	97.45	97.59	98.17	98.05	98.19	94.27	92.58
2.	97.41	97.59	98.27	98.14	98.24	94.14	92.54
3.	97.48	97.57	98.26	98.13	98.21	94.31	92.59
4.	97.41	97.59	98.26	98.11	98.21	94.05	92.39
5.	97.41	97.52	98.18	98.06	98.20	94.25	92.53
6.	97.43	97.57	98.19	98.08	98.17	94.26	92.55
7.	97.45	97.60	98.27	98.10	98.19	94.28	92.53
8.	97.43	97.61	98.29	98.13	98.28	94.02	92.32
9.	97.42	97.63	98.23	98.11	98.22	94.14	92.41
10.	97.40	97.58	98.22	98.12	98.21	94.43	92.72
Mean ± Std	97.43 ± 0.02	97.59 ± 0.03	98.23 ± 0.04	98.11 ± 0.03	98.21 ± 0.03	94.22 ± 0.12	92.51 ± 0.11

For the classifiers trained on the parametrized signals (training procedure SP), the random forests and the SVM with a third degree polynomial kernel excelled and achieved accuracies over 98%. The poorest classification results were obtained for the k-NN classifier, but even this classifier achieved accuracies exceeding 95%.

For the classifiers trained directly on signal samples (training procedure *TDR*), all classifiers but the SVM achieved accuracies better than 97%. Although, the SVM classifier yielded results no worse than 92% correct head gesture recognitions.

We have evaluated the performance of the classifiers and compared whether the accuracies they yield are statistically different. Because we consider more than two classifiers and our data do not have a normal distribution, the Friedman test was used for this purpose [37]. For both the training procedures, i.e., the *SP* (with parametrization of IMU signals that generate 42 dimensional training vectors) and the *TDR* (raw IMU signal samples that generate six dimensional training vectors), we have rejected the null-hypothesis, i.e., we concluded that there are statistically significant differences ($p < 0.05$) between the classifiers. We also compared the chosen seven classifiers (Tables 1 and 2) by conducting the paired Wilcoxon signed rank test. For the *SP* training procedure for all classifiers we have also rejected the null-hypothesis. In the case of the *TDR* training procedure, there were no significant differences between the random forest and the k-NN for $k = 19$ (the null-hypothesis accepted). For other classifiers for the *TDR* procedure we have rejected the null-hypothesis, i.e., the classifiers' performances are statistically equal.

3.1. Results for Datasets Consisting of the Parametrised IMU Signals

For the parametrized signals (the *SP* training procedure), the classifier performances were evaluated for different time window widths T for which the statistical parameters of the signals were calculated. The results obtained for the combined test scenarios $T1$ and $T2$ are shown in Figure 4 for different time window sizes. Also, Figure 5 compares the results of head gesture classifications obtained for two different testing datasets, i.e., the test data for trial participants whose data was used for training the classifiers and the test data from trial participant whose data was not used for training the classifiers. We can conclude that the results shown in Figure 5 show the user independent performance of the system.

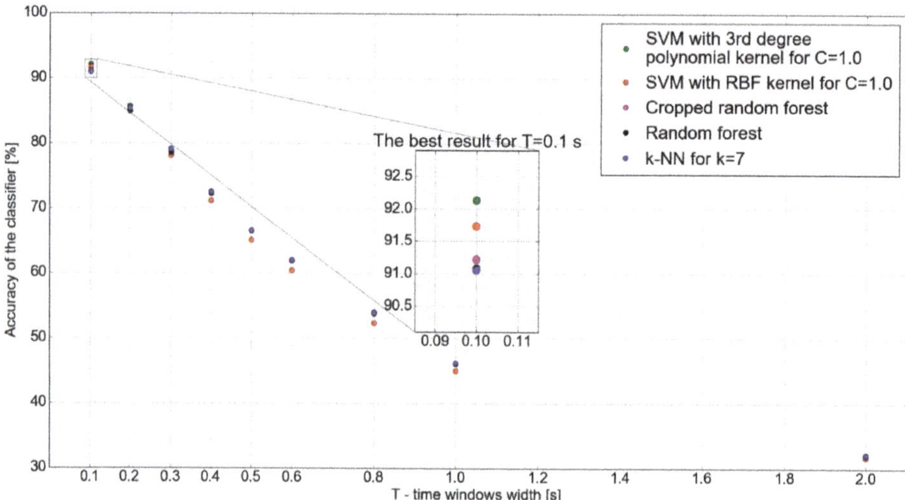

Figure 4. Accuracy of the classifiers for different widths of time window T for the test data.

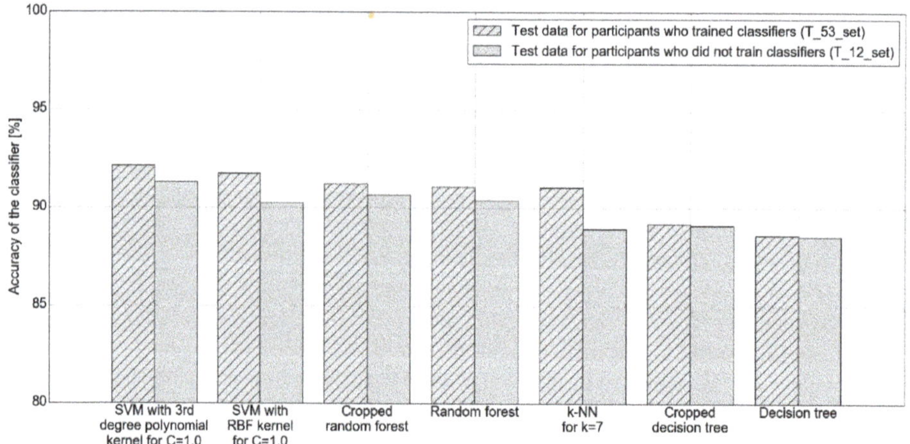

Figure 5. Comparison of the classifiers' accuracy obtained for the test data from participants who trained classifiers on the T_53_set and the test data from individuals who did not record the training data (T_12_set). Classifiers accuracy for the test data—causal system (vectors with 42 parameters).

Interestingly (as illustrated in Figure 4), the best results were obtained for the shortest time window, i.e., $T = 0.1$ s window containing 10 signal samples. An increase in the width of the time window negatively affected the classification accuracy. Note that for the test dataset, regardless of the width of the time window, the best classifier was the SVM with a 3rd degree polynomial kernel (accuracy better than 92%). The best classification accuracy for the k-NN classifier was obtained for

$k = 7$ and was slightly worse than the one achieved by the random forest classifier (both with an accuracy above 91%).

For the shortest time windows, the accuracies of individual classifiers tend to differ, whereas for larger time windows, the performance of all the classifiers converges to similar albeit poorer values. Figure 5 illustrates in more detail how the classifiers performed for the shortest time window ($T = 0.1$ s).

3.2. Results for Datasets Taken Directly from IMU Signal Samples

Subsequently, we evaluated the classifier performances for the testing datasets that were drawn directly from the samples of IMU signals. The achieved classifiers' accuracies for the test datasets are shown in Figure 6.

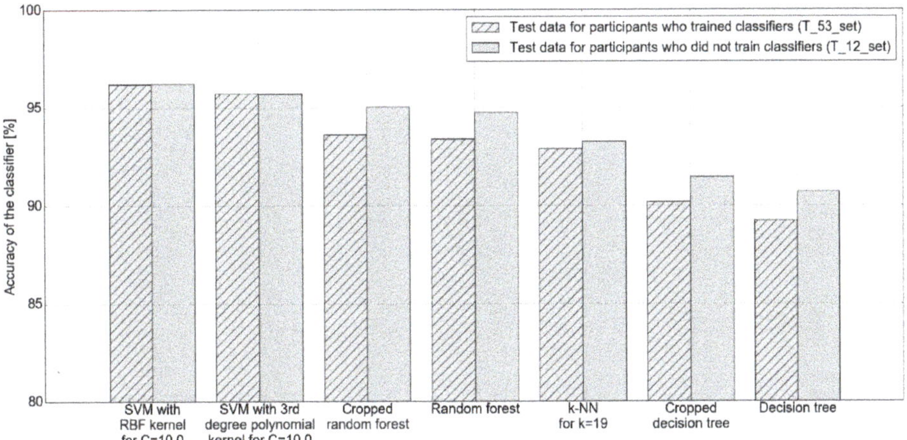

Figure 6. Comparison of the classifiers' accuracy obtained for the test data from participants who trained the classifiers (T_53_set) and the test data from individuals who did not record the training data (T_12_set). The test data collected directly from the accelerometer and gyroscope (vectors with 6 parameters).

Thus, we can note that for the classifiers trained on the IMU signal samples, the best accuracies exceeding 95% were achieved for the SVM classifiers. Also, an important observation is that their performance did not depend on the test data type (T_53_set or T_12_set).

3.3. Person Independent Recognition of Head Gestures

Here we report on the performance of the classifiers for the test data recorded for the 12 trial participants for whom the recorded IMU signals were not used for training the classifiers. In Figures 5 and 6, we compare the recognition rates of head gestures for the new 12 users and the classification results obtained for the test dataset recorded for 53 trial participants. Figure 5 contains the results for the shortest time window $T = 0.1$ s for which 42 statistical parameters were computed and used for training the classifiers, whereas Figure 6 shows the results for the classifiers trained on IMU signal samples, i.e., on six-element vectors.

For the person-independent test datasets, i.e., for the 12 participants who did not take part in training the classifiers, the best head movement recognition results for vectors with 42 parameters were obtained for the SVM and random forests. For all of these classifiers, about 90% accuracy or higher was achieved (Figure 5). Note also that for the person independent test datasets, the results obtained were inferior to those for the T_53_set test data (i.e., person dependent dataset) except for the SVM classifiers which performed equally on the two test datasets (for the *TDR* training procedure).

On the other hand, for the classifiers trained directly on the IMU signal samples, head gesture recognition accuracy exceeded 90% for all classifiers, except the decision tree classifier (see Figure 6). Interestingly, the accuracy for all classifiers except SVM was better for T_12_set, however, the SVM performed best for both T_53_set and T_12_set test datasets.

3.4. Head Gesture Recognition Rates for T1 and T2 Testing Scenarios

We also verified how the classifiers' accuracy depends on the type of testing scenario as defined in Section 2.1. For the causal system, the best results were obtained for test scenario *T1* (see Figure 7). The accuracy of all classifiers exceeded 90%. The results obtained for test scenario *T2* were distinctly inferior. For this testing scenario, the accuracy of the classifiers was below 90%. Again, the best classifiers were the SVM and the random forests.

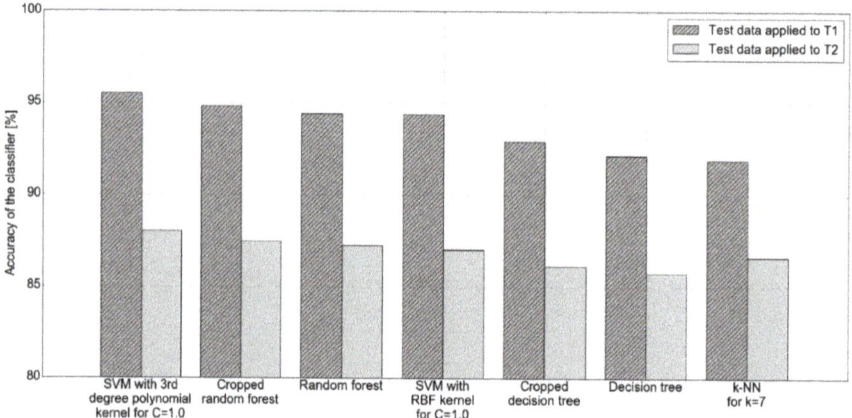

Figure 7. Comparison of the classifiers' accuracy obtained for test scenarios *T1* and *T2* for the test data recorded for individuals who did not record the training data (T_12_set) and the *SP* training procedure (classifiers trained on the parameters of the IMU signals).

For the signal samples recorded directly from the IMU (vectors with six parameters), almost all classifiers' accuracies were above 90%. As is evident in Figure 8, except for the SVM, the best results were obtained for test *T2*. Note that for the SVM classifier, the recognition accuracy for both tests was almost identical and exceeded 95%.

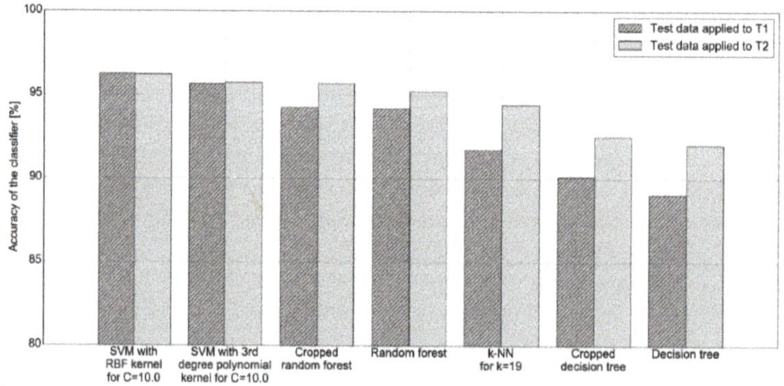

Figure 8. Comparison of the classifiers' accuracy obtained for test scenarios *T1* and *T2*, i.e., for the test data recorder for individuals who did not record the training data (T_12_set). Classifiers accuracy for the test data collected directly from the accelerometer and gyroscope (vectors with six parameters).

We have also verified which head gestures were most often confused with one another for the classifiers which yielded the best results, i.e., for the SVM classifier with a 3rd degree polynomial and regularization parameter $C = 1.0$ for the *SP* training procedure and for the SVM with an RBF kernel and $C = 10.0$ for the *TDR* training procedure. The results are presented as confusion matrices in Tables 1 and 2. These tables show the results obtained for the 12 trial participants who did not take part in training the classifiers.

As shown in Table 3, the highest error rate was reported for pitch (5.5% of the gestures recognised as immobility). Further, considerably high error rates were noted for roll (4.6% of the gestures recognised as immobility). The best recognition rates were obtained for head immobility (92.6%). Interestingly, also for the immobility we have obtained the highest false positive recognitions.

Table 3. The confusion matrix showing inter-gesture recognition errors for the SVM classifier with a 3rd degree polynomial and $C = 1.0$ (the classifier trained on parametrised IMU signals).

	Recognised: Pitch	Recognised: Roll	Recognised: Yaw	Recognised: Immobility	
True: Pitch	92.0%	1.1%	1.4%	5.5%	100%
True: Roll	1.1%	91.7%	2.6%	4.6%	100%
True: Yaw	3.5%	1.8%	91.3%	3.4%	100%
True: Immobility	1.2%	2.8%	3.4%	92.6%	100%

The sensitivity, specificity and F1 scores of the head gesture recognition rates are presented in Table 4. Note that the F1 score is defined as a harmonic average of the sensitivity and positive predictive value. It is a good measure of overall classifier performance.

Table 4. Statistical measures showing detection results of head gestures for the SVM classifier with a 3rd degree polynomial and $C = 1.0$ (the classifier trained on parametrized IMU signals).

	Pitch	Roll	Yaw	Immobility
Sensitivity	92%	92%	91%	93%
Specificity	98%	98%	97%	96%
F1-score	92%	93%	92%	92%

Note that F1-score, sensitivity and specificity rates are similar for different gestures, however, the specificity rates assume the highest values.

For the SVM classifier trained on IMU signal samples (Table 5), the distribution of errors was different. For this case, the immobility was frequently and falsely recognized as yaw or as pitch (4.3% and 3.9% error rates respectively).

Table 5. The confusion matrix showing inter-gesture recognition errors for the SVM classifier with an RBF kernel and $C = 10.0$ (the classifier trained on IMU signal samples).

	Recognized: Pitch	Recognized: Roll	Recognized: Yaw	Recognized: Immobility	
True: Pitch	95.3%	0.3%	0.5%	3.9%	100%
True: Roll	0.3%	96.1%	1.8%	1.8%	100%
True: Yaw	0.6%	1.3%	93.8%	4.3%	100%
True: Immobility	0.5%	0.2%	0.5%	98.8%	100%

Note here that both the sensitivity and specificity rates in Table 6 (classifiers trained on raw IMU signals) outperform the corresponding rates in Table 4 (classifiers trained on parametrized IMU signals).

Table 6. Statistical measures showing detection results of head gestures for the SVM classifier with a 3rd degree polynomial and $C = 1.0$ (the classifier trained on raw IMU signals).

	Pitch	Roll	Yaw	Immobility
Sensitivity	95%	96%	94%	99%
Specificity	99%	99%	99%	97%
F1-score	97%	97%	95%	96%

Overall, however, the gesture recognition rates were better for the SVM classifier trained on IMU signal samples than the classifier trained on the parametrized IMU signals (compare the matrix diagonals in Tables 3 and 5, and also sensitivities, specificities and F1-scores in Tables 4 and 6).

The training times of the classifiers were also evaluated. The computations were performed on an Intel Core i5-7500, 16 GB RAM, Windows 10 64-bit PC. Scripts were written in Python in the Enthought Canopy environment using the Sklearn module. The obtained calculation times required for training the classifiers are presented in Table 7.

Table 7. Training times of the classifiers.

No.	Classifier	Parametrized IMU Signals	Raw IMU Signals
1.	Decision tree	12.6 s	18 s
2.	Decision tree with a minimum of 5 samples in the leaf	11.9 s	17 s
3.	Random forest	9.2 s	52 s
4.	Random forest with a minimum of 5 samples in the leaf	8.8 s	50 s
5.	k-NN for k ∈ {1, 3, 5, 7, 9, 11, 13, 15, 17, 19}	0.6 s	2 s
6.	SVM with an RBF kernel for $C \in \{0.1, 1, 5, 10\}$	5 min, 17 s	from 2 h, 27 min to 8 h, 42 min *
7.	SVM with a 3rd degree polynomial kernel for $C \in \{0.1, 1, 5, 10\}$	2 min, 12 s	from 2 h, 46 min to 9 h, 57 min *

* training times depend on the values of the regularization parameter C.

Depending on the types of datasets (42-element or 6-element vectors), a different size of the training datasets was used, hence the training times of the classifiers varied significantly. For the 42-element training vectors, the longest training time did not exceed 6 min. The training procedure was the fastest for the k-NN classifier. On the other hand, it took several minutes to train the SVM classifier.

For the 6-element training vectors (and larger training datasets), the shortest training times were obtained for the k-NN classifiers. Training of the decision trees (and random forests) took tens of seconds while the longest training times were required for the SVM classifiers (a few hours).

Table 8 shows the time necessary to recognize a head gesture from a single vector by the trained classifiers.

Table 8. Recognition times of a gesture for a single input data pattern.

No.	Classifier	Recognition of Parametrized Signals	Recognition of Raw IMU Signals
1.	Decision tree	~1 ms	~1 ms
2.	Decision tree with a minimum of 5 samples in the leaf	~1 ms	~1 ms
3.	Random forest	~10 ms	~6 ms
4.	Random forest with a minimum of 5 samples in the leaf	~6 ms	~7 ms
5.	k-NN for $k \in \{1, 3, 5, 7, 9, 11, 13, 15, 17, 19\}$	~5 ms	~2 ms
6.	SVM with an RBF kernel for $C \in \{0.1, 1, 5, 10\}$	~10 ms	~5–11 ms
7.	SVM with a 3rd degree polynomial kernel for $C \in \{0.1, 1, 5, 10\}$	~10 ms	~4–10 ms

The computation times required for processing the data by the trained classifiers (for a PC with Intel Core i5-7500 processor) varies from 1 ms for the decision tree classifier to 11 ms for the SVM classifier with an RBF kernel. It should be noted, that for the parametrized approach, we must first compute the statistical parameters; for the shortest time window consisting of 10 signal samples, we need to wait for 0.1 s before we feed these parameters to the classifiers.

4. Discussion

Our primary motivation for the study was to build a hands-free interface for the visually impaired that would facilitate controlling an electronic travel aid built in our project dedicated to help the visually disabled in independent mobility and travel [26]. The main role of the DUO MLX device and the image processing software is to recover a 3D structure of the environment from the captured stereovision images. The sequences of depth images are then sonified and acoustically presented to the visually impaired user. In [38], we have shown that such a non-visual presentation of the environment can help the blind to detect obstacles and find safe walking paths in the surroundings. One of the complications we encountered in the conducted trials is handling of the device by the blind. The use of an additional remote control to manipulate the interface proved to be inconvenient and occupied the other hand of the user who carried a white cane. An example scenario of how the interface might be used is the following. The navigation system generates a series of audio-haptic signals to indicate obstacles or inform about points of interest. Such a system features numerous settings like loudness, sensitivity range, sonification, and haptic activation schemes. The blind user, while standing still, by a series of head gestures, can move within the menu of the system and select the appropriate setting for the encountered environment or a navigation task at hand. System settings are confirmed by synthetized voice messages. It is important to note that a blind person carries a white cane which occupies one hand and the proposed solution does not need to engage the other hand to control the device. Also, after consulting the visually impaired, we plan another application of the interface. The device can play the role of a remote control that can aid the visually impaired in the activities of daily living, e.g., it can be used to control radio/TV volume without a need to seek for the remote of such devices.

Compared with the study reported in [24], we obtained a comparable F1-score of 92% for the parametrized IMU signals. It should be noted, however, that in [24] and in [17], individual gestures were not distinguished, but rather whole body activities, e.g., walking, running, cycling, hence the need for longer time data analysis windows: 5 s, 3 s and 1 s in [24] and 5 s and 12 s in [17]. In our case, we recognize delicate head movements, thereby the shorter the time window, the better the motion recognition. In [17,18], the IMU sensors were attached to the wrist, in [24] two to five sensors were used and they were mounted in different parts of the body, while in [19], IMU was mounted on the head like in our work. The advantage of our study is that 65 people participated in the experiments and our database is more diverse than e.g., in [19] where only five people took part in the trials. None of the works [17–19,24] compared the recognition results of body movements for data derived from the parametrized signals and time domain samples of IMU signals. We also tested the recognition rates from a large pool of different data classifiers, i.e., the decision trees and forests (and their cropped versions), k-NN, and SVMs (with an RBF kernel and a 3rd degree polynomial kernel). Finally, we show that better classification results can be obtained if the classifiers are trained on raw IMU signals rather than on the parametrized signals.

We are aware of the limitations of our study. The participants of the trials were mainly young people recruited from the students. Thus, an open question is on the usability of the system for elder people. Secondly, the trial participants remained seated during data collection. This limitation, for some system applications, should be released. Also, we should underline that our choice of IMU signals parameters, e.g., statistical parameters, was arbitrary and one might hypothesize that a different set of parameters might be composed that would even further improve the recognition performance of the head gestures.

Finally, the proposed system would need to be tested with target groups recruited from persons with physical or sensory disabilities.

5. Conclusions

In the presented study, we have compared two approaches to the classification of head gestures, i.e., training the classifiers on the parametrized signals, and training on direct signal samples recorded from the IMU.

The obtained results of head gestures recognition allow to formulate the following conclusions:

1. Head movements (*roll*, *yaw*, and *pitch*) can be efficiently recognized on the basis of signal recordings from an IMU that is positioned on the user's forehead.
2. The data classifiers trained on IMU signal samples outperformed the classifiers that were trained on a set of statistical parameters derived from the IMU signals. These performance differences were confirmed by running statistical tests. The Friedman test revealed that the classifiers yield statistically different results ($p < 0.05$). Also, the paired Wilcoxon signed-rank test has confirmed statistically valid differences for most of the classifiers (no significant differences between the random forest and the k-NN classifier for $k = 19$ was noted only). Our explanation about high head gesture recognition performance from raw IMU signals is that the six signal channels carry rich enough information for the classifiers to confidently recognize the head gestures. Also, we conclude that the inherent measurement noise (that occurs randomly) is averaged out during the training procedures of the classifiers.
3. The SVM classifier outperformed other classifiers in recognizing the head gestures, and if trained on the IMU signal samples, it achieved a recognition accuracy above 95%.
4. The recognition rates of the head gestures for the person-independent test dataset, i.e., the data taken from 12 persons whose recordings were not used in training the classifiers are comparable to the recognition rates obtained for the data recorded from the individuals who have also recorded the training data (see Figures 5 and 6). Interestingly, if the classifiers were trained on raw IMU signal samples, their recognition performance was generally better for data recorded from participants who did not train the classifiers.
5. The proposed method is suitable for on-line implementations. In Table 8, we show that the computation times required for detection of a gesture (on a PC with Intel Core i5 processor) by pre-trained classifiers do not exceed 11 ms.

We hypothesize that the presented head-gesture-controlled interface can find numerous applications. Firstly, it can offer an alternative communication channel for people with serious physical disabilities, e.g., individuals suffering from serious spinal injuries or stroke that result in tetraplegia. We made a new public benchmark dataset consisting of the training data (from 53 participants) and testing data (from 65 participants). The signal database with description is available for download at [29].

In further studies, on the basis of high head-gesture recognition performance, one could even further improve the recognition accuracy by accumulating classifier recognitions with each arriving IMU signal sample (i.e., with every 100 ms time interval) and achieve close to 100% recognition rates for the entire gesture. Our motivation is to further test the interface in a mobile travel aid system [38] and as a supporting interface in activities of daily living for the visually impaired, e.g., as a remote control system for home appliances. We will also consider the usefulness of such an interface as a potential rehabilitation aid for individuals with neck stiffness.

Author Contributions: A.B.-T.—data curation, formal analysis, methodology, software, visualization, writing—original draft. P.S.—conceptualization, methodology, project administration, supervision, writing—review and editing. All authors have read and agreed to the published version of the manuscript.

Funding: This research was partially funded from the European Union's Horizon 2020 research and innovation programme under grant agreement No 643636 "Sound of Vision".

Acknowledgments: We thank all the participants for taking part in the trials reported in this study.

Conflicts of Interest: The authors declare no conflict of interest.

References

1. International Conferences on Human System Interactions, HIS. Available online: https://ieeexplore.ieee.org/xpl/conhome.jsp?punumber=1002118 (accessed on 29 May 2020).
2. Nielsen, J. *Usability Engineering*; Morgan Kaufmann Publishers Inc.: San Francisco, CA, USA, 1993.
3. Helal, A.; Mounir, M.; Abdulrazak, B. (Eds.) *The Engineering Handbook of Smart Technology for Aging, Disability, and Independence*; John Wiley & Sons, Inc.: Hoboken, NJ, USA, 2008.
4. Strumillo, P.; Pajor, T. A vision-based head movement tracking system for human-computer interfacing. In Proceedings of the IEEE 2012 Joint Conference New Trends in Audio & Video and Signal Processing: Algorithms, Architectures, Arrangements and Applications (NTAV/SPA), Lodz, Poland, 27–29 September 2012.
5. Dumas, B.; Lalanne, D.; Oviatt, S. Multimodal Interfaces: A Survey of Principles, Models and Frameworks. In *Human Machine Interaction. Lecture Notes in Computer Science*; Springer: Berlin/Heidelberg, Germany, 2009; Volume 5440, pp. 3–26.
6. Poryzala, P.; Materka, A. Cluster analysis of CCA coefficients for robust detection of the asynchronous SSVEPs in brain–computer interfaces. *Biomed. Signal Process. Control* **2014**, *10*, 201–208. [CrossRef]
7. Kocejko, T.; Bujnowski, A.; Wtorek, J. Eye mouse for disabled. In Proceedings of the IEEE 2008 Conference on Human System Interaction (HIS), Krakow, Poland, 25–27 May 2008.
8. Krolak, A.; Strumillo, P. Eye-blink detection system for human–computer interaction. *Univers. Access Inf. Soc.* **2011**, *11*, 409–419. [CrossRef]
9. Ascari, R.; Silva, L.; Pereira, R. Personalized gestural interaction applied in a gesture interactive game-based approach for people with disabilities. In Proceedings of the International Conference on Intelligent User Interfaces, Cagliari, Italy, 17–20 March 2020; pp. 100–110.
10. Song, Y.; Luo, Y.; Lin, J. Detection of Movements of Head and Mouth to Provide Computer Access for Disabled. In Proceedings of the IEEE 2011 International Conference on Technologies and Applications of Artificial Intelligence, Chung-Li, Taiwan, 11–13 November 2011.
11. Jia, P.; Hu, H.H.; Lu, T.; Yuan, K. Head gesture recognition for hands-free control of an intelligent wheelchair. *Ind. Robot Int. J.* **2007**, *34*, 60–68. [CrossRef]
12. Solea, R.; Margarit, A.; Cernega, D.; Serbencu, A. Head movement control of powered wheelchair. In Proceedings of the 23rd International Conference on System Theory, Control and Computing, ICSTCC 2019, Sinaia, Romania, 9–11 October 2019; pp. 632–637.
13. Dobrea, M.; Dobrea, D.; Severin, I. A New Wearable System for Head Gesture Recognition Designed to Control an Intelligent Wheelchair. In Proceedings of the 2019 E-Health and Bioengineering Conference (EHB), Iasi, Romania, 21–23 November 2019; pp. 1–5. [CrossRef]
14. Ishizuka, A.; Yorozu, A.; Takahashi, M. Driving Control of a Powered Wheelchair Considering Uncertainty of Gaze Input in an Unknown Environment. *Appl. Sci.* **2018**, *8*, 267. [CrossRef]
15. Matsuzawa, K.; Ishii, C. Control of an electric wheelchair with a brain-computer interface headset. In Proceedings of the IEEE 2016 International Conference on Advanced Mechatronic Systems (ICAMechS), Melbourne, VIC, Australia, 30 November–3 December 2016.
16. Mitra, S.; Acharya, T. Gesture Recognition: A Survey. *IEEE Trans. Syst. Man Cybern. Part C* **2007**, *37*, 311–324. [CrossRef]
17. Yang, J.-Y.; Wang, J.-S.; Chen, Y.-P. Using acceleration measurements for activity recognition: An effective learning algorithm for constructing neural classifiers. *Pattern Recognit. Lett.* **2008**, *29*, 2213–2220. [CrossRef]
18. Maziewski, P.; Kupryjanow, A.; Kaszuba, K.; Czyzewski, A. Accelerometer signal pre-processing influence on human activity recognition. In Proceedings of the IEEE Signal Processing Algorithms, Architectures, Arrangements, and Applications SPA 2009, Poznan, Poland, 24–26 September 2009.
19. Wu, C.W.; Yang, H.Z.; Chen, Y.A.; Ensa, B.; Ren, Y.; Tseng, Y.C. Applying machine learning to head gesture recognition using wearables. In Proceedings of the 2017 IEEE 8th International Conference on Awareness Science and Technology (iCAST), Taichung, Taiwan, 8–10 November 2017; pp. 436–440. [CrossRef]

20. Lorenzi, P.; Rao, R.; Romano, G.; Kita, A.; Irrera, F. Mobile Devices for the Real-time detection of specific human motion disorders. *IEEE Sens. J.* **2016**, *16*, 8220–8227.

21. Baraka, A.; Shaban, H.; Abou El-Nasr, M.; Attallah, O. Wearable Accelerometer and sEMG-Based Upper Limb BSN for Tele-Rehabilitation. *Appl. Sci.* **2019**, *9*, 2795. [CrossRef]

22. He, C.; Kazanzides, P.; Sen, H.T.; Kim, S.; Liu, Y. An Inertial and Optical Sensor Fusion Approach for Six Degree-of-Freedom Pose Estimation. *Sensors* **2015**, *15*, 16448–16465. [CrossRef] [PubMed]

23. Montero Quispe, K.G.; Sousa Lima, W.; Macêdo Batista, D.; Souto, E. MBOSS: A Symbolic Representation of Human Activity Recognition Using Mobile Sensors. *Sensors* **2018**, *18*, 4354. [CrossRef] [PubMed]

24. Allik, A.; Pilt, K.; Karai, D.; Fridolin, I.; Leier, M.; Jervan, G. Optimization of Physical Activity Recognition for Real-Time Wearable Systems: Effect of Window Length, Sampling Frequency and Number of Features. *Appl. Sci.* **2019**, *9*, 4833. [CrossRef]

25. Van der Straaten, R.; Bruijnes, A.K.B.D.; Vanwanseele, B.; Jonkers, I.; De Baets, L.; Timmermans, A. Reliability and Agreement of 3D Trunk and Lower Extremity Movement Analysis by Means of Inertial Sensor Technology for Unipodal and Bipodal Tasks. *Sensors* **2019**, *19*, 141. [CrossRef] [PubMed]

26. Skulimowski, P.; Owczarek, M.; Radecki, A.; Bujacz, M.; Rzeszotarski, D.; Strumillo, P. Interactive Sonification of U-depth Images in a Navigation Aid for the Visually Impaired. *J. Multimodal User Interfaces* **2018**. [CrossRef]

27. Baranski, P.; Strumillo, P. Emphatic trials of a teleassistance system for the visually impaired. *J. Med. Imaging Health Inform.* **2015**, *5*, 1640–1651. [CrossRef]

28. The Producer's DUO MLX. Available online: https://duo3d.com/ (accessed on 29 May 2020).

29. Anna Borowska-Terka. Available online: http://eletel.p.lodz.pl/abterka (accessed on 29 May 2020).

30. Theodoridis, S.; Koutroumbas, K. *Pattern Recognition*, 4th ed.; Academic Press, Inc.: Orlando, FL, USA, 2008.

31. Koronacki, J.; Cwik, J. *Statistical Learning Systems*, 2nd ed.; Academic Publishing House EXIT: Warsaw, Poland, 2015. (In Polish)

32. Random Forests. Available online: https://www.stat.berkeley.edu/~{}breiman/RandomForests/cc_home.htm (accessed on 29 May 2020).

33. Zakariah, M. Classification of large datasets using Random Forest Algorithm in various applications: Survey. *Int. J. Eng. Innov. Technol.* **2014**, *4*, 189–198.

34. Aha, D.W.; Kibler, D.; Albert, M.K. Instance-based learning algorithms. *Mach. Learn.* **1991**, *6*, 37–66. [CrossRef]

35. Poulet, F.; Do, T.N. Mining Very Large Datasets with Support Vector Machine Algorithms. In *Enterprise Information Systems V*; Camp, O., Filipe, J.B.L., Hammoudi, S., Piattini, M., Eds.; Springer: Dordrecht, The Netherlands, 2004; pp. 177–184.

36. Schölkopf, B.; Smola, A.J. *Learning with Kernels*; MIT Press: Cambridge, MA, USA, 2002.

37. Janez, D. Statistical Comparisons of Classifiers over Multiple Data Sets. *J. Mach. Learn. Res.* **2006**, *7*, 1–30.

38. Caraiman, S.; Morar, A.; Owczarek Burlacu, A.; Rzeszotarski, D.; Botezatu, N.; Herghelegiu, P.; Moldoveanu, F.; Strumillo, P.; Moldoveanu, A. Computer Vision for the Visually Impaired: The Sound of Vision System. In Proceedings of the IEEE International Conference on Computer Vision Workshops (ICCVW), Venice, Italy, 22–29 October 2017; pp. 1480–1489.

Article

A Multi-Layer Perceptron Network for Perfusion Parameter Estimation in DCE-MRI Studies of the Healthy Kidney

Artur Klepaczko [1], Michał Strzelecki [1,*], Marcin Kociołek [1], Eli Eikefjord [2] and Arvid Lundervold [2,3,4]

[1] Institute of Electronics, Lodz University of Technology, 90-924 Lodz, Poland; artur.klepaczko@p.lodz.pl (A.K.); MARCIN.KOCIOLEK@P.LODZ.PL (M.K.)
[2] Department of Health and Functioning, Western Norway University of Applied Sciences, 5063 Bergen, Norway; Eli.Bjovad.Eikefjord@hvl.no (E.E.); Arvid.Lundervold@uib.no (A.L.)
[3] Department of Biomedicine, University of Bergen, 5009 Bergen, Norway
[4] Mohn Medical Imaging and Visualization Centre, Department of Radiology, Haukeland University Hospital, 5021 Bergen, Norway
* Correspondence: michal.strzelecki@p.lodz.pl

Received: 28 June 2020; Accepted: 5 August 2020; Published: 10 August 2020

Abstract: Background: Dynamic contrast-enhanced magnetic resonance imaging (DCE-MRI) is an imaging technique which helps in visualizing and quantifying perfusion—one of the most important indicators of an organ's state. This paper focuses on perfusion and filtration in the kidney, whose performance directly influences versatile functions of the body. In clinical practice, kidney function is assessed by measuring glomerular filtration rate (GFR). Estimating GFR based on DCE-MRI data requires the application of an organ-specific pharmacokinetic (PK) model. However, determination of the model parameters, and thus the characterization of GFR, is sensitive to determination of the arterial input function (AIF) and the initial choice of parameter values. Methods: This paper proposes a multi-layer perceptron network for PK model parameter determination, in order to overcome the limitations of the traditional model's optimization techniques based on non-linear least-squares curve-fitting. As a reference method, we applied the trust-region reflective algorithm to numerically optimize the model. The effectiveness of the proposed approach was tested for 20 data sets, collected for 10 healthy volunteers whose image-derived GFR scores were compared with ground-truth blood test values. Results: The achieved mean difference between the image-derived and ground-truth GFR values was 2.35 mL/min/1.73 m^2, which is comparable to the result obtained for the reference estimation method (−5.80 mL/min/1.73 m^2). Conclusions: Neural networks are a feasible alternative to the least-squares curve-fitting algorithm, ensuring agreement with ground-truth measurements at a comparable level. The advantages of using a neural network are twofold. Firstly, it can estimate a GFR value without the need to determine the AIF for each individual patient. Secondly, a reliable estimate can be obtained, without the need to manually set up either the initial parameter values or the constraints thereof.

Keywords: dynamic contrast-enhanced MRI; kidney perfusion; glomerular filtration rate; pharmacokinetic modeling; multi-layer perceptron; parameter estimation

---

1. Introduction

Monitoring the state of kidney functioning has become an increasingly important task in modern medical diagnostics. Failures in renal operation may impair physiological homeostasis, leading to the improper management of electrolytes, acid-base balance perturbation, or the deregulation of arterial blood pressure. Kidneys are responsible for blood filtration, and the removal of water-soluble waste products of metabolism and surplus glucose and other organic substances. Kidney diseases include, among others, acute kidney injury [1], chronic kidney disease (CKD) [2] and various cancers (e.g., renal cell carcinoma). Treatment depends on the pathological condition, and may require life-long dialysis, or kidney removal or transplantation. In any case, precise knowledge about kidney performance is needed for the proper diagnosis, treatment and follow-up prognosis. Especially in the case of CKD, which may be caused by longstanding hypertension or diabetes mellitus, it is important to continuously and precisely monitor renal function, since early detection of the disease allows the prevention of its development to the end stage.

Tissue perfusion is a viable means of preserving cell nutrition, humoral communication and the elimination of waste products over a lifetime [3]. Thus information about local blood flow and the blood filtration taking part in the glomeruli plays a fundamental role in the diagnosis and follow-up of the aforementioned diseases. Routinely, renal diagnosis is accomplished through the creatinine clearance procedure, or by using the more specific iohexol clearance test. These tests lead to determination of the glomerular filtration rate (GFR)—the main indicator of kidney performance. However, neither procedure enables differentiation between left and right kidney function [4].

Dynamic contrast-enhanced (DCE) magnetic resonance imaging (MRI) is an alternative technique available for the assessment of kidney functioning [5–7]. The procedure requires the administration of a contrast agent (CA) intravenously, given as a bolus injection, whose passage through the abdominal arterial system and the capillary bed and tubular systems of the kidneys is measured. In contrast to conventional methods, DCE-MRI enables the estimation of GFR for a single kidney, ultimately at the voxel-level, and simultaneously it provides anatomical characteristics of the renal parenchyma.

The estimation of renal filtration based on the DCE-MR images involves the pharmacokinetic (PK) modeling of the contrast agent, arriving from the abdominal aorta into a kidney artery, the renal parenchyma and the afferent arterioles of the renal corpuscles, and then passing from the glomerular capillaries through the filtration barrier into the Bowman's space and the tubular system, ending up in the renal pelvis, the ureter and then the bladder as a constituent of the urine. There have been numerous PK models proposed in the literature [8], which vary in (1) the number of tissue compartments, (2) the level of incorporating compartment-specific impulse residue functions, (3) the inclusion of the effect of tracer agent leakage from capillaries in cancerous cells, and (4) the way in which the delay and the broadening of peak CA concentration in the renal parenchyma, relative to the abdominal aorta, is modeled. In the case of the normal-state kidney, all recent models exploit a two-compartment approach [9]. In this setting, at any point in time, the contrast agent concentration in the tissue is the sum of concentrations in the intravascular (IV) and extracellular extravascular (EEV) compartments. In the present study, we used the two-compartment filtration model (2CFM) proposed by Tofts et al. [10].

One approach to finding the PK model's parameters (defined below) is to fit the model-based signal intensity time course to the image data using a non-linear least-squares (NLLS) method. Depending on the model, various explicit parameters are used to characterize tracer agent kinetics. The so-called primary or independent parameters include plasma volume, plasma flow, interstitial volume and permeability product [9]. Based on these attributes, one may derive dependent parameters, i.e., the extraction fraction and the volume transfer constant. The latter, denoted as K^{trans}, characterizes regional renal filtration, as it measures the ratio of plasma flow that is transferred from the glomerular capillary bed into the Bowman's space. K^{trans} multiplied by the kidney volume leads to an estimate of the glomerular filtration rate, and is therefore a central descriptor in quantifying kidney function.

In this study, we propose to use a multi-layer perceptron (MLP) network for PK model parameter estimation, instead of the non-linear least-squares curve-fitting method. The rationale behind this concept arises from the fact that the results of the iterative methods may be located in a local and—in general—suboptimal solution. Thus, the outcome depends largely on the initial choice of parameter values. Moreover, one usually has to establish constraints on the expected parameter values, so that the result falls into a physiologically feasible range. Bounds that are too narrow, however, may prevent the fitting algorithm from finding an optimum. Contrary to NLLS, MLP does not require that one provides the initial parameter values; moreover, it ensures the optimal solution when properly trained.

Although MLP networks could be regarded as a legacy method, they are still in use, and prove efficient in complex non-linear regression problems [11,12]. Moreover, according to the universal approximation theorem, an artificial neural network with at least one hidden layer of nonpolynomial activation (such as sigmoid or rectified linear) can model any measurable function that maps one finite-dimensional space to another, provided the hidden layers contain enough units [13]. Therefore, MLP seems to be particularly usefully in its application to the specific problem of PK model parameter estimation in renal DCE-MRI examinations.

Also, the reliability of the fitted PK model parameters is conditioned by—among other factors—the determination of the arterial input function (AIF), i.e., the tracer concentration's time course in the feeding artery. The standard way of annotating AIF consists of estimating it for each patient. However, other approaches are accepted if patient-specific AIF measurements are not possible. Such problems can arise due to data acquisition constraints or the lack of a suitable artery within the imaging field of view. In the alternative strategy, one either utilizes a population-based mean AIF [14] (calculated for a cohort of studies, where it is feasible to do so) or defines its parameterized functional form [15]. In this study, we implicitly follow the population-based approach. The neural network is trained using a pool of AIFs automatically derived for a subset of available DCE images. Then, the GFR for an individual test case is estimated without the need to determine the case-specific AIF.

To sum up, the main contribution of this paper is the establishment of the architecture of an MLP network for estimating parameters of the 2CFM, based solely on image intensity time courses in the kidney region. Additionally, a side-contribution is the introduction of the algorithm for automatic determination of the AIF in DCE-MRI examinations. The latter achievement was needed in order to construct the data set for MLP training, but the algorithm can be equally used in standard approaches to PK modeling.

2. Methods and Materials

2.1. Subjects

A total of 20 DCE-MRI examinations were available for experiments [16]. The data sets were collected from 10 healthy volunteers. Each subject was imaged twice, seven days apart (further, these examinations will be referred to as Session 1 and 2). The acquisition sequence used the standard 3D fast low-angle shot (FLASH) spoiled gradient recalled echo technique with the following parameters: TR = 2.36 ms, TE = 0.8 ms, FA = 20°, in-plane resolution = 2.2×2.2 mm^2, slice thickness = 3 mm, acquisition matrix = 192×192, number of slices = 30. Prior to image acquisition, patients were administered 0.025 mmol/kg of GdDOTA at 3 mL/s flow rate. The contrast agent was injected intravenously. Then, 74 volumetric scans were gathered in 2.3 s time intervals.

Every image series was registered in the time domain using the b-splines method, as implemented in the Insight Toolkit (ITK) library [17]. Specifically, we used Slicer 3D software to accomplish the task. For every study, the same reference frame was used as a fixed volume that every other (moving) volume was matched to. B-splines registration was performed fully automatically, i.e., no fiducial points were marked over tissues of interest. Furthermore, the procedure was launched in a multi-stage configuration. In each stage, various settings of grid size and subsampling rates were used. For a detailed interpretation of these parameters, the reader is referred to the ITK documentation. In short, they allow the registration of images at different scales—starting from coarse matching and then refining the outcome.

A common practice in DCE series analysis is to interpolate the measured signal so as to achieve a higher temporal resolution and capture more precisely the critical phases of the perfusion process, such as, e.g., bolus arrival time or the peak of the signal enhancement [10]. As such, we applied cubic spline interpolation to the observed intensity time courses, and then resampled the interpolated signal curves at 0.4 milliseconds intervals.

As stated in the introduction, analysis of the DCE images leads to estimation of the GFR. Thus, in order to validate the obtained estimates against ground-truth values, volunteers underwent iohexol clearance tests. The measurement was carried out by administrating a dose of 5 mL of iohexol (300 mg I/mL; Omnipaque 300, GE Healthcare), and then acquiring a venous blood sample after 4 h.

No specific diet was imposed on the participating subjects, but they were asked to abstain from alcohol and high-protein meals, avoid exhausting physical exertion, be normally hydrated at least 2 days before examination, and have no caffeine on the examination day. Apart from these restrictions, participants were instructed to retain their ordinary habits concerning the type and amount of consumed food so as to assure stable examination conditions. All subjects gave their informed consent for inclusion before they participated in the study. The study was conducted in accordance with the Declaration of Helsinki, and the protocol was approved by the Regional Committees for Medical Research Ethics Western Norway (REC West 2012/1869).

2.2. Two-Compartment Filtration Model of Kidney Perfusion

In the PK model considered in this study [10], it is assumed that renal tissue is composed of two compartments—an intravascular (IV) one and an extracellular extravascular (EEV) one (Figure 1). The model can be applied either to the whole kidney parenchyma or the kidney cortex only. The assumption which could be violated here is that CA does not flow out from the tubular system within the fitting period.

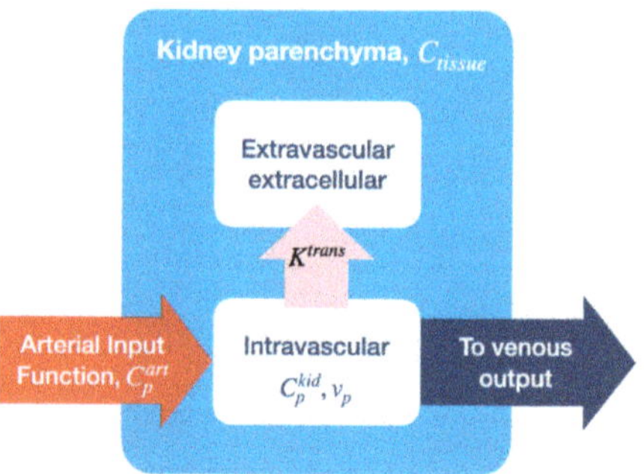

Figure 1. Schematic illustration of the two-compartment filtration model.

The contrast agent's concentration in the renal tissue $C_{tissue}(t)$ is governed by the equation

$$C_{tissue}(t) \; = \; K^{trans} \int_0^t C_p^{kid}(\tau) + v_p C_p^{kid}(t), \tag{1}$$

where C_p^{art} denotes the arterial input function, v_p—plasma volume fraction, and C_p^{kid}—CA concentration in the plasma, given by the convolution

$$C_p^{kid} \; = \; C_p^{art} \otimes g(t) \; = \; \int_0^t C_p^{art}(t - \tau) g(\tau) d\tau. \tag{2}$$

The first term in (1) represents the EEV compartment, whereas Equation (2) models the delay and dispersion of the arterial input upon its arrival to the IV compartment by using the concept of the vascular impulse response function (VIRF), denoted as $g(t)$. In our study we used the delayed exponential VIRF, defined as

$$g(t) = \begin{cases} 0 & t < \Delta \\ \frac{1}{T_g}e^{-\frac{t-\Delta}{T_g}} & t \geq \Delta \end{cases},$$ (3)

with

$$\int_0^\infty g(t)dt = 1.$$ (4)

The variables T_g—the dispersion time constant—and Δ—the delay interval—together with the volume fraction v_p and transfer constant K^{trans} form the complete set of the Tofts model's parameters.

2.3. Arterial Input Function

Although in the proposed approach, the trained neural network enables the determination of GFR without annotation of a patient-specific arterial input function, information about the time course of the contrast agent's concentration in a tissue-feeding artery is required at the training stage. As further described, we use a subset of image-derived AIFs to construct a training data set. Therefore, in this section, we present our custom method for automatic AIF determination.

Most frequently, AIF is determined based on signal averaged over a manually delineated region of interest near to the perfused organ [18,19]. Specifically, in the case of renography it is recommended to locate the AIF region-of-interest (ROI) near the place where the aorta bifurcates into common iliac arteries. This location is in relative proximity to the kidney, and is simultaneously sufficiently far from the upper aorta section, where the signal coming from the blood that enters the imaging slab at high velocity is high, because of the inflow artifact and not because of the contrast agent [19]. Moreover, Cutajar et al. in [18] confirmed the significant dependence of renal perfusion and the filtration coefficients on the size of the chosen AIF ROI. In [20], AIF was determined from voxels chosen semi-automatically on the maximum signal-enhancement maps as the brightest locations in the abdominal aorta, directly below the inlets of the renal arteries. Note that the manual or semi-automatic approaches suffer from poor reproducibility and subjectivism, and they can be time-consuming and can require the availability of well-trained staff. In addition, it may be necessary in such cases to obtain patient-specific hemodynamic parameters in order to adjust the DCE-MRI protocol to the arterial pulse wave, and thus reduce the inflow artifact. Therefore, there were several attempts made to automate determination of the AIF. A common approach employs clustering techniques [21]. On the other hand, in [22], the authors used Kendall's coefficient of concordance. These attempts, however, were either applied in preclinical studies, or to different organs such as the head or neck, but not to the kidney.

We postulate a fully automatic procedure for determining a patient-specific AIF based on the clustering of abdominal aorta voxels' time-series. Our approach consists of three stages, as illustrated in Figure 2. The algorithm starts with delineation of the abdominal aorta region. This task, further decomposed into four steps, is accomplished through the vessel segmentation of one time frame of the DCE-MR image series. The appropriate time frame is selected for maximum signal enhancement in the bottom-central region of the image, near the aorta's bifurcation into common iliac arteries. The peak intensity value in this region corresponds to the time point when the CA fills the whole aorta lumen, thus ensuring the highest contrast between the background tissue and the aorta at its entire length.

Next, the image is blurred by the low-pass Gaussian filter ($\sigma = 3$ voxels) to suppress fine structures that could be misinterpreted as vessels. Such a smoothed image is then submitted to the vessel enhancement routine. In this step, we computed a multi-scale vesselness function in the form proposed by Sato [23], using 5 scales of radius standard deviation ranging from 0.5 to 2.5 mm. As a result, the filtered image contains only the elongated bright structures. Apart from the aorta, it may contain smaller vessels, like renal arteries and some other elongated structures. In order to remove them, the flood fill operation is executed, with the lower threshold set to an empirically found value equal to 12.5% of the maximum image intensity. The upper bound is set to maximum, as we found the aorta to be the brightest structure visible in the time frames selected for segmentation. Similarly, the seed point of the flood fill operation was determined as the brightest point in the central part of the image along the coronal direction.

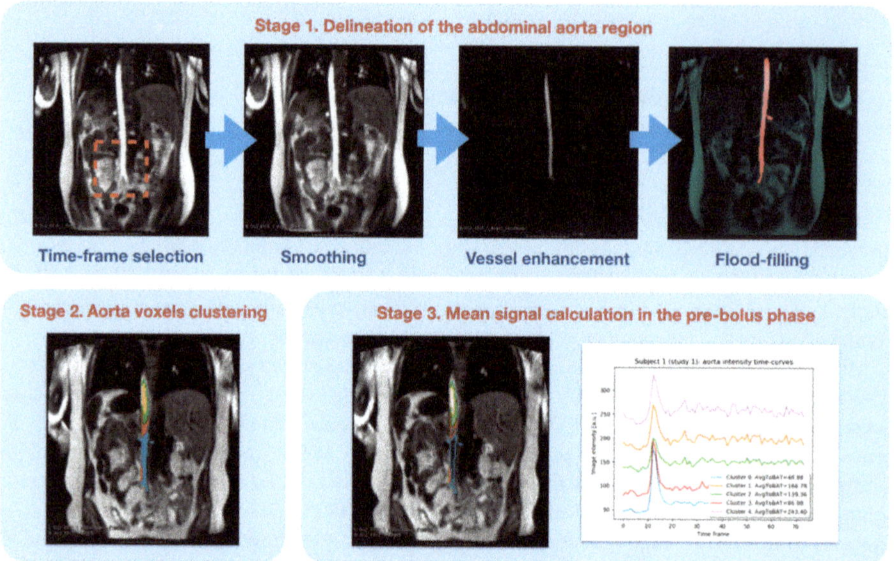

Figure 2. Steps of automatic arterial input function (AIF) determination algorithm.

In the second stage, abdominal aorta voxels are grouped into clusters using their signal intensity time courses as the feature vectors. For this purpose, we employed a k-means clustering algorithm with $k = 5$. We found this value to be a reasonable tradeoff between the need to reflect apparent diversity in the aorta signal dynamics and the need to keep the cluster size sufficiently large, so that each one represents a significant portion of aorta voxels. As an effect of clustering, points whose temporal signal characteristics are similar are joined. Eventually, we select one of the clusters as the AIF ROI. The selection is based on the minimum mean signal value calculated in the pre-bolus phase. We assume that the best AIF candidate region has the lowest signal in this initial phase of bolus passage, as it is free from the inflow-enhancement artifact.

2.4. PK Model Fitting

There are several optimizers utilized in the NLLS curve-fitting problems. Downhill Simplex, Levenberg–Marquardt or non-linear conjugate gradients belong to the most frequently cited algorithms [24]. In the DCE-MRI context, a popular solver is based on the gradient expansion method implemented in the IDL Software [7]. Often, in order to ensure that the resulting parameter set falls into a range of reasonable physiological values, it is also necessary to restrain the optimization procedure to a specific solution subspace. In such a case, the Trust Region Reflective technique can be considered as the algorithm of choice as it enables optimization with constraints. All the above-mentioned methods are iterative, and require initiation with a starting point representing a feasible solution. As such, there is a risk of getting stuck in a local minimum of the objective function, usually defined as the mean squared error of the model residuals.

Therefore, we propose to solve the curve-fitting problem using an artificial neural network (ANN); specifically, a fully-connected multi-layer perceptron structure. ANNs have been successfully employed in a variety of optimization scenarios, including the fitting of complex, multi-parametric relationships to the observed noisy data [25]. The neural network training process is configured to explore a wide range of feasible parameter values, potentially covering the whole solution space. For every parameter candidate set, there is a corresponding CA concentration time course generated based on the PK model equation. Thus, the network incorporates a priori knowledge about the underlying mechanism of signal generation. Then, the network's response to an observed, patient-specific CA time-course is estimated in the generalized context of tracer kinetics, making the fitting process more robust to noise, measurement errors and local minima.

The first step in the design of an MLP network is the choice of the number of hidden layers. As formulated by the already recalled universal approximation theorem, one sufficiently large hidden layer should enable the accurate approximation of the modeled function. In cases of complex problems, however, two layers with lower numbers of units may be easier to train, and could better acquire a general pattern that links input data vectors with estimated function parameters. We have observed such behavior in the current study. One-hidden-layer structures turned out to be more prone to overfitting, and their training process was less liable to converge. Therefore, in the rest of the paper we concentrate on the description of the three-layer architecture, i.e., consisting of two hidden layers and one output layer. However, in the Results section we provide example outcomes of training two-layer structures to support the above observations. On the other hand, increasing the network depth to three hidden layers did not improve training efficacy, while making the task of optimizing the network architecture more complicated. Hence, the results of these computations are omitted in the experiments description.

The overall concept of a three-layer perceptron network operating as an estimator of the Tofts model parameters is shown in Figure 3. The network design is represented by weight vectors $w = [w_1, w_2]$, and v for two hidden layers and the output layer neurons respectively. The symbols $f(\cdot)$ and $h(\cdot)$ denote the activation functions of perceptrons in a given layer. In our implementation we used the same activation for both hidden layers, which was the rectified linear unit function, i.e., $h(x) = \max(x, 0)$. For the output layer we used linear activation. The detailed structure of implemented MLP is presented in the Results section.

In principle, the network inputs accept a vector $C_{tissue}(t)$ of CA concentrations in the subsequent time frames of a given DCE series. Transformation from the image signal intensity $S(t)$ to concentration $C_{tissue}(t)$ is given in the Appendix A. In the following, the time variable t will be skipped where possible, for the sake of clarity.

There are two modes of network operation. In the *recall* mode, a trained network recognizes the pattern in the observed data and predicts the possible values of the PK model's parameters. Formally, the network realizes the transformation $\hat{\theta} = f(v, h(w, C_{tissue}))$, which approximates the unknown mapping $\hat{\theta} = \mathcal{F}(C_{tissue})$ from the observation space to the parameter space. It is assumed that this mapping exists and is unique. In the *training* mode, the network is fed by a large and versatile collection of input vectors, and simultaneously it is presented with an expected parameter set θ for each individual input example. The difference between these true parameter values and the output drives the process of network weights adjustment. Specifically, we applied the mean squared error (MSE) metric as the network loss function. Apart from MSE, during the network weights adjustment, the mean absolute percentage error (MAPE) was also calculated for both the training and validation samples. The definitions of the MSE and MAPE metrics are given in Appendix A.

The weights assigned to random values are updated after each iteration using the error back-propagation algorithm [26]. The loss function (dependent on inter-layer weights) was optimized using the stochastic gradient descent algorithm [27]. The network was implemented in a Python script using Keras library with the Tensorflow backend [28].

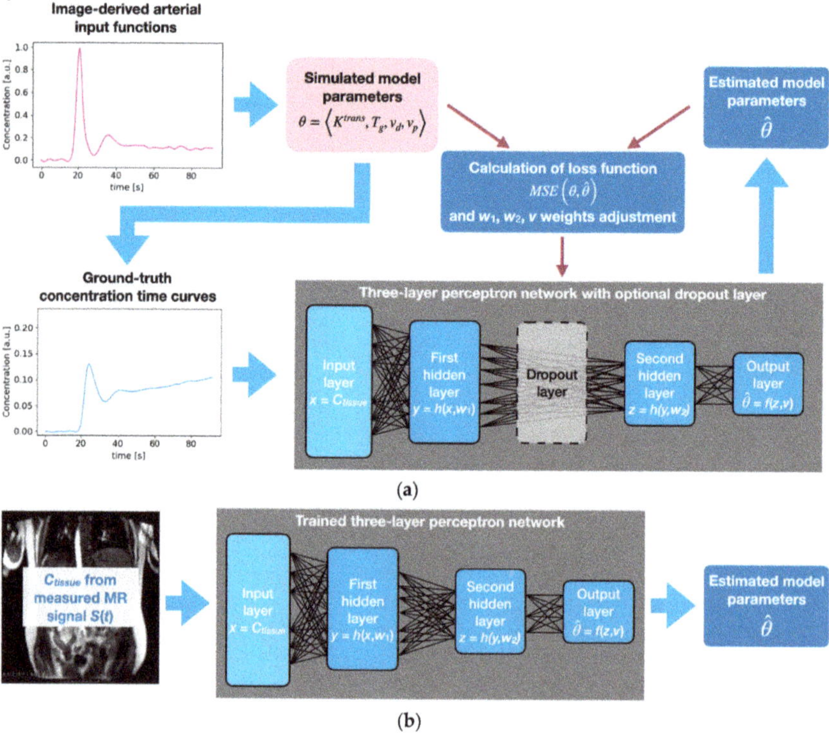

Figure 3. Overview of a pharmacokinetic (PK) model parameter fitting using an artificial neural network operating in the training (**a**) and recall (**b**) modes. The three-layer perceptron structure is shown only conceptually—the number of hidden units was adjusted during the network design process. The number of output units was always set to 4, i.e., the number of PK model parameters. All layers are fully connected, as in the conventional multi-layer perceptron (MLP) structures, except dropout mechanism was switched on, in which case random connections between hidden layers were removed during training to prevent network overfitting.

The crucial step in the above-described procedure is in the preparation of the training set. On the one hand it must be sufficiently large and varied for the network to gain the generalization capabilities. On the other hand, the input signal intensity vectors must be accompanied by the true parameter values, which allow the observation of a given C_{tissue} in response to a specific arterial input function. Moreover, since the available data set size is relatively small, we repeat this procedure using the leave-one-out approach. Having a data set of 20 examinations, we invoked calculations 20 times, each time excluding another subject AIF from the training stage. Hence, a training set in a given repetition was established via the 3-step procedure described below.

Step 1. Selecting a subset of 19 realistic arterial input functions.

Step 2. Establishing ranges of model parameter values. Table 1 presents the assumed parameter limits along with their corresponding probability distributions. These limits were fixed with respect to the published values for normal subjects [7,10,29]. Training input samples were generated using the model equation and parameters sampled from these distributions. According to the Tofts model [10], there are four such parameters: K^{trans}, v_p, T_g and Δ. The latter two are the parameters of the vascular impulse response function, whose sum defines the so-called mean residence time (MRT). Since [7,10] report only the value of MRT, we sampled T_g and MRT, whereas Δ was calculated as

$$\Delta = \mathrm{MRT} - T_g \tag{5}$$

Step 3. For a given AIF, we sample the model parameters θ and calculate C_{tissue} according to Equation (1). Sampling is repeated 10^4 times for each AIF. As a result, there were $19 \times 10^4 < C(t), \theta >$ pairs available in each leave-one-out repetition. The calculated C_{tissue} curves were probed at time steps adjusted to match the temporal resolution of the network input vector.

Table 1. Parameter limits and types of probability distributions used for sampling training data set for ANN (artificial neural network).

Parameter	Min	Max	Mean	SD	Probability Distribution
K^{trans} [min^{-1}]	0	–	0.25	0.1	Normal
v_p [mm^3]	0.2	0.8	–	–	Uniform
Δ [s]	1	3.5	–	–	Uniform
MRT [s]	–	–	5.5	0.7	Normal
T_g [s]	0.02	–	–	–	Normal

The simulated tracer concentration time curves were optionally corrupted by the Gaussian noise to increase their variability and make them more realistic. Data corruption was performed by adding to each simulated CA concentration a value drawn from the normal distribution, with mean = 0, and standard deviation was adjusted to a current time step, i.e., $\sigma_{noise} = s \cdot [C_{tissue}(t) - min(C_{tissue}(t))]$. The scale factor s was experimentally set to 0.025 to achieve reasonable deflection from the modeled CA concentration time curves. Then, the data set was partitioned into training and test sets in the proportion 7:3. Moreover, 30% of the training data vectors were randomly set apart in each epoch for validation purposes.

The last issue which needs to be fixed is the establishing of a proper network architecture. During experiments, it occurred that only one hidden layer cannot accurately encapsulate the non-linear relationship between the PK model's parameters and the variety of CA concentration time-curves. The two-layer structure ensured a significant decrease in the loss function value, while the usage of more than two hidden layers did not offer any further improvement in this respect. Eventually, the number of perceptrons in each layer had to be established. We tested 12 configurations, listed in Table 2. Apart from the various cardinalities of neurons in hidden layers, we also considered inclusion of additional dropout layers between them to overcome the problem of overfitting [30]. For every configuration and each leave-one-out fold, the training was run for a fixed number of 100 epochs.

Table 2. Number of neurons in the tested two-layer perceptron structures.

Configuration No.	First Hidden Layer	Dropout Rate *	Second Hidden Layer	Output Layer
#1	20	0.0	10	
#2		0.2		
#3	26	0.0	13	
#4		0.2		
#5	30	0.0	15	4
#6		0.2		
#7	34	0.0	17	
#8		0.2		
#9	40	0.0	20	
#10		0.2		
#11	50	0.0	25	
#12		0.2		

* Dropout rate = 0.0 means that effectively there was no dropout mechanism between hidden layers.

3. Results

We compared the effectiveness of the postulated ANN-based approach to optimizing a PK model's parameters in reference to the algebraic curve-fitting method, configured to utilize the Trust Region Reflective algorithm [31]. Apart from the direct comparison of particular model parameters, we also estimated single- and two-kidney GFRs. The image-derived estimates were then compared against reference values measured with iohexol clearance tests.

We also validated the designed algorithm for the determination of the arterial input function. In this experiment, we calculated the GFR scores using the Trust Region Reflective method for AIFs annotated manually, and using the proposed automated approach.

For every study we performed manual segmentation of the whole kidney, as well as cortex and medulla. This step was accomplished by two professionals experienced in medical image analysis, using our custom software annotation tool. The rate of agreement between the respective kidney regions delineated by both experts was estimated by the Jaccard coefficient, and was equal to 0.93 on average. For a given study, kidney segments were outlined only in a time frame corresponding to the maximum signal enhancement in the perfusion phase, which ensured the largest contrast between cortex, medulla and pelvis. Depending on the study, this maximal enhancement was observed in frames numbered from 12 to 18. Then, the segmentations were applied to all other time frames reflected in the parameter estimation procedure, so as to determine mean signal time courses in the perfused renal tissue, and the selected frames became fixed reference frames in the b-spline registration algorithm.

3.1. Assessment of Automatic AIF Determination Method

Figure 4 visualizes the locations of the arterial input function ROIs, as identified by the proposed automatic method and manual selection. The presented examples correspond to the two cases, where the best and the worst model fits were obtained. As shown, in comparison to manual AIF annotations, automatically found ROIs occupy many more voxels that are free from inflow artifacts. As a result, fitted intensity time curves better encapsulate image signal changes. For example, in Figure 4c,e, the curve is visible in the region of maximum signal enhancement. In the case of automatically determined AIF, the curve approximates well the measured signal samples, whereas in the case of manual annotations, the model overestimates signal values. The reliability of the GFR measurements based on automatically determined AIFs is visualized in Figure 5b, and can be compared to the Bland–Altman plots shown in Figure 5a, obtained with manually annotated AIFs. Since the automatically determined AIFs seemed to improve the accuracy and precision of GFR assessment, this allowed us to use them in the generation of the training set for the neural network.

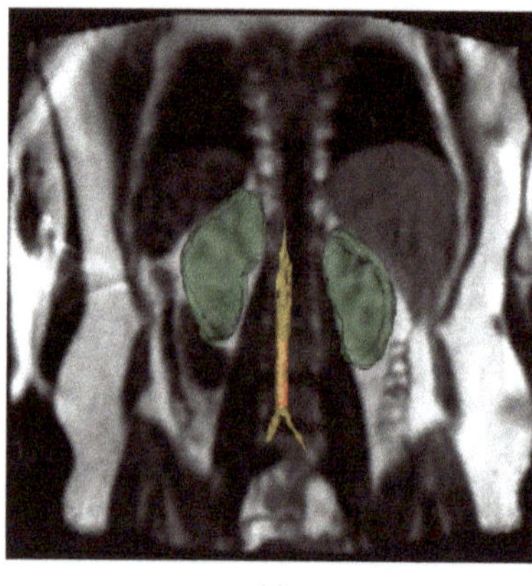 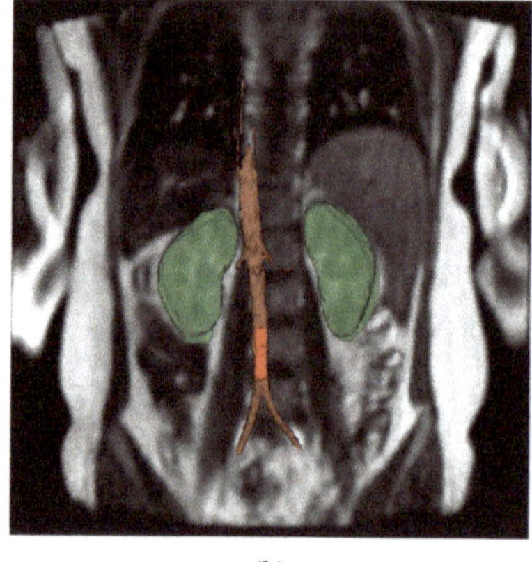

(a)   (b)

Figure 4. *Cont.*

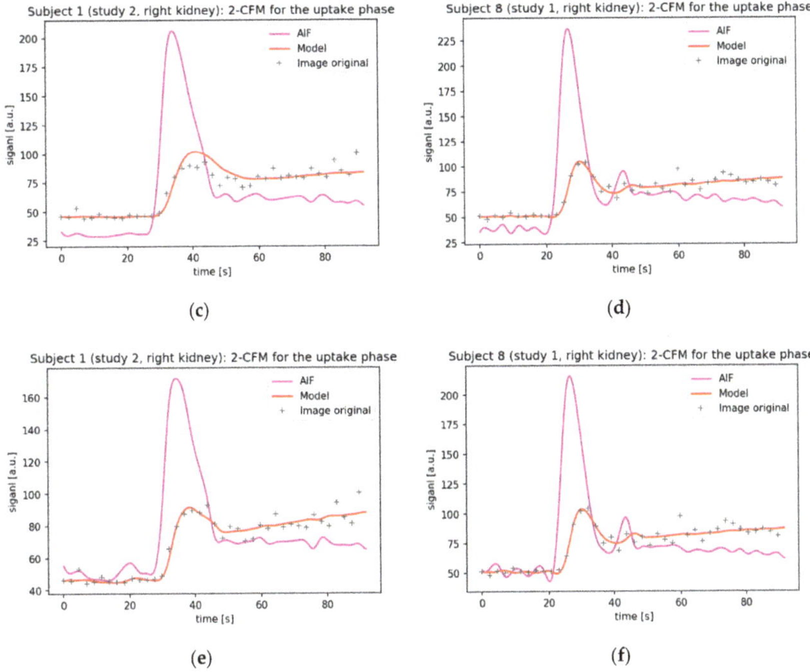

(c)

(d)

(e)

(f)

Figure 4. Examples of automatically determined arterial input function region-of-interests (ROIs) (**a**,**b**). For comparison, manually delineated ROIs placed directly above aorta bifurcation into iliac arteries are shown as red rectangles. The corresponding fitted model curves are shown in (**c**,**d**) for manual ROIs and in (**e**,**f**) for the automatic ones.

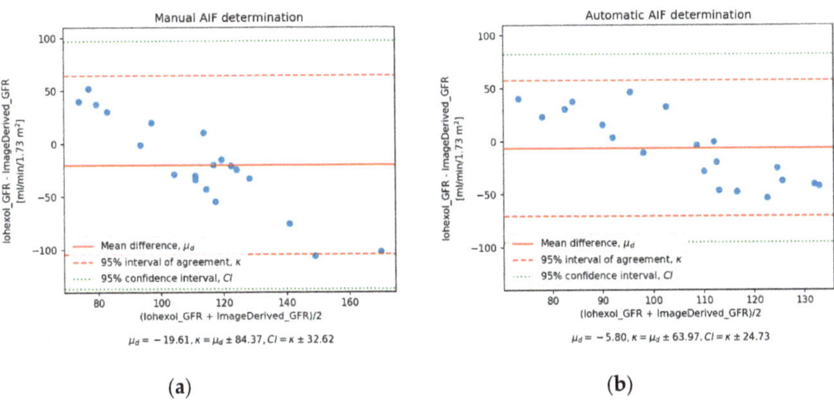

(a)

(b)

Figure 5. Bland–Altman plots of agreement for manually (**a**) and automatically (**b**) determined AIF ROIs. Measurements were evaluated against normality using Shapiro–Wilk test. The obtained p-values were equal to 0.52 (iohexol-based glomerular filtration rate (GFRs)), 0.29 (image-derived GFRs with manual ROIs) and 0.12 (for image-derived GFRs with automatically determined ROIs), thus disallowing us to reject the null hypothesis of the data with a normal distribution.

3.2. Selecting Network Configuration

In the first phase of experiments, we aimed to select an architecture for the neural network suitable to solving the problem of fitting the 2CFM to the DCE-MRI data. We began with deciding upon the number of hidden layers and their activation units. Figure 6 presents the learning curves for three configurations of two-layer architectures, whereas the plots shown in Figure 7 correspond to the three-layer network configurations listed in Table 2. Although the selected learning curves were obtained only for one of the leave-one-out repetitions, they are representative for all the subjects in the study. It can be seen that in all cases, the loss function initially decreases rapidly, and converges to a value in the range 0.01–0.13 depending on the network architecture and the existence of noise in the training data. However, for configurations without the dropout layer, it is observable that the MSE value in a validation subset heavily fluctuates until the very end of the training process. It must be noted that for the single-hidden-layer configurations (Figure 6), the problem with the convergence of the loss function also persists in the largest tested network, with 50 hidden units, despite even inserting a dropout layer before the output nodes. This observation, combined with potential overfitting (which manifests in larger errors in a validation sample, with respect to a training set), substantiates the use of two hidden layers. In the latter case, the loss function stabilizes after around 40 iterations for 'dropout' configurations with higher numbers of neurons (architectures number 6, 8, 10 and 12). It is especially pronounced if the training data was corrupted by noise.

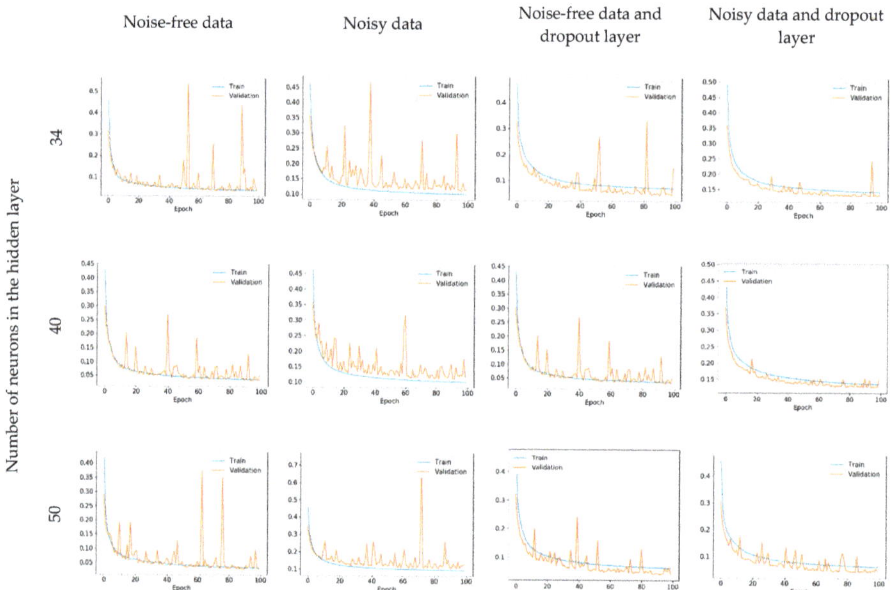

Figure 6. Learning curves for various two-layer MLP configurations—training set scores in blue, validation set scores in orange. The plots in the first two and second two columns correspond to architectures without and with the dropout layer, respectively. The curves in columns 1 and 3 were obtained for the noise-free training data, whereas in columns 2 and 4 for the data corrupted with Gaussian noise.

Apparently, increased data variability due to the presence of noise, along with the inclusion of a dropout layer, ensures better loss function convergence and helps prevent the network from overfitting. As can be observed in Figure 7 (second column), the loss function scores on the validation set start to surpass the values obtained for the training set quite early in the learning process. This overfitting dynamic is further confirmed by the MAPE scores obtained after 100 learning epochs (see Figure 8). Without the dropout mechanism, the error remains higher for the validation set than for the training one, for almost every network configuration.

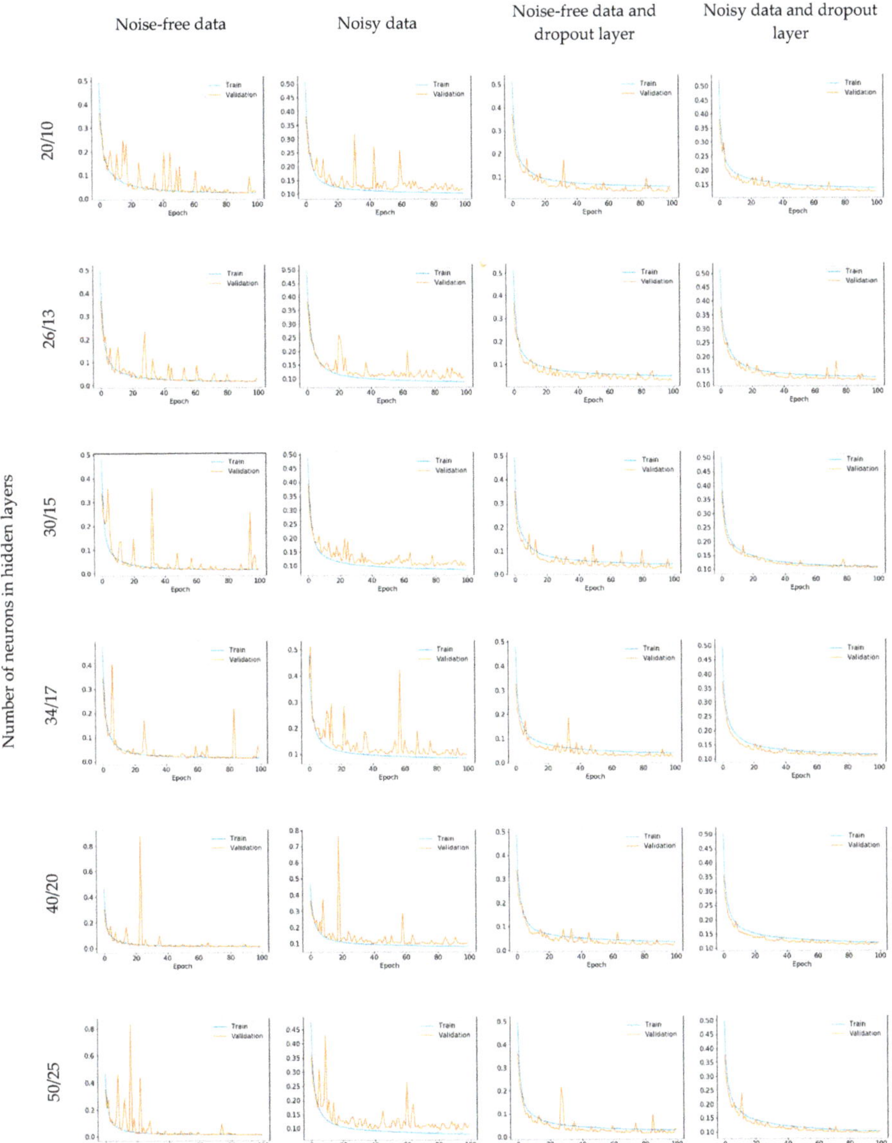

Figure 7. Learning curves for various three-layer MLP configurations—training set scores in blue, validation set scores in orange. The plots in the first two and two second two columns correspond to architectures without and with the dropout layer, respectively. The curves in columns 1 and 3 were obtained for the noise-free training data, whereas in columns 2 and 4 for the data corrupted with Gaussian noise.

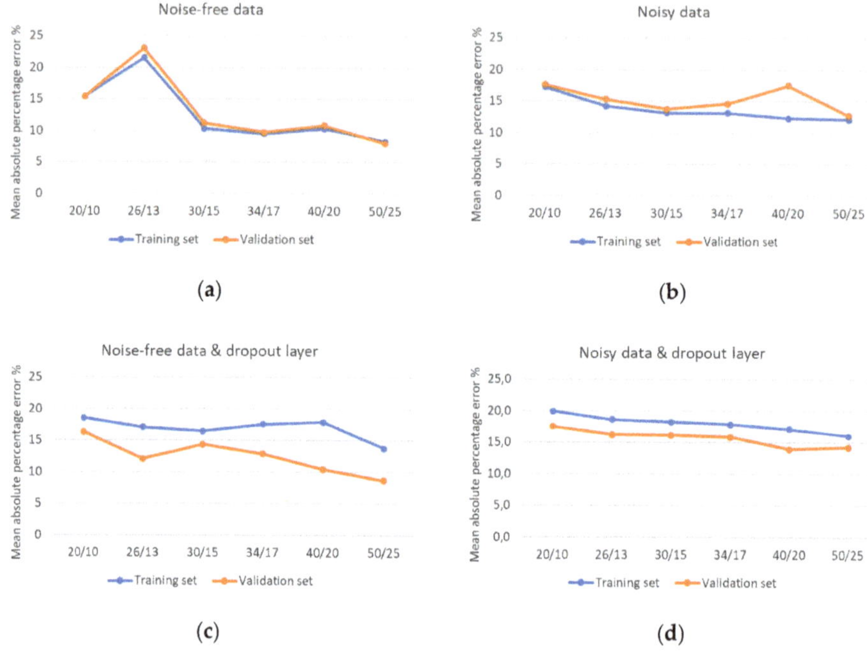

Figure 8. (**a–d**): Mean absolute percentage error obtained for various network architectures and with respect to existence of noise in the training and validation data sets.

While the above analysis settles upon the usage of the dropout layer and the noisy training vectors, the question regarding the sizes of the hidden layers must still be resolved. On one side, increasing the number of neurons leads to the lowering of the MSE and MAPE scores. On the other hand, however, it escalates the risk of overfitting. Figure 9 presents the distributions of the Δ parameter estimates against the true values in the test set. Regression analysis shows good agreement between the predicted and the actual parameter estimates. The performed t-test and its resulting large p-values indicate that one cannot reject the hypothesis of equality between the group means. However, the overfitting effect manifests itself in the case of noise-free data, and it appears deeper for the more complex MLP architectures. The network tends to predict parameter values close to some discrete levels. Although the determination coefficients R^2 are higher in these cases, clearly the predicated outcomes do not conform to the actual data distribution. This tendency is not observed in the case of noisy data; however, changing the number of neurons in the first hidden layer from 40 to 50 did not bring significant improvement in the R^2 score (0.85 versus 0.86), and simultaneously, as shown in Figure 8d, it slightly increased the mean absolute percentage error for the validation set (from 13.9% to 14.3%).

Therefore, for the rest of the experiments we chose configuration #10, with 40 and 20 perceptrons in the first and second hidden layers, respectively, and with a dropout layer in between, with a dropout rate = 0.2. This setting ensures the lowest mean absolute percentage error, relatively high R^2 scores for all estimated parameters (Figure 10), and that the network model's capacity is appropriate for the data set's complexity.

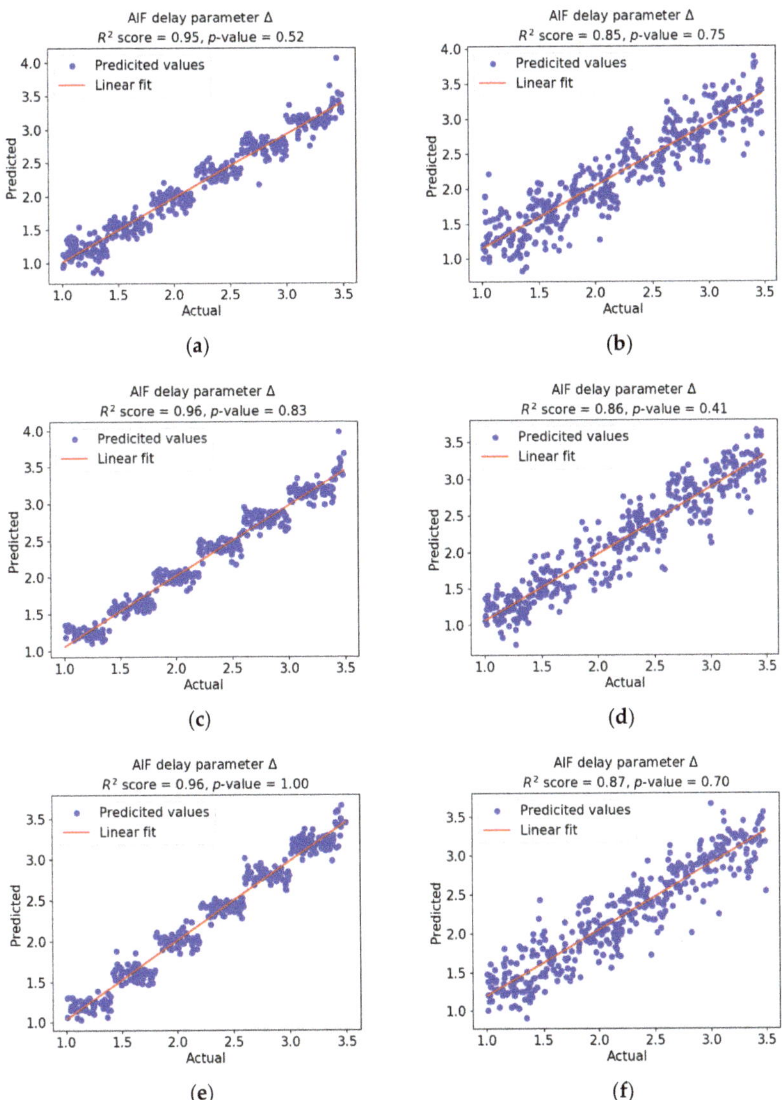

Figure 9. Predicted against actual values of the Δ parameter in the test set for three configurations of the MLP structure with the dropout layer. The number of neurons in the hidden layers: 30/15 (**a**,**b**), 40/20 (**c**,**d**), 50/25 (**e**,**f**). Plots in the left and right columns represent, correspondingly, noise-free and noisy data.

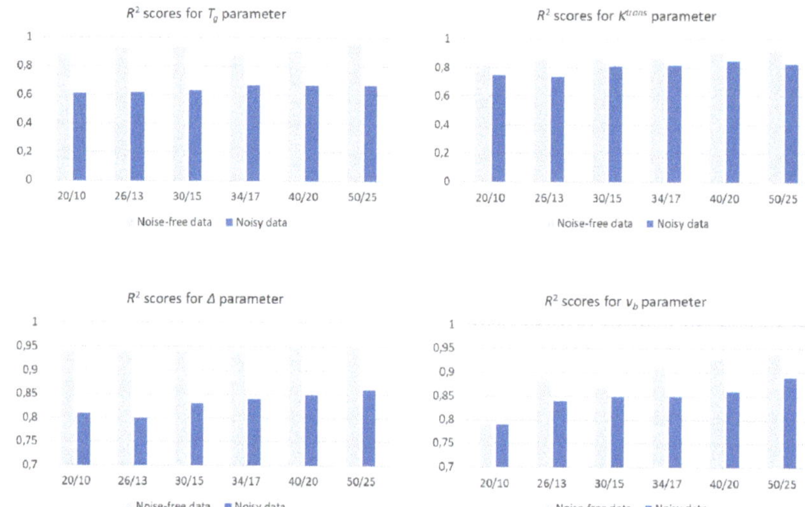

Figure 10. Determination coefficients for two-compartment filtration model (2CFM) model parameters obtained for various network configurations (test set).

3.3. Evaluation of MLP-Based Model Fitting Reliability

The second phase of experiments consisted of using the trained MLP network to estimate 2CFM model parameters, as well as the GFR values for the 20 DCE-MRI examinations available in this study. The statistics of the parameter estimates are summarized in Table 3, whereas the calculated GFR values are collected in Table 4. The reliability of GFR estimation was also evaluated against the ground-truth using the Bland–Altman plot, shown in Figure 11. When compared to Figure 5b, it emerges that the mean biases from the reference values are similar for both optimization methods—MLP and NLLS curve-fitting. However, the limits of agreement κ and confidence intervals are narrower for the neural network. Moreover, our ANNs ensure the better balance of the estimated values around the mean. The NLLS method, on the other hand, tends to overestimate the measured quantity, especially in the lower range of values. On the other hand, in the case of ANN, GFR was estimated without explicit determination of AIF for a given subject. It proves that the designed network was capable of extracting the general characteristics of the underlying perfusion process.

From the analysis of Table 3, it emerges that MLP determined K^{trans} and v_p parameter values close to those of the reference method. The box plots presented in Figure 12a,b visually illustrate that the ranges of estimated K^{trans} and v_p values partially overlap, and their medians are similar. The result of the t-test gives evidence that the v_p results, as estimated by MLP and NLLS, are not statistically different (p-value > 0.3), which effectively shows that both methods can be used interchangeably for blood volume fraction assessment. In the case of K^{trans}, the difference in parameter estimates becomes statistically significant (p-value < 10^{-3}), although the medians remain relatively close with respect to the overall data variability. Estimates of T_g and Δ vary more significantly (cf. Figure 12c,d). The evidence against the null hypothesis of there being equal means in the MLP- and NLLS-based calculations is even stronger than for the K^{trans} parameter (p-value < 10^{-6}). Apparently, the variation in the vascular impulse response function manifests itself in the CA concentration time courses in a way which is less evident in comparison to the other two PK model parameters. As such, the neural network could not precisely infer the relationship between the concentration time curves and the expected T_g and Δ values. This effect is consistent with the observed R^2 scores calculated on the test set upon MLP training. The VIRF timing factors are remarkably lower than the blood volume fraction and transfer constant.

Table 3. Statistics of the estimated values of the two-compartment filtration model parameters.

Parameter	MLP Estimates				NLLS Estimates			
	Min	Max	Mean	Std. Dev.	Min	Max	Mean	Std. Dev.
K^{trans} [min^{-1}] ($\times 10^{-2}$)	0.40	0.50	0.45	0.02	0.39	0.60	0.51	0.03
v_p [mm^3]	0.33	0.91	0.62	0.08	0.35	0.86	0.67	0.09
Δ [s]	1.10	1.69	1.34	0.15	1.00	4.97	2.46	0.92
T_g [s]	4.68	5.35	4.95	0.18	1.00	4.70	1.84	0.82

Table 4. Juxtaposition of GFR values estimated by MLP and NLLS methods based on DCE-MRI data acquired in two examination sessions in comparison with ground-truth scores measured with iohexol clearance test [ml/min/1.73 m^2].

Subject Number	MLP Estimates						NLLS Estimates						Ground-Truth (Total)
	Session 1			Session 2			Session 1			Session 2			
	Left Kidney	Right Kidney	Total	Left Kidney	Right Kidney	Total	Left Kidney	Right Kidney	Total	Left Kidney	Right Kidney	Total	
1	48	70	118	56	59	115	48	62	110	69	75	144	107
2	52	62	114	56	71	127	52	67	119	34	48	82	98
3	40	35	75	44	42	86	41	25	66	67	69	136	90
4	43	49	92	46	58	104	49	55	103	72	68	140	93
5	42	45	87	41	44	85	19	34	53	49	41	90	94
6	33	32	65	51	54	105	32	33	65	47	75	122	103
7	50	46	96	55	47	101	72	65	137	71	83	154	112
8	41	48	89	41	47	88	38	34	72	44	42	86	96
9	49	55	104	54	65	119	55	69	124	70	79	149	119
10	64	72	136	52	53	105	58	54	112	63	89	152	112

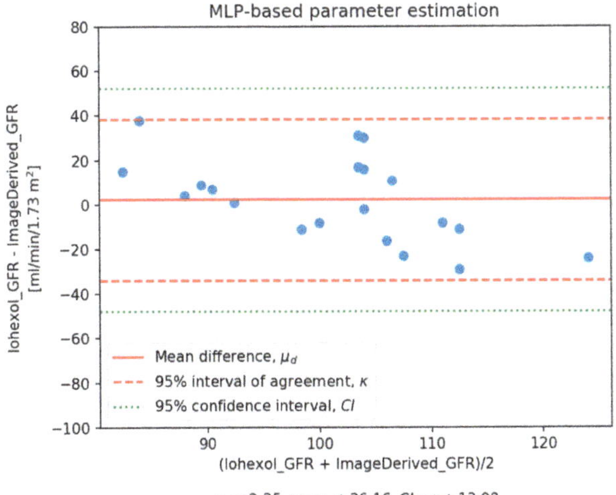

Figure 11. Bland–Altman plot of agreement for MLP-based GFR estimation method. Normal distribution of the measurements was confirmed using Shapiro–Wilk test. The obtained *p*-values were equal to 0.52 (iohexol-based GFRs) and 0.97 (image-derived MLP-based GFRs).

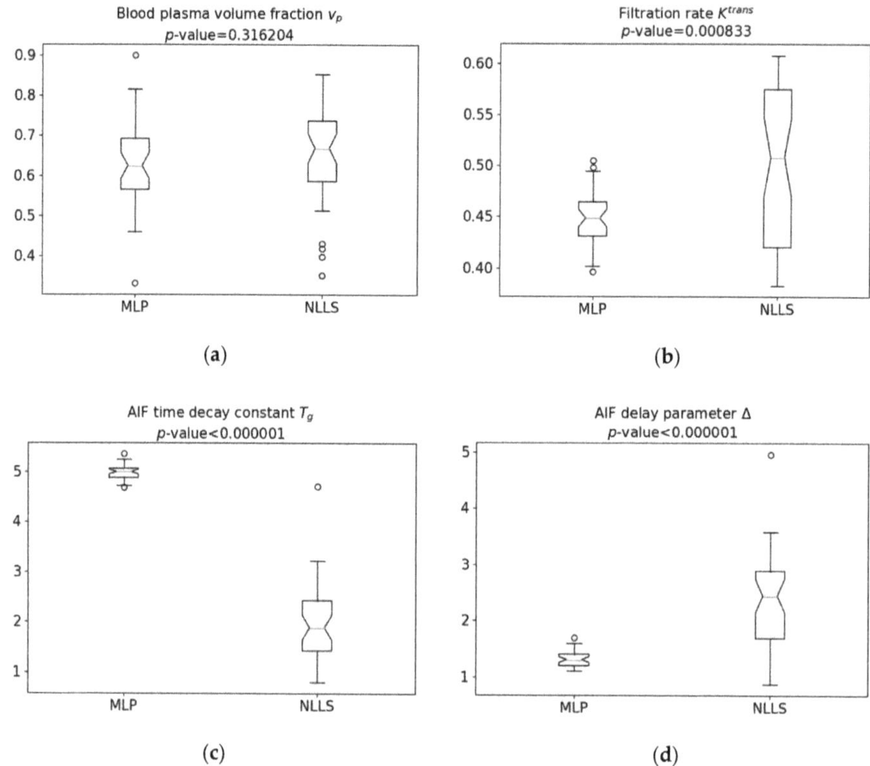

Figure 12. (a–d): Box plots showing distribution of 2CFM parameter values estimated by MLP and NLLS methods. Provided *p*-values were obtained for the *t*-test under the null hypothesis that mean estimated values are the same for both methods.

4. Discussion and Conclusions

The main goal of this study was to evaluate the possibility of estimating the parameters of the two-compartment filtration model using an artificial neural network. As shown in the Results section, especially in Table 4, overall the task was accomplished successfully. The estimated single- and two-kidney estimates of glomerular filtration rate fall into the physiologically feasible range, and are close to the reference values calculated with the Trust Region Reflective algorithm (a non-linear least-squares curve-fitting method), as well as those measured by blood tests. Comparison with the latter method, which serves as a gold standard in clinics, appears particularly optimistic. The mean difference between the corresponding measurements is to μ_d = 2.35 mL/min/1.73 m^2, with the agreement interval κ = μ_d ± 36.16 mL/min/1.73 m^2. This interval is even narrower than the one obtained for the NLLS method (κ = μ_d ± 63.97 mL/min/1.73 m^2), which signals the greater precision of the designed MLP structures. This effect was achieved thanks to the usage of dropout layers between the hidden ones, and the training of the network with data vectors purposely corrupted with noise.

Although the trained MLP network performs relatively well in predicting K^{trans}, v_p and Δ parameters, it actually fails in estimating accurate values for the VIRF decay time constant T_g. The corresponding R^2 determination coefficients calculated for the test sets do not exceed the value of 0.64, which is remarkably less than for the other three parameters (0.85–0.92). T_g, however, does not affect the calculation of GFR, which depends only on K^{trans}. In contrast with the NLLS procedure, a neural network does not actually fit a modeled curve to the input data. A failure in predicting a given parameter value does not necessarily cause an error in another parameter's estimation.

It is instructive to compare the results obtained in this study to similar measurements presented in [16], partly conducted on the same subjects. The scores presented therein were obtained using a conventional curve-fitting approach, and were divided with respect to examination session. The mean differences and limits of agreement reported in their study were $\mu_d = 1.5$ mL/min/1.73 m^2 and $\kappa = \mu_d \pm$ 43.2 mL/min/1.73 m^2 (Session 1), and $\mu_d = 6.1$ mL/min/1.73 m^2 and $\kappa = \mu_d \pm 31.9$ mL/min/1.73 m^2 (Session 2). Hence, our MLP-based method estimates true GFR values at a comparable level, demonstrating either slightly better (in the case of Session 1) or poorer (Session 2) precision. Apart from different parameter estimation methods, the observed discrepancies may be caused by other factors, including various post-processing algorithms, e.g., image registration and segmentation. Above all, a different pharmacokinetic model has been utilized, namely the Sourbron's two-compartment separable model [7].

Regarding the PK models themselves, the most common of them are often claimed to be too simple to properly represent patient-specific biophysiological processes. Then, even the most precise parameter estimation may not lead to accurate GFR calculation. There have been several attempts to develop more advanced models, including multi-compartmental [32] or patient-specific models [33], as well as those with a modified functional form of CA retention in the renal tissue compartments [34]. As such, the reliability of our proposed MLP-based method for DCE-MRI processing should be further examined with respect to various mathematical models of kidney perfusion.

Another constraint that narrows the scope of the conclusions that can be derived from this study is that it presents methodological achievements tested solely on healthy subjects. Although we believe that the proposed methods of ANN-based PK model parameter estimation and AIF determination will manifest its applicability to diseased kidneys, this paper focuses exclusively on algorithms per se and numerical image processing, rather than clinical applications. Such an approach is common in numerous works published in the biomedical engineering domain, which postulate novel processing techniques, including those devoted to renal perfusion measurement [7,10,16,35]. The reasoning behind this approach states that the technical quality of the designed solution must be verified under well-defined and controlled conditions for a statistically representative data sample. Such conditions are ensured by a cohort of healthy volunteers. Besides, in the current study, the inclusion of abnormal cases would not necessarily provide additional insight into the behavior of the automatic determination of AIF. In the case of diseased kidneys, it should run in an unchanged manner, since tissue lesion is independent of the blood flow dynamics in abdominal aorta. On the other hand, taking into account renal impairments would involve experimenting with various pharmacokinetic models of perfusion, and it would remarkably extend the scope of this study.

To conclude, the performed experiments show the potential of neural networks as an alternative computational framework to the standard curve-fitting procedure in the context of quantitative perfusion estimation. The overall agreement with the gold standard method obtained by ANN was comparable to the results obtained using the non-linear least-squares approach. Moreover, GFR can be estimated without explicit determination of AIF for a given patient, which constitutes the major advantage of the proposed approach. Instead, AIFs must be found only for a limited number of studies included in the training data set. The described algorithm for AIF ROI determination allows one to determine it in an automatic manner, reducing the effect of the inflow artifact. It simultaneously increases the reliability and precision of the 2CFM model's parameter estimation. Hence, no additional process is required in assessing patient-specific hemodynamic parameters by utilizing either clinical examinations (MR angiography [36,37] or ultrasound [38]) or non-clinical methods (e.g., photopletysmographic arterial pulse wave measurement [39]) that would otherwise be necessary to properly trigger the DCE-MRI sequence.

Author Contributions: Conceptualization, A.K.; Data curation, E.E. and A.L.; Investigation, A.K. and M.S.; Methodology, A.K.; Resources, E.E. and A.L.; Software, A.K. and M.K.; Supervision, M.S. and A.L.; Validation, E.E.; Visualization, M.K.; Writing—original draft, A.K. All authors have read and agreed to the published version of the manuscript.

Funding: The paper was partially supported by the Polish National Science Centre grant no. 2014/15/B/ST7/05227.

Conflicts of Interest: The authors declare no conflict of interest.

Appendix A

Appendix A.1. Transforming Signal to Concentration

Assuming the gradient echo sequence, C_{tissue} in a given time step t is derived from the observed signal $S(t)$ using the transformation

$$C_{tissue}(t) = \frac{R_1(t) - 1/T_{10}}{r_1} \tag{A1}$$

where T_{10} is the longitudinal relaxation time before arrival of the tracer agent bolus, r_1 is the compartment-specific longitudinal relaxivity constant and

$$R_1(t) = -\frac{1}{TR} \ln\left[\frac{1 - a(t)b}{1 - a(t)b \cos\alpha}\right] \tag{A2}$$

with

$$a(t) = \frac{S(t)}{S_0} \tag{A3}$$

$$b = \frac{\left(1 - e^{-R_{10}TR}\right)}{1 - \cos\alpha\, e^{-R_{10}TR}}. \tag{A4}$$

In the equations above, α and TR are the MRI sequence parameters of flip angle and repetition time, S_0 is the signal baseline, i.e., before arrival of the CA bolus, and R_{10} is the relaxation constant equal to $1/T_{10}$. Note that in the case of the tissue signal, the conversion function is valid only under the assumption of equal longitudinal relaxivities in the EEV and IV compartments.

Appendix A.2. Assessment of MLP Training Process

The *MSE* metric used in our study as the loss function driving the MLP training procedure and the *MAPE* score used to validate training efficiency are defined as follows

$$\text{MSE} = \frac{1}{NM} \sum_{i=0}^{N-1} \sum_{j=0}^{M-1} \left(\hat{\theta}_{ij} - \theta_{ij}\right)^2 \tag{A5}$$

$$\text{MAPE} = \frac{1}{NM} \sum_{i=0}^{N-1} \sum_{j=0}^{M-1} \left|\frac{\hat{\theta}_{i,j} - \theta_{i,j}}{\theta_{i,j}}\right| \tag{A6}$$

where $\theta_{i,j}$, $\hat{\theta}_{i,j}$ denote the actual and estimated value of a j-th parameter for an i-th data training vector, whereas M and N are the total numbers of parameters and data vectors, correspondingly. Since the model parameters have different scales, it was necessary to scale their values to the same range so as to balance their contribution to the overall loss function.

References

1. Gameiro, J.; Fonseca, A.J.; Jorge, S.; Lopes, A.J. Acute Kidney Injury Definition and Diagnosis: A Narrative Review. *J. Clin. Med.* **2018**, *7*, 307. [CrossRef]
2. Thomas, R.; Kanso, A.; Sedor, R.J. Chronic kidney disease and its complications. *Prim. Care* **2008**, *35*, 329–344. [CrossRef] [PubMed]
3. Vander, A.J.; Sherman, J.H.; Luciano, D.S. *Human Physiology. The Mechanism of Body Function*, 8th ed.; McGraw-Hill Publishing: New York City, NY, USA, 2001.

4. Stevens, L.A.; Levey, A.S. Measured GFR as a Confirmatory Test for Estimated GFR. *J. Am. Soc. Nephrol.* **2009**, *20*, 2305–2313. [CrossRef] [PubMed]

5. Hackstein, N.; Heckrodt, J.; Rau, W.S. Measurement of single-kidney glomerular filtration rate using a contrast-enhanced dynamic gradient-echo sequence and the Rutland-Patlak plot technique. *J. Magn. Reson. Imaging* **2003**, *18*, 714–725. [CrossRef] [PubMed]

6. Annet, L.; Hermoye, L.; Peeters, F.; Jamar, F.; Dehoux, J.P.; Van Beers, B.E. Glomerular filtration rate: Assessment with dynamic contrast enhanced MRI and a cortical compartment model in the rabbit kidney. *J. Magn. Reson. Imaging* **2004**, *20*, 843–849. [CrossRef] [PubMed]

7. Sourbron, S.P.; Michaely, H.J.; Reiser, M.F.; Schoenberg, S.O. MRI-measurement of perfusion and glomerular filtration in the human kidney with a separable compartment model. *Investig. Radiol.* **2008**, *43*, 40–48. [CrossRef] [PubMed]

8. Cárdenas-Rodríguez, J.; Li, X.; Whisenant, J.G.; Barnes, S.; Stollberger, R.; Gore, J.C.; Yankeelov, T.E. The basic principles of dynamic contrast-enhanced magnetic resonance imaging. In *MR and CT Perfusion and Pharmacokinetic Imaging. Clinical Applications and Theory*; Bammer, R., Ed.; Wolters Kluwer: Philadelphia, PA, USA, 2016; Chapter 23; pp. 332–347.

9. Sourbron, S.; Lee, T.-Y. Pharmacokinetic models for dynamic contrast-enhanced computed tomography and magnetic resonance imaging. In *MR and CT Perfusion and Pharmacokinetic Imaging. Clinical Applications and Theory*; Bammer, R., Ed.; Wolters Kluwer: Philadelphia, PA, USA, 2016; Chapter 29; pp. 412–430.

10. Tofts, P.S.; Cutajar, M.; Mendichovszky, I.A.; Peters, A.M.; Gordon, I. Precise measurement of renal filtration and vascular parameters using a two-compartment model for dynamic contrast-enhanced MRI of the kidney gives realistic normal values. *Eur. Radiol.* **2012**, *22*, 1320–1330. [CrossRef]

11. Ohn, I.; Kim, Y. Smooth Function Approximation by Deep Neural Networks with General Activation Functions. *Entropy* **2019**, *21*, 627. [CrossRef]

12. Liu, L.; Ma, D.; Azar, A.T.; Zhu, Q. Neural Computing Enhanced Parameter Estimation for Multi-Input and Multi-Output Total Non-Linear Dynamic Models. *Entropy* **2020**, *22*, 510. [CrossRef]

13. Goodfellow, I.; Bengio, Y.; Courville, A. *Deep Learning*; The MIT Press: Cambridge, MA, USA, 2016.

14. Shukla-Dave, A.; Lee, N.; Stambuk, H.; Wang, Y.; Huang, W.; Thaler, H.T.; Patel, S.G.; Shah, J.P.; Koutcher, J.A. Average arterial input function for quantitative dynamic contrast enhanced magnetic resonance imaging of neck nodal metastases. *BMC Med. Phys.* **2009**, *9*, 4. [CrossRef]

15. Parker, G.J.M.; Roberts, C.; Macdonald, A.; Buonaccorsi, A.; Cheung, S.; Buckley, D.L.; Jackson, A.; Watson, Y.; Davies, K.; Jayson, G.C. Experimentally-Derived Functional Form for a Population-Averaged High-Temporal-Resolution Arterial Input Function for Dynamic Contrast-Enhanced MRI. *Magn. Reson. Med.* **2006**, *56*, 993–1000. [CrossRef] [PubMed]

16. Eikefjord, E.; Andersen, E.; Hodneland, E.; Hanson, E.A.; Sourbron, S.; Svarstad, E.; Lundervold, A.; Rørvik, J.T. Dynamic contrast-enhanced MRI measurement of renal function in healthy participants. *Acta Radiol.* **2017**, *58*, 748–757. [CrossRef] [PubMed]

17. Johnson, H.J.; McCormick, M.M.; Ibanez, L. *The ITK Software Guide: Design and Functionality*, 4th ed.; Kitware: Clifton Park, NY, USA, 2015.

18. Dujardin, M.; Luypaert, R.; Vandenbroucke, F.; Van der Niepen, P.; Sourbron, S.; Verbeelen, D.; Stadnik, T.; de May, J. Combined T1-based perfusion MRI and MR angiography in kidney: First experience in normals and pathology. *Eur. J. Radiol.* **2009**, *69*, 542–549. [CrossRef] [PubMed]

19. Cutajar, M.; Mendichovszky, I.A.; Tofts, P.S.; Gordon, I. The importance of AIF ROI selection in DCE-MRI renography: Reproducibility and variability of renal perfusion and filtration. *Eur. J. Radiol.* **2010**, *74*, e154–e160. [CrossRef]

20. Shi, L.; Wang, D.; Liu, W.; Fang, K.; Wang, Y.X.; Huang, W.; King, A.D.; Heng, P.A.; Ahuja, A.T. Automatic detection of arterial input function in dynamic contrast enhanced MRI based on affinity propagation clustering. *J. Magn. Reson. Imaging* **2014**, *39*, 1327–1337. [CrossRef]

21. Yin, J.; Yang, J.; Guo, Q. Automatic determination of the arterial input function in dynamic susceptibility contrast MRI: Comparison of different reproducible clustering algorithms. *Neuroradiology* **2015**, *57*, 535–543. [CrossRef]

22. Kim, J.; Im, G.H.; Yang, J.; Choi, D.; Lee, W.J.; Lee, J.H. Quantitative dynamic contrast-enhanced MRI for mouse models using automatic detection of the arterial input function. *NMR Biomed.* **2012**, *25*, 674–684. [CrossRef]

23. Yoshinobu, S.; Shin, N.; Hideki, A.; Thomas, K.; Guido, G.; Shigeyuki, Y.; Ron, K. *3D Multi-Scale Line Filler for Segmentation and Visualization of Curvilinear Structures in Medical Images*; Troccaz, J., Grimson, E., Mösges, R., Eds.; Proc. CVRMed-MRCAS'97, LNCS; Springer: Berlin/Heidelberg, Germany, 1997; pp. 213–222.

24. Bammer, R. *MR and CT Perfusion and Pharmacokinetic Imaging. Clinical Applications and Theory*; Wolters Kluwer: Philadelphia, PA, USA, 2016.

25. Materka, A.; Mizushina, S. Parametric Signal Restoration Using Artificial Neural Networks. *IEEE Trans. Biomed. Eng.* **1996**, *43*, 357–372.

26. Witten, I.H.; Frank, E. *Data Mining. Practical Machine Learning Tools and Techniques*, 2nd ed.; Morgan Kaufman, Elsevier: Amsterdam, The Netherlands, 2005.

27. Bottou, L. On-line Learning and Stochastic Approxmiations. In *On-Line Learning in Neural Networks*; Saad, D., Ed.; Cambridge University Press: Cambridge, UK, 1998; Chapter 2; pp. 9–42.

28. Chollet, F.; Allison, K.; Wicke, M.; Bileschi, S.; Bailey, P.; Gibson, A.; Allaire, J.J. "Keras". 2015. Available online: https://keras.io (accessed on 7 August 2020).

29. Tsushima, Y.; Blomley, M.J.; Kusano, S.; Endo, K. Use of contrast-enhanced computed tomography to measure clearance per unit renal volume: A novel measurement of renal function and fractional vascular volume. *Am. J. Kidney Dis.* **1999**, *33*, 754–760. [CrossRef]

30. Srivastava, N.; Hinton, G.; Krizhevsky, A.; Sutskever, I.; Salakhutdinov, R. Dropout: A Simple Way to Prevent Neural Networks from Overfitting. *J. Mach. Learn. Res.* **2014**, *15*, 1929–1958.

31. Yuan, Y. Recent advances in trust region algorithms. *Math. Program.* **2015**, *151*, 249–281. [CrossRef]

32. Lee, V.S.; Rusinek, H.; Bokacheva, L.; Huang, A.J.; Oesingmann, N.; Chen, Q.; Kaur, M.; Prince, K.; Song, T.; Kramer, E.L.; et al. Renal function measurements from MR renography and a simplified multicompartmental model. *Am. J. Physiol. Renal Physiol.* **2007**, *292*, F1548–F1559. [CrossRef] [PubMed]

33. Tipirneni-Sajja, A.; Loeffler, R.B.; Oesingmann, N.; Bissler, J.; Song, R.; McCarville, B.; Jones, D.P.; Hudson, M.; Spunt, S.L.; Hillenbrand, C.M. Measurement of glomerular filtration rate by dynamic contrast-enhanced magnetic resonance imaging using a subject-specific two-compartment model. *Physiol. Rep.* **2016**, *4*, e12755. [CrossRef] [PubMed]

34. Chen, B.; Zhang, Y.; Song, X.; Wang, X.; Zhang, J.; Fang, J. Quantitative Estimation of Renal Function with Dynamic Contrast-Enhanced MRI Using a Modified Two-Compartment Model. *PLoS ONE* **2014**, *9*, e105087. [CrossRef]

35. Zoellner, F.G.; Sance, R.; Rogelj, P.; Ledesma-Carbayo, M.J.; Rørvik, J.; Santos, A.; Lundervold, A. Assessment of 3D DCE-MRI of the kidneys using non-rigid image registration and segmentation of voxel time courses. *Comput. Med. Imaging Graph.* **2009**, *33*, 171–181. [CrossRef]

36. Klepaczko, A.; Szczypiński, P.; Strzelecki, M.; Stefańczyk, L. Simulation of phase contrast angiography for renal arterial models. *Biomed. Eng. OnLine* **2018**, *17*, 41. [CrossRef]

37. Klepaczko, A.; Szczypiński, P.; Dwojakowski, G.; Strzelecki, M.; Materka, A. Computer simulation of magnetic resonance angiography imaging: Model description and validation. *PLoS ONE* **2014**, *9*, e93689. [CrossRef]

38. Ciccone, M.M.; Iacoviello, M.; Gesualdo, L.; Puzzovivo, A.; Antoncecchi, V.; Doronzo, A.; Monitillo, F.; Citarelli, G.; Paradies, V.; Favale, S. The renal arterial resistance index: A marker of renal function with an independent and incremental role in predicting heart failure progression. *Eur. J. Heart Fail.* **2014**, *16*, 210–216. [CrossRef]

39. Gircys, R.; Kazanavicius, E.; Maskeliunas, R.; Damasevicius, R.; Wozniak, M. Wearable system for real-time monitoring of hemodynamic parameters: Implementation and evaluation. *Biomed. Signal. Proc. Control* **2020**, *59*, 101873. [CrossRef]

Article

Deep Instance Segmentation of Laboratory Animals in Thermal Images

Magdalena Mazur-Milecka *, Tomasz Kocejko * and Jacek Ruminski *

Department of Biomedical Engineering, Faculty of Electronics, Telecommunications and Informatics, Gdansk University of Technology, 80-233 Gdansk, Poland
* Correspondence: magdalena.milecka@pg.edu.pl (M.M.-M.); tomasz.kocejko@pg.edu.pl (T.K.); jacek.ruminski@pg.edu.pl (J.R.)

Received: 9 July 2020; Accepted: 25 August 2020; Published: 28 August 2020

Abstract: In this paper we focus on the role of deep instance segmentation of laboratory rodents in thermal images. Thermal imaging is very suitable to observe the behaviour of laboratory animals, especially in low light conditions. It is an non-intrusive method allowing to monitor the activity of animals and potentially observe some physiological changes expressed in dynamic thermal patterns. The analysis of the recorded sequence of thermal images requires smart algorithms for automatic processing of millions of thermal frames. Instance image segmentation allows to extract each animal from a frame and track its activity and thermal patterns. In this work, we adopted two instance segmentation algorithms, i.e., Mask R-CNN and TensorMask. Both methods in different configurations were applied to a set of thermal sequences, and both achieved high results. The best results were obtained for the TensorMask model, initially pre-trained on visible light images and finally trained on thermal images of rodents. The achieved mean average precision was above 90 percent, which proves that model pre-training on visible images can improve results of thermal image segmentation.

Keywords: instance segmentation; deep learning; computer vision

---

1. Introduction

Laboratory animals' behavior analysis is often used in studies of stress, anxiety, depression or neurodegenerative diseases [1]. The automation of this analysis has undoubtedly many advantages, among which objectivity and standardization are one of the most desirable. As far as the position and motion analysis are easy and commonly used parameters [2], the action or behavior recognition are quite difficult to automatize. Existing systems for laboratory animals' behavior analysis are restricted by, for example, the number or the color of the objects [2]. The spectrum of behavior analysis performed by the available systems is usually limited to exploration, rest or grooming. All these behaviors are detected based on simple object parameters representing the position, speed, direction of the movement or the shape parameters of the observed rodent [3]. Objective detection and analysis of more complex behaviors would open up new possibilities. An important parameter of animals' health and well-being is its social behavior. It ensures survival of the species and is the specific characteristics of a single individual. Futhermore, deviations and abnormalities of social actions may be related to stress, fear or illness [1]. Social behaviors are divided into three categories: aggressive, defensive and neutral [4]. An example of aggressive behavior is an attack, bite or aggressive grooming. However, gentle grooming can also be a form of defense, it can also be confused with other behavior—climbing. All these insignificant differences make the classification of complex behaviors an extremely difficult task, mainly performed by human observers.

The presence of a man who, after all, is a rodent's natural enemy, probably adversely affects the results of behavior research [5]. This can be solved by using camcorders and analyze the behavior from the recordings. An additional advantage of this solution is the ability to slow down or stop the action, certainly helpful during fast animal movements, e.g., fights. Nevertheless, the need for a good lighting for cameras is a very unfavorable and stressful factor for nocturnal animals such as rodents [6,7]. This is where technology that offers alternative imaging comes to the rescue. Thermal camcorders record the surface temperature and can work in limited light and even in complete darkness. In addition, they provide data of the object's surface temperature distribution, proved to be useful in early disease diagnosis [8,9]. Changes in the rodent's surface temperature can be an indicator of his health condition, behavioral changes, anxiety or stress [10,11]. In paper [12] authors use mouse surface temperature distribution as an object identifier to recover tracking algorithm identity swap after close contact.

The automatic analysis of rodents from video recordings requires the identification and tracking of each animal in a scene. The distinction between the object and the background in animal tracking visual systems is most often based on difference between frames, color threshold or color matching [13]. A clear temperature difference in thermal imaging greatly simplifies foreground-background segmentation. Usually, the Otsu's thresholding supported by morphology is enough [2,14]. Figure 1b demonstrates the result of Otsu's thresholding on thermal image. In a cluttered environment there is also a need to segment the object from other parts of the foreground or to distinguish individuals from each other. For small connections of the animals' bodies, methods such as marker-controlled watershed segmentation technique are sufficient [15]. However, the solution to the segmentation problem for the objects that overlap significantly is not easy and most image analysis methods fail. Figure 1c shows the results of rodent segmentation using watershed technique with markers build from the low gradient regions. There are no clear differences between the images of both objects and no border between them is visible. For these reasons a significant number of systems analyze only one individual at a time [2,13]. That is why the correct segmentation between objects is necessary for further analysis.

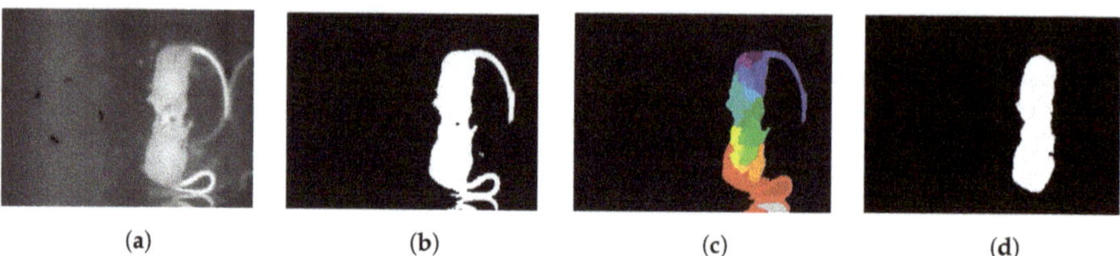

(a)   (b)   (c)   (d)

Figure 1. Results of the most popular algorithms used for segmentation of (a) original image by: (b) Otsu's thresholding, (c) watershed with gradient regions markers, (d) U-Net—semantic segmentation.

Many segmentation techniques have been proposed in literature. However, the methods based on deep-learning have been recently found to provide the best segmentation results in many computer vision tasks. Segmentation techniques based on deep learning may be divided into two approaches: (i) semantic segmentation—pixel classification-based approach and (ii) instance segmentation–object detection and classification based approach.

Segmentation using (i) does not discriminate between instances. In paper [16] we have used popular methods of semantic segmentation—U-Net [17] and V-Net [18] to separate two objects. Algorithms correctly segmented animals in close contact, but they were not able to draw the boundary between the objects at the time of body overlapping. Figure 1d shows the result of U-Net architecture trained on 200 images on five epochs, 2000 steps each. The animals are not separated and are recognized together as one object.

In (ii) each object is detected, represented as a separate segment and labeled with the class name. Instance segmentation is one of the most challenging computer vision tasks. It was introduced by Hariharan [19] and then popularized by COCO [20]. Instance segmentation algorithms classify and localize each object's bounding box while also precisely segment each instance. Both those subtasks can be performed as a single- or two-stage detection process.

1.1. Two-Stage Methods

Treating segmentation as an extension of the object detection is adopted by most of two-stage methods [21–24]. They are designed according to the dominant paradigm for instance segmentation which is: detect-then-segment. First, it detects an object with a box and then segments each object using the box as a guide.

Mask R-CNN [23] is one of a state of the art two-stage method of instance segmentation. It extends Fast [25] and Faster R-CNN [26] methods by adding a branch for predicting an object mask in parallel with the existing branch for bounding box recognition [23]. This method detects object bounding boxes, and then crops and segments them to find the objects.

Many approaches based on Mask R-CNN also have dominated leaderboards of recent segmentation challenges.

In 2019 Mask Scoring R-CNN algorithm was introduced in [24]. This method focuses on scoring the predicted instance masks. The model learns an Intersection-over-Union (IoU) score for each mask and improves segmentation results by rewarding more accurate predictions during COCO AP evaluation.

The method named PANet presented in [27] improves Mask R-CNN in three steps by:

1. creating bottom-up path augmentation,
2. adaptive feature pooling,
3. augmenting mask prediction with fully connected layers.

The results outperform Mask R-CNN architecture but are re-implemented by the PANet authors.

Another extension of the Mask R-CNN model, this time with a cascade architecture, is the Hybrid Task Cascade (HTC) [28]. It improves the segmentation accuracy by incorporating cascade and multi-tasking at each stage and explores more contextual information from the spatial context.

State-of-the-art two-stage methods achieve outstanding results. However, due to the cascade strategy, these methods are usually time- and memory-consuming. That is why recently, performing instance detection and segmentation in a single-stage architecture [29–32] has become more and more popular, even though it does not yet perform as good as the two-stage methods.

1.2. Single-Stage Methods

Though single-stage methods simplify the procedure by removing the re-pooling step, usually they cannot produce masks that are as accurate as the two-stage methods. Some of those methods works in a per-pixel prediction way, similar to the semantic segmentation.

The algorithm Fully Convolutional One-Stage detection (FCOS) [33] has fully eliminated pre-defined anchor boxes, thus avoids complicated computation such as calculating boxes overlapping during training.

EmbedMask proposed in [32] unifies both segmentation- and proposal-based methods and takes the advantages of both. It is built on top of the detection models and thus has strong detection capabilities, just like the proposal-based methods. It applies extra embedding modules to generate embeddings for pixels and proposals, what enables EmbedMask to generate high-resolution masks without missing details from repooling, like segmentation-based methods.

TensorMask is an example of a dense sliding-window instance segmentation method [34]. The idea is to use structured 4D tensors to represent masks over a spatial domain. It performs multiclass classification in parallel to mask prediction. Authors show that this algorithm yields similar results to Mask R-CNN.

In this paper, we are mainly interested in the detection of each rodent in reference to each other and in reference to the background objects. Therefore, we focus on instance segmentation approach and more specifically deep instance segmentation architectures represented by two different approaches: one- and two-stage methods. In particular, we addressed the problem of animals instance segmentation in close physical contact on thermal images. The aim of this study was also to verify whether thermal data bit depth reduction or re-scaling has an effect on the results.

The paper is organized as follows: Section 2 presents the general description of methods and materials. Results are introduced in Section 3 and discussed in Section 4. The last section concludes the work.

2. Methods

In this section we describe the methods for instance segmentation adopted by us for thermal image of laboratory animals. For this purpose we use thermal database of rats introduced in [35]. The database consists of 300 min of social behavior tests recordings. Every single test was carried out on two healthy male rats of the Wistar strain at the age of 12–16 weeks kept in a plexiglass cage (dimensions: 35 length × 45 width × 46 cm height) for about 17 min in dimly illuminated room with temperature about 22 Celsius degrees. Animals were not accustomed to the cage earlier, it was an unfamiliar environment. There were no food, drink nor bedding material in the cage. The image sequences were recorded by FLIR A320G camera situated 120 cm above the cage with the spatial resolution of 320 × 240 pixels, 60 fps and 16-bit image representation. The results of semantic segmentation on the same dataset are presented in paper [16].

The principles for the care and use of laboratory animals in research, as outlined by the Local Ethical Committee, were strictly followed and all the protocols were reviewed and approved by the Committee.

2.1. Data Preprocessing

Thermal data were measured and stored with a resolution of 16-bit. The reduction of bit depth to 256-level standard images caused data loss. However, some data, e.g., background temperature, were redundant. In order to minimize data loss and investigate the effect of a temperature range selection, we proposed thermal frames processing to achieve five different images. This procedure is described in details in paper [16] and presented in Figure 3 in brief. First, the entire range of recorded raw thermal data (Figure 3a) was scaled to 256-level gray image (Figure 3b) and called orig image, as these were the original raw thermal data re-scaled to 8 bits. Then the gray-scale image was created from only three selected thermal ranges: animal body range (Figure 3c) and two thermal ranges (Figure 3d,e), all marked as ch1, ch2 and ch3 respectively. Minimal temperature value of the object was assumed to be equal to the threshold temperature between the background and the object calculated by the Otsu method, marked as a red line in the Figure 3a and called $maxOtsu_{thres}$. The green and blue dashed lines named as T_2 and T_3 respectively defined one- and two-thirds of the temperature interval from $maxOtsu_{thres}$ to the maximal value for all data. The ch1 range was selected between $maxOtsu_{thres}$ and the maximal temperature value. The ch2 image was created for the temperature range from T_2 to the maximal temperature value, and analogously, the lower limit of the ch3 image was T_3 value and the upper limit was the maximum temperature. Re-scaling 16-bit raw thermal data representation to 8-bit image caused loss of accuracy. Therefore, the last type of images was an image of the entire range of recorded data saved as a 16 bit image (16-bit). In [16] we proved that proper range selection improved the results of semantic segmentation.

2.2. Training Models

In this study we chose one single- and one two-stage architectures to compare the instance segmentation methods on thermal images of laboratory rodent. The decision was made for the Mask R-CNN [23] and the TensorMask [34].

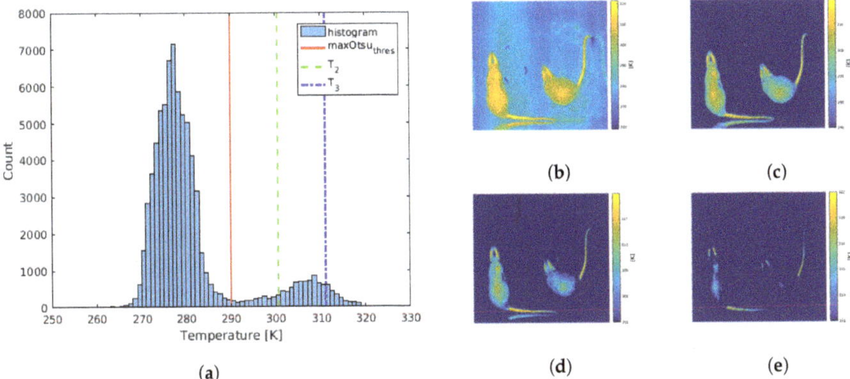

Figure 3. Range selection: (**a**) histogram of one frame raw thermal data with values of: minimal rodent's surface thermal value named $maxOtsu_{thres}$, marked with red solid line; the minimum values for images *ch2* and *ch3* marked as green (T_2) and blue (T_3) lines respectively, (**b**) orig image of the whole thermal range, (**c**) ch1 image of the first range: $maxOtsu_{thres}$ to T_{max}, (**d**) ch2 image of the second range: T_2 to T_{max}, (**e**) ch3 image of the third range: T_3 to T_{max}.

2.3. Learning Configurations

We made two training and one testing manually segmented datasets. The only difference between training sets was their size: 200 and 500 images respectively. The smaller set was used for the transfer-learning training, where we used pre-trained network: Mask R-CNN with a ResNet-Feature Pyramid Network (FPN) backbone 50 layers deep trained on the COCO dataset [36], and similarly TensorMask with a ResNet-FPN backbone 50 layers deep [37], both implementations from Detectron2 v0.1.2.

The 500 images were used for training the whole architectures from random initialization. In this training the number of training iterations had to be increased so models could converge. However, even despite the very large number of iterations, the TensorMask architecture sometimes had convergence problems: the loss function increased over time. In such cases we have used normalization techniques according to [38], and replaced Frozen Batch Normalization with Group Normalization [39]. Group Normalization's accuracy was insensitive to batch sizes [39] and allowed for successful TensorMask model training from scratch.

The testing set consisted of 50 images not included in the training sets. Images in all data-sets depicted two rodents during physical contact of varying degrees: from a small part of the body contact (e.g., nose-to-nose) to the body overlapping and covering.

For every training model we performed three independent training sessions. Json files with ground-truth rat regions were manually created for every training and testing set using the VGG Image Annotator (VIA) [40]. The regions were marked on a zoomed image by a set of points forming a closed polygon.

To compare the trained and commonly available models, we tested the inference using weights pre-trained on MS COCO Dataset [20], Citiscapes [41] and LVIS [42] and not trained on our database. We have also used two different Python3 implementations of Mask R-CNN (Mask R-CNN2.1 [36], Detectron2 v0.1.2 [37]) and one for TensorMask Detectron2 v0.1.2 [37].

In total, we tested 24 different model configurations for pre-trained learning and 22 for training from scratch.

For both models and both training methods with the best segmentation results 3-fold cross-validation was performed. The first fold was made for all the model configurations. The second and third folds used the whole testing data-set from the first fold in their training sets and replaced the training set with 50 different images selected from the training set of the first fold.

All experiments were performed using NVIDIA DGX-1 Station with Ubuntu 18.

2.4. Evaluation Metrics

To evaluate the results we are using three different detection metrics for bounding box- (bbox) and segmentation-level (segm): mean Average Precision (mAP), AP under Intersection-over-Union thresholds of 0.5 (AP_{50}) and 0.75 (AP_{75}).

The mean average precision (mAP) is actually the standard for the quantitative evaluation of object detection and instance segmentation methods [43,44]. The general definition for the Average Precision is finding the area under the precision-recall curve. The precision and recall are defined as follows:

$$Precision = \frac{TP}{TP+FP}; \quad Recall = \frac{TP}{TP+FN} \tag{1}$$

where: TP = True Positive, FP = False Positive and FN = False Negative.

The precision measures how accurate the prediction was, and recall measures how good all the positives were found.

In object detection and instance segmentation it is important to appropriately define a true positive and other types of results. As a current standard, the IoU metric was used. It was defined as the intersection between the predicted segment (bbox) and the actual segment (bbox) divided by their union:

$$IoU = \frac{Area\ of\ Overlap}{Area\ of\ Union} \tag{2}$$

In this study the actual (ground truth) segment was manually drawn by the authors on each reference image.

A prediction is considered to be True Positive if the IoU value is higher than the threshold (e.g., IoU > 0.5), and False Positive otherwise. However, different thresholds could be used and the precision recall curve could be obtained. Then, the area under the curve could be calculated as:

$$AP = \int_0^1 p(r)dr, \tag{3}$$

where: p(r) is the precision-recall curve.

In practice, in a majority of current papers about instance segmentation, the interpolated curve is calculated in ranges of IoU threshold values from 0.5 to 0.95 with a step size of 0.05. This method was also used in this paper. The mAP metric calculates the mean of AP across all of the IoU thresholds. We also used AP at fixed IoUs: IoU = 0.5 (AP_{50}) and IoU = 0.75 (AP_{75}). Which means, that a prediction is considered as True Positive if its IoU is >0.5 and >0.75 respectively. Thus, the higher the threshold, the greater the requirement for the accuracy of matching the prediction to the ground-truth. The standard metrics used in this study allow direct comparison of future methods with the approach used in this paper.

We have evaluated bounding box AP for object detection and segmentation AP for instance segmentation. The metrics are measured for all five types of images (orig, ch1, ch2, ch3, 16-bit) and training parameters shown in Table 1.

Table 1. Training parameters that achieved the best results in training.

Transfer Learning				Training from Scratch			
Mask R-CNN		TensorMask		Mask R-CNN		TensorMask	
Batch	Epochs	Batch	Epochs	Batch	Epochs	Batch	Epochs
2	1000	2	2000	4	16,000	2	**100,000**
4	1000	4	1000	4	24,000	4	100,000
4	**2000**	4	**2000**	8	16,000		
		8	2000	8	24,000		
				2	100,000		

3. Results

The best results among all the training parameter combinations were achieved by the values presented in Table 1.

The TensorMask's convergence problems during training from scratch forced the use of a large iteration number, much larger than for Mask R-CNN architecture. In order to compare the effectiveness of both architectures, we performed additional Mask R-CNN training with optimal parameters for TensorMask (2 batches, 100,000 epochs). Values in bold indicate parameters that have been selected for cross-validation.

In Tables 2 and 3 we compare Mask R-CNN testing results for pre-trained and from scratch learning respectively. Analogous measurements for the TensorMask architecture are presented in Tables 4 and 5.

Table 2. Testing results for Mask R-CNN pre-trained learning and 200 images.

	mAP	AP_{50}	AP_{75}	mAP	AP_{50}	AP_{75}
orig	bbox			segm		
2 batch, 1000 epochs	88.641	100	98.99	89.007	100	98.99
4 batch, 1000 epochs	**89.025**	100	**100**	**89.444**	100	**100**
4 batch, 2000 epochs	88.781	100	99	89.068	100	99
ch1	bbox			segm		
2 batch, 1000 epochs	87.332	100	100	89.289	100	100
4 batch, 1000 epochs	86.942	100	100	**89.715**	100	100
4 batch, 2000 epochs	**90.205**	100	100	89.655	100	100
ch2						
2 batch, 1000 epochs	85.252	100	**99**	87.323	100	100
4 batch, 1000 epochs	85.847	100	98.98	88.256	100	100
4 batch, 2000 epochs	**87.915**	100	98.98	**88.893**	100	100
ch3						
2 batch, 1000 epochs	63.719	99.114	73.409	52.901	93.649	59.722
4 batch, 1000 epochs	69.028	99.417	80.98	55.181	97.284	60.683
4 batch, 2000 epochs	**73.908**	**99.952**	**87.824**	**65.687**	**98.902**	**83.376**
16-bit	bbox			segm		
2 batch, 1000 epochs	88.128	100	99.01	**89.604**	100	100
4 batch, 1000 epochs	89.965	100	100	89.194	100	100
4 batch, 2000 epochs	**90.549**	100	100	89.417	100	100

Table 3. Training results for Mask R-CNN from scratch learning and 500 images.

	mAP	AP$_{50}$	AP$_{75}$	mAP	AP$_{50}$	AP$_{75}$
orig	bbox			segm		
4 batch, 16,000 epochs	81.811	100	98.307	85.586	100	100
4 batch, 24,000 epochs	82.567	100	100	86.864	100	100
8 batch, 16,000 epochs	81.658	100	100	84.728	100	98.584
8 batch, 24,000 epochs	84.465	100	100	86.224	100	100
2 batch, 100,000 epochs	**86.695**	100	100	**89.569**	100	100
ch1						
4 batch, 16,000 epochs	82.085	100	100	83.666	100	100
4 batch, 24,000 epochs	84.368	100	100	85.821	100	100
8 batch, 16,000 epochs	81.426	100	100	84.038	100	100
8 batch, 24,000 epochs	85.153	100	98.772	86.389	100	100
2 batch, 100,000 epochs	**86.564**	100	100	**87.208**	100	100
ch2						
4 batch, 16,000 epochs	80.629	100	98.95	80.896	100	100
4 batch, 24,000 epochs	82.979	100	100	**84.834**	100	100
8 batch, 16,000 epochs	80.983	100	100	79.931	100	100
8 batch, 24,000 epochs	**83.091**	100	100	84.428	100	100
2 batch, 100,000 epochs	82.55	100	97.55	82.397	100	100
ch3						
4 batch, 16,000 epochs	60.947	97.766	73.004	48.183	93.325	47.4
4 batch, 24,000 epochs	**67.739**	98.113	79.816	**58.782**	**96.882**	**69.92**
8 batch, 16,000 epochs	65.355	97.874	74.09	54.855	96.873	64.968
8 batch, 24,000 epochs	66.014	98.4	78.156	57.452	95.887	63.699
2 batch, 100,000 epochs	66.369	**98.504**	**84.891**	56.116	94.891	67.155
16-bit						
4 batch, 16,000 epochs	82.826	100	98.842	84.561	100	100
4 batch, 24,000 epochs	84.29	100	100	86.111	100	100
8 batch, 16,000 epochs	83.493	100	97.82	85.313	100	100
8 batch, 24,000 epochs	**85.596**	100	100	**86.607**	100	100
2 batch, 100,000 epochs	84.862	100	100	85.877	100	100

Table 4. Training results for TensorMask pre-trained learning and 200 images.

	mAP	AP$_{50}$	AP$_{75}$	mAP	AP$_{50}$	AP$_{75}$
orig	bbox			segm		
2 batch, 2000 epochs	90.646	100	98.01	89.82	100	100
4 batch, 1000 epochs	**91.264**	100	**99.01**	89.877	100	99.01
4 batch, 2000 epochs	90.916	100	**99.01**	**90.158**	100	100
8 batch, 2000 epochs	90.672	100	98.96	89.824	100	100

Table 4. Cont.

	mAP	AP$_{50}$	AP$_{75}$	mAP	AP$_{50}$	AP$_{75}$
ch1						
2 batch, 2000 epochs	89.091	100	100	89.808	100	**100**
4 batch, 1000 epochs	89.624	100	100	89.547	100	99.01
4 batch, 2000 epochs	**90.299**	100	100	**90.171**	100	99.01
8 batch, 2000 epochs	90.052	100	100	90.087	100	99.01
ch2						
2 batch, 2000 epochs	88.319	100	100	**89.496**	100	100
4 batch, 1000 epochs	88.17	100	100	87.949	100	100
4 batch, 2000 epochs	87.912	100	98.931	89.335	100	100
8 batch, 2000 epochs	**88.828**	100	100	89.264	100	100
ch3						
2 batch, 2000 epochs	74.37	99.933	88.433	65.062	**99.933**	80.743
4 batch, 1000 epochs	71.325	99.834	84.895	60.251	99.602	69.757
4 batch, 2000 epochs	75.031	99.961	**91.558**	64.849	99.423	79.397
8 batch, 2000 epochs	**76.159**	**100**	91.406	**65.307**	99	**83.105**
16-bit						
2 batch, 2000 epochs	91.248	100	100	**90.249**	100	100
4 batch, 1000 epochs	90.443	100	98.951	89.857	100	98.951
4 batch, 2000 epochs	**91.609**	100	100	90.199	100	100
8 batch, 2000 epochs	90.939	100	100	90.171	100	98.99

Table 5. Training results for TensorMask from scratch learning and 500 images.

	mAP	AP$_{50}$	AP$_{75}$	mAP	AP$_{50}$	AP$_{75}$
orig	bbox			segm		
2 batch, 100,000 epochs	**78.235**	**100**	92.41	79.615	**100**	**95.535**
4 batch, 100,000 epochs	76.413	99.99	**95.435**	**80.371**	99.99	91.678
ch1						
2 batch, 100,000 epochs	**78.862**	100	**98.307**	77.724	100	90.687
4 batch, 100,000 epochs	77.725	100	94.535	**81.259**	100	**95.188**
ch2						
2 batch, 100,000 epochs	64.42	99.18	69.516	62.307	96.206	71.867
4 batch, 100,000 epochs	**75.275**	**99.833**	**90.342**	**74.431**	**99.833**	**89.674**
ch3						
2 batch, 100,000 epochs	44.557	84.367	40.402	13.097	60.133	0.012
4 batch, 100,000 epochs	**55.982**	**95.072**	**57.009**	**32.752**	**89.025**	**16.88**
16-bit						
2 batch, 100,000 epochs	**78.112**	99.99	**97.29**	80.198	99.99	94.435
4 batch, 100,000 epochs	76.296	**100**	90.662	**81.267**	**100**	**94.887**

Figure 4 presents the exemplary testing results for two images (Figures 4a,b) which constituted quite a serious challenge for segmentation. For a better illustration, the contact area was zoomed and manually segmented with a yellow line in Figures 4c,d. The examples given include animals that overlapped one another. Figures 4e–h present the results of pre-trained learning for Mask R-CNN and TensorMask respectively. The instance segmentation made by those architectures trained from scratch are shown in Figures 4i–l.

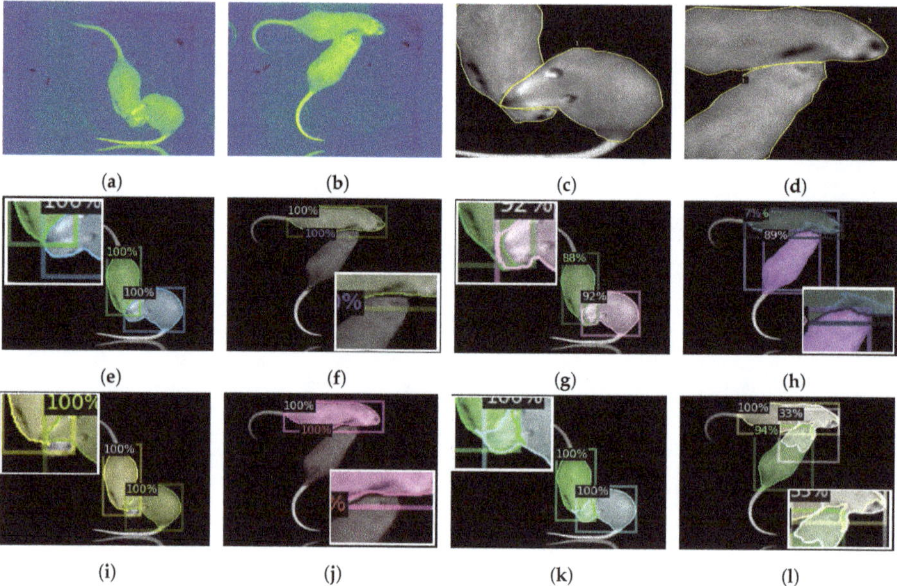

Figure 4. Examples of segmentation and detection prediction made for (**a**,**b**) the original images; (**c**,**d**) proper segmentation zoomed and marked with yellow lines. Results for: (**e**,**f**) pre-trained Mask R-CNN; (**g**,**h**) pre-trained TensorMask; (**i**,**j**) random initialized Mask R-CNN; (**k**,**l**) random initialized TensorMask.

Table 6 demonstrates the results of the inference made for *orig* images using commonly available models pre-trained only on MS COCO Dataset, Citiscapes and LVIS. We also used two different implementations of Mask R-CNN pre-trained on the COCO dataset.

Table 6. Segmentation results for commonly available models.

	mAP	AP$_{50}$	AP$_{75}$	mAP	AP$_{50}$	AP$_{75}$
Mask R-CNN	bbox			segm		
MS COCO—implementation [37]	4.14	13.39	1.4	5.36	14.44	1.05
MS COCO—implementation [36]	7.62	13.5	7.62	-	-	-
Citiscapes	0.06	0.3	0	0.01	0.09	0
LVIS	0.71	3.45	0.04	0	0	0
TensorMask						
MS COCO—implementation [37]	1.69	3.41	1.29	3.06	6.31	2.6

The results of 3-fold cross-validation for the parameters of four batch size with 2000 epochs and two batch size with 100,000 epochs for pre-trained and trained from random initialization models respectively are presented in Table 7. The results are presented in the form of box and segmentation mAP. The folds of model training for three different training and testing datasets were repeated twice. Presented result values are the best selected repetition.

Table 7. 3-fold cross-validation mean Average Precision (mAP) results for pre-trained (four batch, 2000 epochs) and trained from scratch (two batch, 100,000 epochs) models of Mask R-CNN and TensorMask.

mAP	Mask Pre-Trained		Mask From Scratch		TensorMask Pre-Trained		TensorMask From Scratch	
	bbox	segm	bbox	segm	bbox	segm	bbox	segm
orig								
1. fold	88.8	89.1	83.6	85.9	90.9	90.2	78.2	79.6
2. fold	**92.0**	89.5	84.7	87.6	91.9	**90.5**	73.9	70.6
3. fold	89.7	89.2	87.7	88.0	90.5	90.2	70.6	76.6
average	90.2	89.3	*85.3*	*87.2*	**91.1**	**90.3**	74.2	75.6
ch1								
1. fold	90.2	89.7	84.6	87.2	90.3	90.2	78.9	77.7
2. fold	88.4	89.1	88.3	88.9	**91.3**	**90.4**	78.7	77.8
3. fold	89.6	89.4	88.3	89.0	90.2	89.8	75	74.4
average	89.4	89.4	*87.1*	*88.4*	**90.6**	**90.1**	77.5	76.6
ch2								
1. fold	87.9	88.9	82.6	82.4	87.9	**89.3**	64.4	62.3
2. fold	84.7	86.4	86.6	88.2	89.1	89.0	66	57.2
3. fold	**89.5**	88.1	85.0	87.7	88.3	88.9	65.7	54.3
average	87.4	87.8	*84.7*	*86.1*	**88.4**	**89.1**	65.4	57.9
ch3								
1. fold	73.9	**65.7**	66.4	56.1	**75.0**	64.8	44.6	13.1
2. fold	63.0	51.9	65.6	57	70.9	61.9	40.2	11.2
3. fold	72.6	62.9	66.6	62.4	71.3	62.5	36.6	11.2
average	69.8	60.2	*66.2*	*58.5*	**72.4**	**63.1**	40.5	11.8
16-bit								
1. fold	90.5	89.4	84.9	85.9	91.6	90.2	78.1	80.2
2. fold	89.2	89.4	85.8	88.5	**92.5**	**90.8**	76.8	77.6
3. fold	90.0	89.1	86.6	87.8	90.6	89.9	71.5	73.9
average	89.9	89.3	*85.8*	*87.4*	**91.6**	**90.3**	75.5	77.2

4. Discussion

The objective of the study was to compare two different instance segmentation approaches and two different learning methods for the purpose of experimental animals segmentation on thermal images. Various methods for re-scaling thermal data to a standard image were also compared.

In paper [38] authors showed that the model trained from random initialization, if it only has proper training parameters, can get results similar to the pre-trained models, however, it needs more iterations to converge. In our experiments training Mask R-CNN from scratch needed about 16 times the number of epochs (16,000 or 24,000) (Tables 4 and 5) than for the pre-trained model (1000 or 2000) to achieve similar results (Tables 2 and 3). Increasing the number of epochs to 100,000 improved the results only for the orig and ch1 images, but not significantly. TensorMask architecture required a much larger number of epochs for the random initialization training (100,000), and still did not achieve results comparable to the pre-trained model (see Tables 4 and 5).

Mask R-CNN training from scratch neither is time nor data expensive. Only a slightly larger number of images and epochs allow it to achieve all-layers training results comparable to the pre-trained models.

The values in Tables 2 and 4 indicate that the best segmentation results for the pre-trained models can be obtained for the 16-bit and ch1 image followed by the orig for Mask R-CNN and also 16-bit with orig this time followed by ch1 using TensorMask. Ch2 images achieved only slightly lower results than the top ones. The bbox and segmentation mAP values for both architectures were close to 90 percent, with a slight TensorMask advantage. Detection for ch3 in both cases achieved the best mAP above 70, while the segmentation best mAP was equal to 65 percent. If we look at the results for all combinations of training parameters, not only the best ones, it can be clearly seen that TensorMask achieves better results for various training models.

The results of training from scratch show the opposite trend (see Tables 3 and 5). This time it is Mask R-CNN that achieves better results for all images (mAP above 80 percent for almost all images) under various criteria, suggesting that this model trained from scratch catch up not only by chance for a single metric. The differences were especially visible for the ch3 image, for which the best TensorMask segmentation mAP was equal only to 32.752 percent.

The pre-trained model shows generally similar results for instance segmentation and detection for all images except ch3. The image ch3 contains a very narrow range of only the highest animals' surface thermal values (Figure 3a,e), so it is deprived of a significant part of the body area. Such an object is detectable, but difficult to correctly segment if there is no information about the object's boundaries. That is why the mAP for segmentation is much smaller than for detection.

The data in Tables 3 and 5 show that models trained from scratch are likely to perform segmentation more accurately than the detection (except ch3). Detection and segmentation accuracy for the pre-trained models are more similar.

In the Figure 4 we have presented the results of the detection and segmentation for challenging views with the critical regions zoomed in white frames. The image in Figure 4c is difficult to segment, because the snout of one rodent covers the snout of the other one over the entire width of the body. However, there is a small element of the snout that belongs to the animal at the bottom and in this camera view is not connected to the rest of the body. One of animals in Figure 4d has its snout hidden under the body of the other rat, which, as a matter of fact, is not so rare. Here, the cooler water mark left on the fur creates a line, that can be mistaken as a continuation of the body boundary. In this particular way it was segmented by the pre-trained TensorMask (Figure 4h), as a result of which a small area of the body on the border was miss-assigned to the wrong object. It also marked both individuals as the object, however with low probability (7%). The Mask R-CNN architecture set the boundaries more precisely (Figure 4f,j) but left a few pixels' gap between the objects. TensorMask trained from scratch segmented the rodents' body areas similarly to the pre-trained model, (Figure 4l), and here also an additional object—a combination of both body parts—was detected.

The separated object from the Figure 4c caused the Mask R-CNN the most problems. The pre-trained model (Figure 4e) considered the separated part of the animal's mouth as an element of the other rat's body. In addition, as the only one, it inaccurately determined the bbox of the detected object. Mask R-CNN trained from scratch did not assign this small part of the snout to any object, but correctly detected the bbox (Figure 4i). The pre-trained TensorMask (Figure 4g) classified the problematic snout partly to both individuals—this line of segmentation seems to be the closest to the correct one. TensorMask architecture trained from scratch assigned both animal snouts to both objects at once (Figure 4k).

Although the segmentation results were not always satisfactory, it should be remembered that difficult cases were presented here. For the vast majority of images, the prediction was very similar to ground-truth images and succeeded in its mission of rodent segmentation, in contrast to the semantic segmentation algorithms presented in paper [16]. The results of U-Net segmentation (see Figure 1d) show that the boundary between objects disappears during close contact, although it was previously visible during small connection.

The results of segmentation made by the commonly available models (see Table 6) are much worse than those trained on the target images, and do not exceed 8 mAP. The Citiscapes and LVIS datasets achieved extremely low values—bellow 0.75 mAP. During the evaluation, the correctness of class assignment is also taken into account. Although the COCO data-set has a "mouse" class [20], however, rat objects were not assigned to it, so in this case the assignment correctness was zero.

The 3-fold cross-validation results presented in Table 7 are very similar. The top segmentation and detection results (marked as bold in Table 7) on average was achieved by the pre-trained TensorMask. Both pre-trained models show greater accuracy than those trained from scratch, which is consistent with the previous results (see Tables 2–5). It is probably related to a much smaller training set comparing to the COCO set [20]. As far as a random initialized models are concerned, the difference between the both models is significant. Mask R-CNN (marked as bold italic in Table 7) obtains mAP even 10 percent higher than TensorMask. It is possible that TensorMask needs more data and/or epochs number to achieve a Mask R-CNN-like results.

5. Conclusions

The deep instance segmentation algorithms are able to distinguish between two individuals in close contact where overlaps may appear. The segmentation mAP almost reach value of 90 percent. The detection results are slightly higher and usually oscillate around 90 percent. The top results are achieved by the single-stage (TensorMask) pre-trained network, however, two-stage method (Mask R-CNN) works better than single-stage when trained from scratch. Training Mask R-CNN model from random initialization is neither time nor data consuming. It can be used for training non-standard images, however, keeping in mind that networks trained from scratch focuses more on segmentation than detection. In turn, the cost of training TensorMask is large, and still does not achieve results similar to the pre-trained version. The pre-trained TensorMask model shows the best performance. The research indicates that thermal images can be successfully analyzed by architectures pre-trained on standard images. However, some layers of the network must be trained on thermal images, because the models pre-trained only on publicly available databases achieve very poor segmentation results and even worse in detection.

The thermal data conversion from different thermal data range does not improve the quality of segmentation. The results for 16-bit (*16-bit*) and 8-bit (orig) image representation as well as for image deprived of the background (ch1) are comparable. Results of images with narrower thermal range (ch2) do not differ much from the others. Unlike the results for ch3 image, where only the information about the warmest parts of the body is visible. However, the bounding box mAP values of this images for pre-trained model suggest that detection for limited-data images is possible.

The main contributions of this paper are the following:

- the adopted deep instance segmentation algorithms have been experimentally verified for the laboratory rodents detection from thermal images,
- it was shown that laboratory rodents can be accurately detected (and separated from each other) from the thermal images using the Mask R-CNN and TensorMask models,
- the obtained results demonstrated that the adopted TensorMask model, pre-trained using visible light images and trained with thermal sequences gave the best results with the mean average precision (mAP) greater that 90,
- it was verified that thermal data conversion from narrower raw thermal range does not improve the quality of segmentation,
- single-stage pre-trained networks achieves better results than two-stage pre-trained models, however two-stage methods seem to work better than single-stage when trained from scratch,
- network pre-training using visible light images improves the segmentation results for thermal images.

The conclusion of this work is that segmentation algorithms can be used to segment experimental animals in thermal images. Depending on the needs, one can customize architectures, learning methods or image types for the best performance. Instance segmentation methods work better than the semantic segmentation methods. The presented approach will work well in social behaviour tests, but not only that. It can be used wherever identification and tracking of experimental animals is required, especially in numerous groups.

In the future it is worth training both architectures with a more diverse thermal database. Increasing the amount of training data can also improve the results, especially for the TensorMask.

Author Contributions: Conceptualization, M.M.-M. and J.R.; data curation, M.M.-M.; formal analysis, M.M.-M. and J.R.; funding acquisition, J.R.; investigation, M.M.-M.; methodology, M.M.-M. and J.R.; project administration, M.M.-M. and J.R.; resources, M.M.-M.; software, M.M.-M. and T.K.; supervision, J.R.; validation, M.M.-M., J.R. and T.K.; visualization, M.M.-M.; writing—original draft preparation, M.M.-M.; writing—review and editing, J.R. and T.K. All authors have read and agreed to the published version of the manuscript.

Funding: This work has been partially supported by Statutory Funds of Electronics, Telecommunications and Informatics Faculty, Gdansk University of Technology

Conflicts of Interest: The authors declare no conflict of interest.

References

1. Lezak, K.; Missig, G.; Carlezon, W.A., Jr. Behavioral methods to study anxiety in rodents. *Dialogues Clin. Neurosci.* **2017**, *19*, 181–191. [PubMed]
2. Franco, N.H.; Gerós, A.; Oliveira, L.; Olsson, I.A.S.; Aguiar, P. ThermoLabAnimal—A high-throughput analysis software for non-invasive thermal assessment of laboratory mice. *Physiol. Behav.* **2019**, *207*, 113–121. [CrossRef] [PubMed]
3. Junior, C.F.C.; Pederiva, C.N.; Bose, R.C.; Garcia, V.A.; Lino-de-Oliveira, C.; Marino-Neto, J. ETHOWATCHER: Validation of a tool for behavioral and video-tracking analysis in laboratory animals. *Comput. Biol. Med.* **2012**, *42*, 257–264. [CrossRef] [PubMed]
4. Grant, E.; Mackintosh, J. A comparison of the social postures of some common laboratory rodents. *Behaviour* **1963**, *21*, 246–259.
5. Kask, A.; Nguyen, H.P.; Pabst, R.; von Hoorsten, S. Factors influencing behavior of group-housed male rats in the social interaction test—Focus on cohort removal. *Physiol. Behav.* **2001**, *74*, 277–282. [CrossRef]
6. Aslani, S.; Harb, M.; Costa, P.; Almeida, O.; Sousa, N.; Palha, J. Day and night: diurnal phase influences the response to chronic mild stress. *Front. Behav. Neurosci.* **2014**, *8*, 82. [CrossRef]
7. Roedel, A.; Storch, C.; Holsboer, F.; Ohl, F. Effects of light or dark phase testing on behavioural and cognitive performance in DBA mice. *Lab. Anim.* **2006**, *40*, 371–381. [CrossRef]
8. Manzano-Szalai, K.; Pali-Schöll, I.; Krishnamurthy, D.; Stremnitzer, C.; Flaschberger, I.; Jensen-Jarolim, E. Anaphylaxis Imaging: Non-Invasive Measurement of Surface Body Temperature and Physical Activity in Small Animals. *PLoS ONE* **2016**, *11*, e0150819. [CrossRef]
9. Etehadtavakol, M.; Emrani, Z.; Ng, E.Y.K. Rapid extraction of the hottest or coldest regions of medical thermographic images. *Med. Biol. Eng. Comput.* **2019**, *57*, 379–388. [CrossRef]
10. Jang, E.; Park, B.; Park, M.; Kim, S.; Sohn, J. Analysis of physiological signals for recognition of boredom, pain, and surprise emotions. *J. Phys. Anthropol.* **2015**, *34*, 1–12. [CrossRef]
11. Tan, C.; Knight, Z. Regulation of Body Temperature by the Nervous System. *Neuron* **2018**, *98*, 31–48. [CrossRef] [PubMed]
12. Sona, D.; Zanotto, M.; Papaleo, F.; Murino, V. Automated Discovery of Behavioural Patterns in Rodents. In Proceedings of the 9th International Conference on Methods and Techniques in Behavioral Research, Wageningen, The Netherlands, 27–29 August 2014.
13. Koniar, D.; Hargaš, L.; Loncová, Z.; Simonová, A.; Duchoň, F.; Beňo, P. Visual system-based object tracking using image segmentation for biomedical applications. *Electr. Eng.* **2017**, *99*, 1349–1366. [CrossRef]
14. Fleuret, J.; Ouellet, V.; Moura-Rocha, L.; Charbonneau, É.; Saucier, L.; Faucitano, L.; Maldague, X. A Real Time Animal Detection And Segmentation Algorithm For IRT Images In Indoor Environments. *Quant. InfraRed Thermogr.* **2016**, 265–274.
15. Kim, W.; Cho, Y.B.; Lee, S. Thermal Sensor-Based Multiple Object Tracking for Intelligent Livestock Breeding. *IEEE Access* **2017**, *5*, 27453–27463. [CrossRef]
16. Mazur-Milecka, M.; Ruminski, J. Deep learning based thermal image segmentation for laboratory animals tracking. *Quant. InfraRed Thermogr. J.* **2020**, 1–18. [CrossRef]

17. Ronneberger, O.; Fischer, P.; Brox, T. U-Net: Convolutional Networks for Biomedical Image Segmentation. In *International Conference on Medical Image Computing and Computer-Assisted Intervention*; Springer: Cham, Switzerland, 2015; pp. 234–241.

18. Milletari, F.; Navab, N.; Ahmadi, S. V-Net: Fully Convolutional Neural Networks for Volumetric Medical Image Segmentation. In Proceedings of the IEEE 2016 Fourth International Conference on 3D Vision (3DV), Stanford, CA, USA, 25–28 October 2016; pp. 565–571.

19. Hariharan, B.; Arbelaez, P.; Girshick, R.B.; Malik, J. Simultaneous Detection and Segmentation. In *European Conference on Computer Vision*; Springer: Cham, Switzerland, 2014; pp. 297–312.

20. Lin, T.; Maire, M.; Belongie, S.J.; Bourdev, L.D.; Girshick, R.B.; Hays, J.; Perona, P.; Ramanan, D.; Dollár, P.; Zitnick, C.L. Microsoft COCO: Common Objects in Context. In *European Conference on Computer Vision*; Springer: Cham, Switzerland, 2014; pp. 740–755.

21. Dai, J.; He, K.; Sun, J. Instance-aware Semantic Segmentation via Multi-task Network Cascades. In Proceedings of the IEEE Conference on Computer Vision and Pattern Recognition, Boston, MA, USA, 7–12 June 2015; pp. 3150–3158.

22. Chen, L.; Hermans, A.; Papandreou, G.; Schroff, F.; Wang, P.; Adam, H. MaskLab: Instance Segmentation by Refining Object Detection with Semantic and Direction Features. *arXiv* **2017**, arXiv:1712.04837.

23. He, K.; Gkioxari, G.; Dollár, P.; Girshick, R.B. Mask R-CNN. In Proceedings of the IEEE International Conference on Computer Vision, Venice, Italy, 22–29 October 2017; pp. 2961–2969.

24. Huang, Z.; Huang, L.; Gong, Y.; Huang, C.; Wang, X. Mask Scoring R-CNN. In Proceedings of the IEEE Conference on Computer Vision and Pattern Recognition, Long Beach, CA, USA, 16–20 June 2019; pp. 6409–6418.

25. Girshick, R.B. Fast R-CNN. In Proceedings of the IEEE International Conference on Computer Vision, Santiago, Chile, 7–13 December 2015; pp. 1440–1448.

26. Ren, S.; He, K.; Girshick, R.; Sun, J. Faster R-CNN: Towards Real-Time Object Detection with Region Proposal Networks. *arXiv* **2015**, arXiv:1506.01497.

27. Liu, S.; Qi, L.; Qin, H.; Shi, J.; Jia, J. Path Aggregation Network for Instance Segmentation. In Proceedings of the IEEE Conference on Computer Vision and Pattern Recognition, Salt Lake City, UT, USA, 18–23 June 2018; pp. 8759–8768.

28. Chen, K.; Pang, J.; Wang, J.; Xiong, Y.; Li, X.; Sun, S.; Feng, W.; Liu, Z.; Shi, J.; Ouyang, W.; et al. Hybrid Task Cascade for Instance Segmentation. In Proceedings of the IEEE Conference on Computer Vision and Pattern Recognition, Long Beach, CA, USA, 16–20 June 2019; pp. 4974–4983.

29. Yao, J.; Yu, Z.; Yu, J.; Tao, D. Single Pixel Reconstruction for One-stage Instance Segmentation. *arXiv* **2019**, arXiv:1904.07426.

30. Bolya, D.; Zhou, C.; Xiao, F.; Lee, Y.J. YOLACT: Real-time Instance Segmentation. In Proceedings of the IEEE International Conference on Computer Vision, Seoul, Korea, 27 October–2 November 2019; pp. 9157–9166.

31. Xiang, C.; Tian, S.; Zou, W.; Xu, C. SAIS: Single-stage Anchor-free Instance Segmentation. *arXiv* **2019**, arXiv:cs.CV/1912.01176.

32. Ying, H.; Huang, Z.; Liu, S.; Shao, T.; Zhou, K. EmbedMask: Embedding Coupling for One-stage Instance Segmentation. *arXiv* **2019**, arXiv:cs.CV/1912.01954.

33. Tian, Z.; Shen, C.; Chen, H.; He, T. FCOS: Fully Convolutional One-Stage Object Detection. In Proceedings of the IEEE International Conference on Computer Vision, Seoul, Korea, 27 October–2 November 2019; pp. 9627–9636.

34. Chen, X.; Girshick, R.B.; He, K.; Dollár, P. TensorMask: A Foundation for Dense Object Segmentation. In Proceedings of the IEEE International Conference on Computer Vision, Seoul, Korea, 27 October–2 November 2019; pp. 2061–2069.

35. Mazur-Milecka, M.; Ruminski, J. Automatic analysis of the aggressive behavior of laboratory animals using thermal video processing. In Proceedings of the IEEE Conference of the Engineering in Medicine and Biology Society, EMBC, Seogwipo, Korea, 11–15 July 2017; pp. 3827–3830.

36. Abdulla, W. Mask R-CNN for Object Detection and Instance Segmentation on Keras and TensorFlow. 2017. Available online: https://github.com/matterport/Mask_RCNN (accessed on 1 February 2020).

37. Wu, Y.; Kirillov, A.; Massa, F.; Lo, W.Y.; Girshick, R. Detectron2, 2019. Available online: https://github.com/facebookresearch/detectron2 (accessed on 1 May 2020).

38. He, K.; Girshick, R.; Dollar, P. Rethinking ImageNet Pre-Training. In Proceedings of the 2019 IEEE/CVF International Conference on Computer Vision (ICCV), Seoul, Korea, 27–28 October 2019; pp. 4917–4926.
39. Wu, Y.; He, K. Group Normalization. *Int. J. Comput. Vis.* **2019**. [CrossRef]
40. Dutta, A.; Zisserman, A. The VIA Annotation Software for Images, Audio and Video. In Proceedings of the 27th ACM International Conference on Multimedia, Nice, France, 21–25 October 2019; ACM: New York, NY, USA, 2019. [CrossRef]
41. Cordts, M.; Omran, M.; Ramos, S.; Rehfeld, T.; Enzweiler, M.; Benenson, R.; Franke, U.; Roth, S.; Schiele, B. The Cityscapes Dataset for Semantic Urban Scene Understanding. In Proceedings of the IEEE Conference on Computer Vision and Pattern Recognition, Las Vegas, NV, USA, 27–30 June 2016; pp. 3213–3223.
42. Gupta, A.; Dollár, P.; Girshick, R. LVIS: A Dataset for Large Vocabulary Instance Segmentation. In Proceedings of the IEEE Conference on Computer Vision and Pattern Recognition, Long Beach, CA, USA, 16–20 June 2019; pp. 5356–5364.
43. Lu, Y.; Lu, C.; Tang, C. Online Video Object Detection Using Association LSTM. In Proceedings of the 2017 IEEE International Conference on Computer Vision (ICCV), Venice, Italy, 22–29 October 2017; pp. 2363–2371.
44. Oksuz, K.; Cam, B.C.; Akbas, E.; Kalkan, S. Localization Recall Precision (LRP): A New Performance Metric for Object Detection. In *Computer Vision—ECCV 2018*; Ferrari, V., Hebert, M., Sminchisescu, C., Weiss, Y., Eds.; Springer International Publishing: Cham, Switzerland, 2018; pp. 521–537.

Article

MRU-NET: A U-Shaped Network for Retinal Vessel Segmentation

Hongwei Ding [1,2], Xiaohui Cui [1,2,*], Leiyang Chen [1,2] and Kun Zhao [1,2]

[1] School of Cyber Science and Engineering, Wuhan University, Wuhan 430000, China; hwding@whu.edu.cn (H.D.); cly_edu@whu.edu.cn (L.C.); zhaokun@whu.edu.cn (K.Z.)

[2] Key Laboratory of Aerospace Information Security and Trusted Computing, Ministry of Education, Wuhan 430000, China

* Correspondence: xcui@whu.edu.cn

Received: 11 August 2020; Accepted: 25 September 2020; Published: 29 September 2020

Abstract: Fundus blood vessel image segmentation plays an important role in the diagnosis and treatment of diseases and is the basis of computer-aided diagnosis. Feature information from the retinal blood vessel image is relatively complicated, and the existing algorithms are sometimes difficult to perform effective segmentation with. Aiming at the problems of low accuracy and low sensitivity of the existing segmentation methods, an improved U-shaped neural network (MRU-NET) segmentation method for retinal vessels was proposed. Firstly, the image enhancement algorithm and random segmentation method are used to solve the problems of low contrast and insufficient image data of the original image. Moreover, smaller image blocks after random segmentation are helpful to reduce the complexity of the U-shaped neural network model; secondly, the residual learning is introduced into the encoder and decoder to improve the efficiency of feature use and to reduce information loss, and a feature fusion module is introduced between the encoder and decoder to extract image features with different granularities; and finally, a feature balancing module is added to the skip connections to resolve the semantic gap between low-dimensional features in the encoder and high-dimensional features in decoder. Experimental results show that our method has better accuracy and sensitivity on the DRIVE and STARE datasets (accuracy (ACC) = 0.9611, sensitivity (SE) = 0.8613; STARE: ACC = 0.9662, SE = 0.7887) than some of the state-of-the-art methods.

Keywords: retinal blood vessel image; computer-aided diagnosis; U-shaped neural network; residual learning; semantic gap

---

1. Introduction

Retinal blood vessel images are often used by doctors as a window to observe the response of various diseases, such as hypertension, coronary heart disease, and diabetes, and vessel abnormalities can reflect the severity of these diseases. Retinal blood vessels are the only blood vessels in the human body that can be obtained without trauma. The fundus camera can directly take images of fundus blood vessels and can respond clearly to microvessels and lesions. Therefore, the research of fundus images has important medical application value. In order to make an effective diagnosis of the disease, it is necessary to accurately segment blood vessels in the fundus. However, because the retinal blood vessel image may have problems such as lesions, uneven lighting, noise, and low contrast between small blood vessels and the background, it is difficult to completely segment the fundus blood vessel image. Therefore, fundus blood vessel image segmentation has become a hot topic at home and abroad.

In recent years, many scholars have conducted extensive and in-depth research on automatic segmentation of fundus blood vessel images and have proposed many segmentation methods. These algorithms can be divided into supervised segmentation and unsupervised segmentation based on whether a gold standard image is required. Unsupervised fundus vascular segmentation does not require a gold standard image as a segmentation standard. For example, Dash et al. [1] proposed a morphological-based fundus image segmentation technology, which can use Kirsch edge detection method to identify retinal vein from retinal image. The advantage of the method is that it can identify the vasculature devoid of any tiny information, but the disadvantage is that the segmentation of microvascular is insufficient. A matching filtering method designed by Zhang et al. [2] used a multi-scale second-order Gaussian derivative filter and performs filtering in the direction score domain to obtain the maximum response. This method has a good effect in dealing with complex vessels, but there is a phenomenon of error segmentation. Abdallah et al. [3] employed a multi-scale tracking method based on Heisson matrix feature vectors and gradient information. This method can extract blood vessels, especially small blood vessels, at different resolutions. Although many scholars have conducted in-depth research on the unsupervised segmentation method, its segmentation effect still needs to be improved. Compared with the unsupervised segmentation method, the supervised segmentation method can segment the fundus blood vessel images more effectively. For supervised fundus blood vessel image segmentation, before being able to use end-to-end deep learning methods for feature learning, researchers must manually extract image features based on prior knowledge of the image. Orlando et al. [4], for example, adopted multi-scale linear detector response and two-dimensional wavelet responses on the green channel as features, used structured output support vector machine to train parameters of a fully connected conditional random field model, and then used the trained conditional random field to segment blood vessels. This method can effectively improve the performance index, but there will be error segmentation in the bright central reflection area. Wang et al. [5] proposed a supervised learning method based on multi-feature and multi-classifier fusion to segment retinal vessels. Four types of features were extracted, and the results of decision tree and AdaBoost classifier were fused together to make a joint decision. This method can effectively improve the accuracy and sensitivity of segmentation, but it needs to be improved compared with the deep learning method. Although the above method has achieved certain effects, the extraction of artificial features requires a wealth of prior knowledge, and it has a large subjectivity for feature extraction, so it may cause problems such as insufficient microvessel segmentation and incorrect segmentation.

The core idea of deep learning is to extract the main representation features from the original data through a series of nonlinear transformations. This feature is multi-level and multi-angle, which also makes the features extracted by deep learning method have stronger generalization and expressive ability, which is exactly what the image segmentation process needs. Convolutional neural network, as an effective deep learning model of image segmentation, has been widely concerned by researchers. Gu et al. [6] proposed a context encoder network (CE-Net) to capture more high-level information, to save spatial information, and to be used for 2D medical image segmentation. Although this method has certain effects on various medical data sets, its accuracy still needs to be improved. Zhou et al. [7] Proposed a neural network structure, UNET + +, for semantic and instance segmentation. By using nested structure and improved hop connection structure, the segmentation performance of the model is effectively improved. The disadvantage of the model is that the parameters increase greatly compared with the original model. Soomro et al. [8] employed a fundus vessel segmentation method based on deep convolutional neural network (CNN). Firstly, image enhancement was performed using fuzzy logic and image processing strategy, and then, an encoding and decoding CNN model with skip connection structure was proposed for retinal blood vessel segmentation.This method can effectively improve the contrast between blood vessel and background, but there is over-segmentation in the segmentation process. Kumawat et al. [9] adoped an improved local phase unit (ReLPU), which is an efficient and trainable convolution layer. When the ReLPU layer is used at the top of the U-Net segmentation network, it can effectively improve the segmentation performance of the U-Net model. The disadvantage of this method is that it cannot segment the microvessels effectively. Tamim et al. [10] proposed an improved Multi-Layer Perceptron (MLP) neural network that can automatically extract and distinguish blood vessels and background pixels in fundus images. This method has strong robustness, but the model is more complex. Cheng et al. [11] used an improved U-Net model, that is, adding dense block structures to the original model, which can effectively improve the accuracy, but the segmentation sensitivity needs to be improved. Francia et al. [12] proposed a double U-Net model structure of fundus image segmentation method. This method uses two U-Net models; the second U-Net model adds residual structure. By adding the information flow extracted from the first U-Net model to the second model, information loss can be effectively avoided. Pan et al. [13] developed an improved U-Net model for retinal vascular segmentation. This method adds the Resnet module to the original structure, which solves the problem that the traditional U-Net model cannot deepen. However, the segmentation results show that the method cannot effectively segment the peripheral blood vessels. Li et al. [14] used a new Resnet module to improve the U-Net. By adding more Resnet modules, the network layer can be deepened, so the model can extract features better. However, with the deepening of the network, the model becomes more complex.

Based on the above problems, such as discontinuous segmentation, difficult segmentation of microvesselss and complex segmentation models, we propose a MRU-Net deep learning model structure, i.e., a multi-scale residual U-shaped network model. The main work of our paper is as follows:

- In MRU-Net, we add a multi-scale feature fusion module in the transition phase between the encoder and decoder and extract the feature information with different granularities using multi-scale dilated convolutions, so as to improve the model's understanding of context information and to improve the model's segmentation ability for microvessels.
- In MRU-Net, we add a feature balance module in the skip connection between the encoder and decoder to solve the problem of possible semantic gaps between low-dimensional features in the encoder and high-dimensional features in the decoder.

2. Proposed Method

As shown in Figure 1, the retinal vessel segmentation framework proposed in this paper is divided into two stages, namely training stage and testing stage. The overall process are as follows:

- Data preprocessing: The original retinal blood vessel image cannot be effectively segmented due to factors such as uneven illumination, low noise, and low contrast. Therefore, image preprocessing is required before training to ensure the maximum possible increase in contrast between the retinal blood vessels and the background, thereby effectively improving the segmentation effect.
- Data expansion: Supervised segmentation training requires manual segmentation of images as labels, so data acquisition is difficult, resulting in insufficient training data for deep learning training. To solve this problem, we randomly divided each preprocessed image data into many smaller image blocks to achieve the purpose of expanding the data set.
- Model training: The image block is divided into a training set and validation set according to the ratio of 9:1. The training set is used as the input of U-Net model, and the corresponding gold standard image block is used as the label. The gradient descent method is used to train the model. According to the training effect of the training set and validation set, the model is continuously optimized until the model reaches the optimum.
- Effect test: Firstly, preprocess the test data in the same way; secondly, orderly segment the test image; then, input the segmented image blocks into the trained U-Net model to obtain the corresponding segmentation results; and finally, the segmentation results of each image block are combined to obtain a complete retinal blood vessel segmentation image.

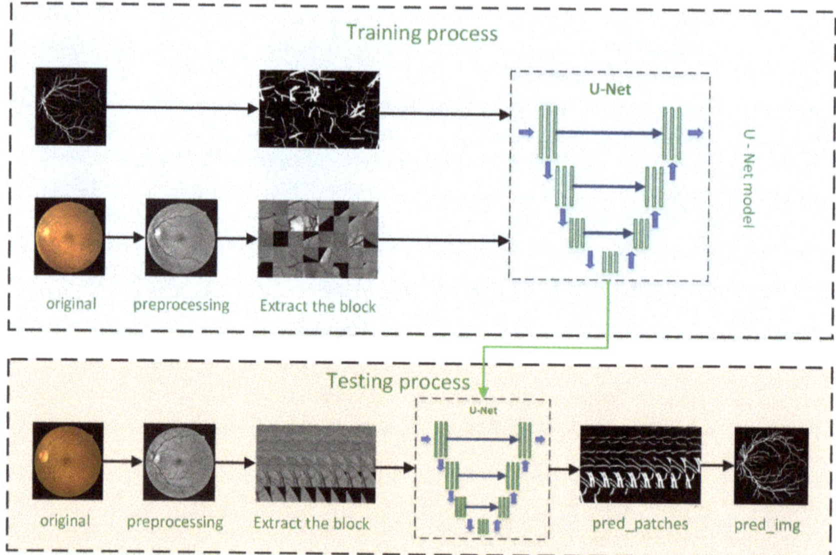

Figure 1. Overview of the architecture framework: the upper part shows the training process of the model; the lower part shows the test process of the model.

2.1. Image Preprocessing

2.1.1. Data Set

We used two datasets for validation experiments: the DRIVE dataset and STARE dataset.

The DRIVE dataset was derived from the Dutch Diabetic Retinopathy Screening Project [15], in which 400 subjects were aged between 25 and 90 years. The DRIVE dataset randomly selected 40 retinal blood vessel images, of which 20 were used as the training set and 20 were used as the training set. Each image is 584×565 in size, and each image has a corresponding expert manual segmentation result and mask image.

The STARE dataset is from the University of California, San Diego [16]. It contains 20 retinal blood vessel images with a size of 700×605, 10 healthy images, and 10 pathological images, and each image is provided with a corresponding expert manual segmentation result image.

Figure 2 is a data example of the two datasets, and the three columns are original fundus blood vessel image, manual segmentation of blood vessels, and masks. Since the masks are not given by STARE, we manually created masks among them.

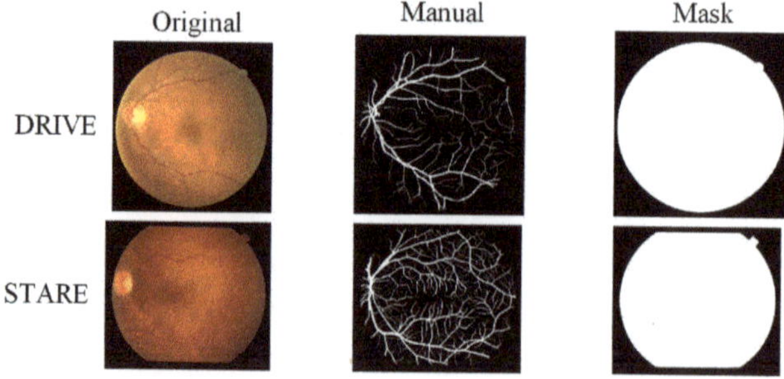

Figure 2. Data example: the images from left to right are the original image, the manually segmented label, and the mask image.

2.1.2. Image Enhancement

Original retinal blood vessel image data are three-channel image data. Due to the optical effect when taking pictures, contrast between the blood vessels and the background in the fundus image is low, so it cannot effectively distinguish the fundus blood vessels from the background image. Image enhancement preprocessing can effectively enhance the contrast between retinal blood vessels and the background. The image enhancement preprocessing process is divided into the following four types:

- Through the contrast experiment, the green channel of the retinal blood vessel image has a higher contrast, so the fundus blood vessel image of the green channel is selected for experiments.
- The green channel image is processed for data standardization, and the conversion formula is as follows:

$$x = \frac{x_i - x_{min}}{x_{max} - x_{min}} \tag{1}$$

  where X_i is the current pixel value, x_{min} is the minimum pixel value, and x_{max} is the maximum pixel value.
- Limited contrast histogram (CLAHE) is used to equalize the normalized image, thereby increasing the contrast between the blood vessel and the background image and making the image easier to segment.
- Gamma adaptive correction is performed according to different pixel characteristics of blood vessels and background images, thereby suppressing uneven light and centerline reflection in the fundus image, and the gamma value was set to 1.2.

The top-down images in Figure 3 are DRIVE and STARE data images, respectively. From left to right: the first column is the original image, and columns 2–5 are the green channel image, the normalized image, the Clahe equalized image, and the gamma locally adaptive correction image, respectively.

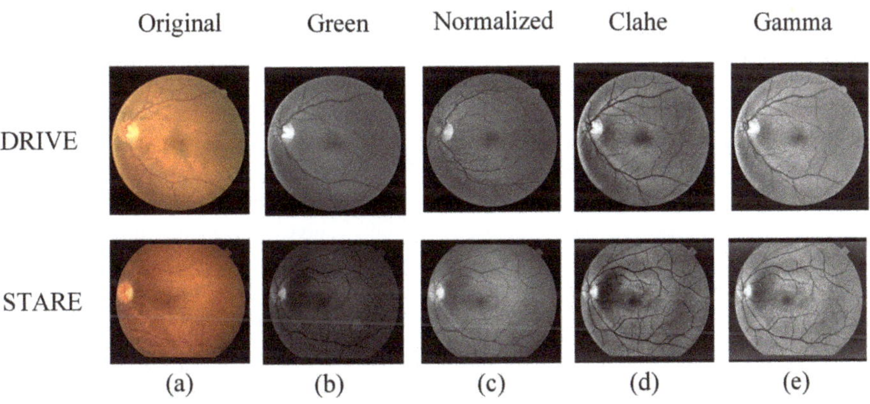

Figure 3. Examples of parts of the image enhancement process: (**a**) original image, (**b**) green channel, (**c**) normalized, (**d**) Clahe processing results, and (**e**) gamma processing results.

2.1.3. Image Expansion

Due to the complexity of manual segmentation of retinal blood vessels, the number of existing images is very scarce, and deep learning is a complex model structure trained based on a large amount of data. When the number is small, the model will be overtrained, which will lead to overfitting. In this paper, the original image data is randomly divided into smaller image blocks to increase the amount of training data.

We randomly cut the original image into 48 × 48 image blocks. The original high-dimensional pixel image segmentation into many low-dimensional pixel images has two main advantages. On the one hand, the high-dimensional image segmentation into more low-dimensional image blocks can effectively play a role in expanding the data, thereby avoiding overfitting caused by a small amount of data. On the other hand, compared with high-dimensional images, the complexity of model design based on low-dimensional image blocks is lower because the U-Net model of low-dimensional images does not need to design deeper layers to extract complex high-dimensional image features and it also effectively reduces the loss of image information caused by the deepening of the model. Figure 4 is an example of several image blocks randomly extracted from the preprocessed DRIVE fundus image.

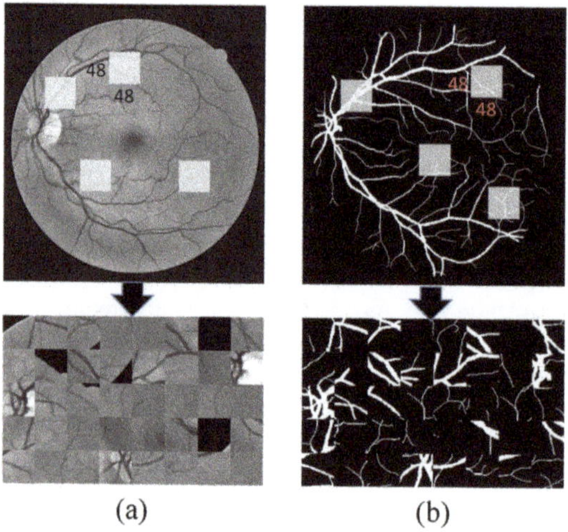

Figure 4. Samples of random segmentation: (**a**) randomly extract 48 × 48 patch image blocks from the preprocessed image and (**b**) the corresponding label from the gold standard image.

2.2. MRU-Net Structure

U-Net [17] has a good effect in the field of medical image segmentation, and many scholars have made different improvements on the segmentation of U-Net network in medical images. The limitation of U-Net and its variants is that. in order to learn more abstract feature information in an image, continuous convolution and pooling operations will be performed, but resolution of image features will be reduced with the deepening of the network. The intuitive expression is that, as the network level deepens, the detailed information in the image will be seriously lost. Based on this, we designed a new U-Net structure, namely MRU-Net. As showned in Figure 5, the proposed model is mainly composed of three modules: a feature encoder–decoder module, a multi-scale feature fusion module, and a skip connection module.

2.2.1. Encoder–Decoder Structure

In the traditional U-Net structure, each block in the encoder and decoder contains two convolution layers and a pooling layer, but frequent convolution and pooling will cause the image to lose more semantic information. As the depth of the network deepens, the phenomenon of gradient disappearance may occur. In order to avoid the gradient disappearing and speeding up network convergence, the traditional convolution module is replaced by the Resnet module. In addition, the convolution operation with a step size of 2 is used to replace the traditional pooling operation in order to reduce the loss of image information as much as possible. The feature map obtained by the convolution operation has the same dimensions as the feature map obtained by the traditional pooling method.

Figure 6 shows the proposed Res-Block structure. Among them, dropout is used to randomly inactivate some neurons to prevent overfitting during model training; Batch Normalization (BN) is used to normalize the activation value of hidden layer neurons, which can prevent the gradient vanishing due to noise in the retina and can improve the expression ability of the model; each convolutional layer in the paper uses linear correction units for feature extraction. Relu can effectively reduce the complexity of the network and increase the convergence rate of the network. Its formula is as follows:

$$F(x) = \begin{cases} 0 & x \leq 0 \\ x & x > 0 \end{cases} \tag{2}$$

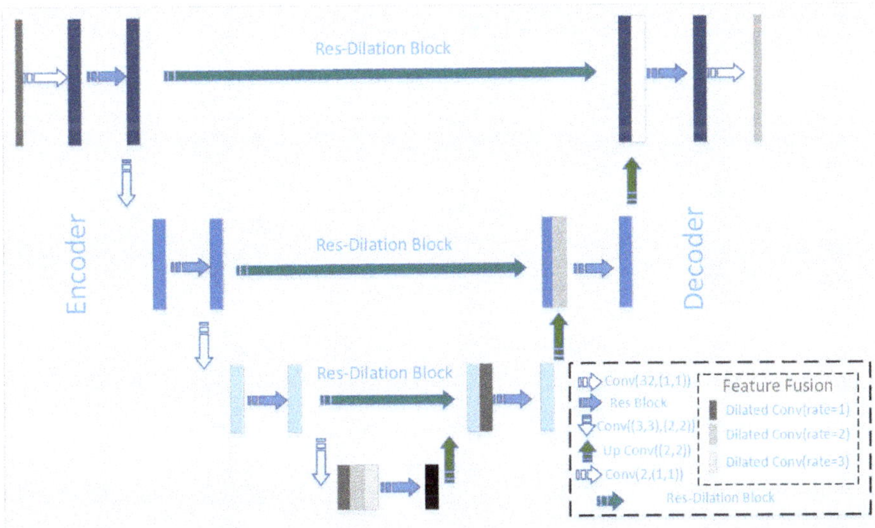

Figure 5. Improved U-shaped neural network (MRU-Net) structure: the structure of the MRU-Net model is described in the lower right corner of the figure, including the methods used for encoder–decoder, feature fusion, and skip connection.

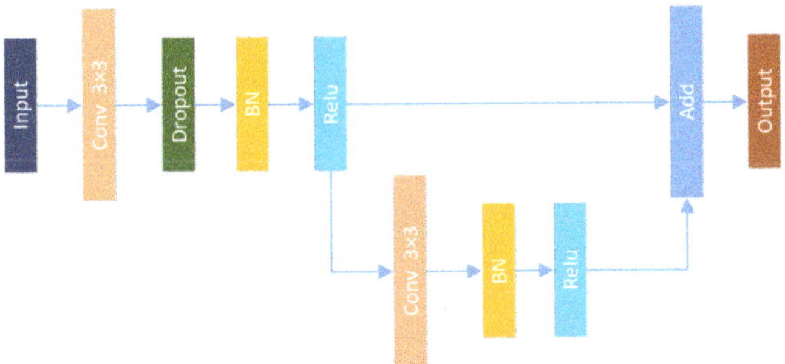

Figure 6. Res-Block structure: each convolution process includes three processes: dropout, BN, and relu.

2.2.2. Feature Fusion Module

Medical image segmentation is different from traditional scene segmentation and has higher requirements for fine-grained segmentation. The retinal blood vessel image is more complicated, and the difference is large in different parts, such as the main part of the blood vessel and microvessel part. We propose a multiscale dilated convolution [18] fusion method. The fusion method mainly relies on different dilation rates to provide multiple effective fields of view, thereby detecting segmented objects of different sizes. In mathematics, the dilated convolution calculation under two-dimensional signals is shown as follows:

$$y[i] = \sum_k x[i + rk]w[k] \tag{3}$$

Among them, x is the input feature map, w is the filter, and r is the expansion rate, which determines the stride of sampling the input signal. It is equivalent to convoluting input x with upsampled filters produced by inserting $r - 1$ zeros between two consecutive flter values along each spatial dimension. The schematic diagram of the dilated convolution is shown in Figure 7.

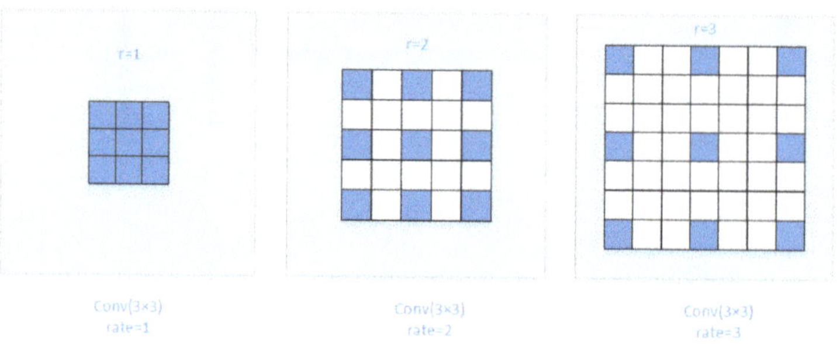

Figure 7. Schematic diagram of dilated convolution.

Context semantic information is mainly determined by the size of the receiving domain. If the receiving domain can provide more abundant information, then more context information can be used. Pooling operations are usually used to increase the receptive field to extract image features at different scales, but the pooling operation also brings loss of image semantic information. In order to overcome this shortcoming and to effectively extract image features of different dimensions, we have adopted the dilated convolution operation mentioned above. Dilated convolution can expand the receptive field arbitrarily without the need to introduce additional parameters and can use the context information of the image, so it is very suitable for multi-scale image segmentation tasks.

Our proposed multi-scale feature fusion module is shown in Figure 8. Generally, the convolution of a larger receptive domain can extract more abstract features of a large object, while the convolution of a small receptive domain is better for small objects. We use three branches to receive the semantic information in the encoder module. Firstly, the dilated rate in the dilated convolution is set to 1, 2, and 3 to expand the receptive field, thereby extracting feature information of different scales in the encoder module; then, the image semantic features extracted from different dilated rates are combined; and finally, in order to reduce the parameters and computational complexity, the Conv (1×1) convolution operation is used to reduce the channel dimension of the feature map to 1/3 of the original dimension.

2.2.3. Res-Dilated Block

One of the main features of the U-Net model is the addition of a skip connection structure between the encoder and decoder. Decoding is a process of recovering coding sampling, in which there will inevitably be loss of information. The addition of a skip connection structure can supplement the original information in the coding structure to the decoding structure, which helps to supplement the lost semantic information.

Because the U-Net model is a deep structure, as the depth of the model increases, the extraction of image feature information will become more abstract. Ibtehaz et al. [19] have shown that, if skip connection is directly used to merge low-dimensional image information and high-dimensional image information, a semantic gap may be generated due to the large difference between image features, which will affect the segmentation effect. Based on this, in the skip connection, we add the Res-Dilated module that combines the residual network and dilated convolution. On the one hand, the high-dimensional representation information of the image is extracted, and on the other hand, the detailed representation information in the receptive field extraction image is increased. The Res-Dilated structure is shown in Figure 9.

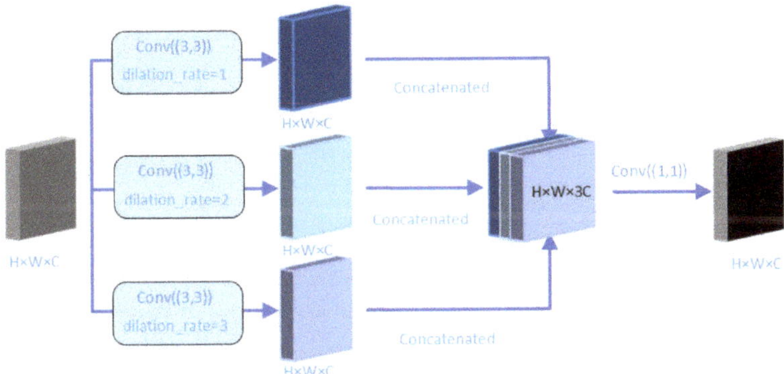

Figure 8. Feature fusion: first, different dilated convolution rates are used for feature extraction, and then, a 1 × 1 convolution operation is used to reduce the feature map, thereby reducing computational complexity.

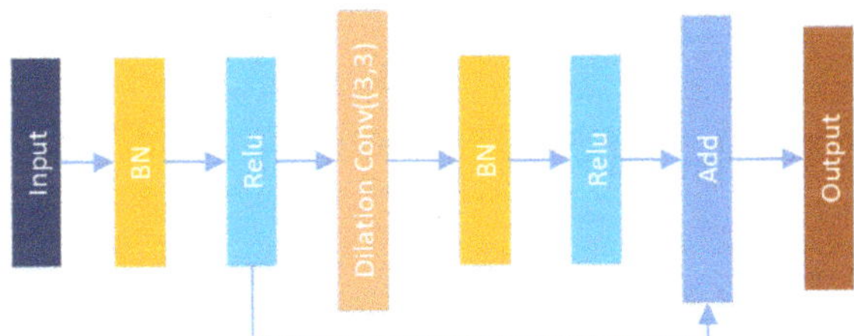

Figure 9. Res-Dilated block: adding an dilated convolution structure to Resnet.

2.2.4. Loss Function

In U-Net model training, the loss function we use is a binary cross-entropy loss function. The formula of the loss function is designed as follows:

$$L = -\frac{1}{n} \sum_{i=1}^{n} \hat{y}_i log y_i + (1 - \hat{y}_i) log(1 - y_i) \tag{4}$$

Among them, $\hat{y}_i$ is the expected output, that is, the real data label, and the value is $\hat{y}_i \in \{0,1\}$; y_i is the actual output, and the value is $y_i \in [0,1]$. In the U-Net network training process, in order to improve its training performance and to obtain better segmentation results, we use Adaptive Moment Estimation (Adam) with the Nesterov momentum term as the optimization algorithm of the model training process. Compared with traditional optimization algorithms, the Adam optimizer has the advantages of high computing efficiency, small memory consumption, and adaptive adjustment of the learning rate. It can also better process noise samples and has a natural annealing effect.

3. Experiment

We used DRIVE and STARE as datasets, where DRIVE contained 40 images (20 training and 20 test images) and STARE contained 20 images (10 training and 10 test images). We extracted the training image with 48×48 pixels, and DRIVE randomly extracts about 190,000 image blocks while STARE randomly extracts about 181,500 image blocks. The extracted image blocks were used as the input of the model, 90% of them were used for training data, and 10% of them were used as validation data. In the test set, the image blocks with pixel values of 48×48 were moved with the height and width of 5 pixels to extract the image blocks in the test set. The number of training epochs was 20, and the batch size was 64. The environment of this experiment was set up on a Linux system. Network construction and experimental tests were implemented based on Keras and TensorFlow. The GPU configuration was GTX 1080Ti. In the experiment, the parameter quantity of the model was 942,020, and the parameter size was about 3.59 M. The training time of each epoch was about 47 s, the total training time was about 15.6 min, and the total test time was about 53 s.

3.1. Evaluation Indicators

For retinal blood vessel image segmentation, the pixels in the image were actually divided into a blood vessel image and a background image. In order to make a qualitative evaluation of the experimental results of this article, we used several general retinal vessel segmentation performance evaluation indicators:

- Sensitivity (SE): The ratio of the total number of correctly segmented blood vessel pixels to the total number of manually segmented blood vessel pixels.
- Accuracy (ACC): The ratio of the total number of correctly segmented blood vessels and background pixels to the pixels of the entire image.
- precision: The ratio of the actual blood vessel pixels in the segmented blood vessel pixels.
- F1 value: The result of combining sensitivity and precision. When the F1 value is high, the method is more effective.
- Area Under Curve (AUC): The area under the receiver operating characteristic (ROC) curve. The larger the value, the better the segmentation effect.

$$Sensitivity = \frac{T_P}{T_P + F_N} \tag{5}$$

$$Accuracy = \frac{T_P + T_N}{T_P + T_N + F_P + F_N} \tag{6}$$

$$Precision = \frac{T_P}{T_P + F_P} \tag{7}$$

$$F1 = 2 \times \frac{Precision \times Sensitivity}{Precision + Sensitivity} \tag{8}$$

Among them, T_P, T_N, F_P, and F_N respectively represent true positive, true negative, false positive, and false negative.

$$Jaccard = \frac{|G \cap P|}{|G \cup P|} \tag{9}$$

where G is the set of ground truth pixels and P is the set of predicted pixels.

3.2. Experimental Results

3.2.1. Subjective Assessment Results

We performed experiments on two databases: DRIVE and STARE. Several sets of images were randomly selected from the test results of the two databases. Figures 10 and 11 show the intuitive effects of the method in this paper. In the figure, the DRIVE data image and the STARE data image are respectively from top to bottom. Among them are (a) preprocessed images, (b) gold standard images manually segmented by experts, (c) images obtained by segmentation using U-Net models, and (d) images segmented by our proposed method.

DRIVE analysis: the first and third lines in Figure 10 are two normal retinal images. On the whole, our proposed method achieves better results than the U-Net model, especially in some details, our method has better segmentation results. The first line in the figure marks some details of the differences. It can be seen that the result of U-Net model segmentation has a problem of blood vessel rupture, and our proposed method solves this problem well; as can be seen from the third line of the figure, compared with the U-Net model, our method has a better effect on the segmentation of microvessels and the small vessels have been completely retained. The images in the second and fourth rows of the figure have noise caused by the lesion. The U-Net model is sensitive to noisy data, resulting in a poor segmentation effect. The segmentation effect proposed in this paper also has a more effective segmentation effect in the face of noisy data.

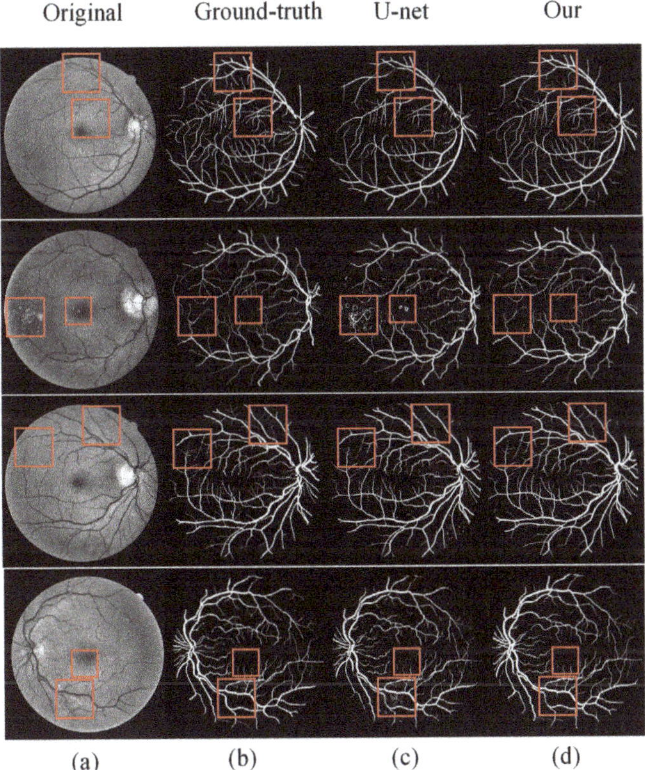

Figure 10. Comparison of DRIVE segmentation results: (**a**) original image, (**b**) ground-truth, (**c**) U-Net segmentation results, and (**d**) MRU-Net segmentation results.

STARE analysis: the three lines of image data in Figure 11 indicate the comparison of the segmentation effects of the U-Net model and the method in this paper. In the first and third rows, U-Net has segmentation breakage and missing vessel segmentation problems; the second line of the retinal blood vessel image is more complicated, and there are many microvessels. Our proposed method also has good results in the face of complex blood vessel images. From this, we can see that our proposed method have a better solution to the problem of microvessel and blood vessel rupture and can perform more coherent and effective segmentation of microvessel.

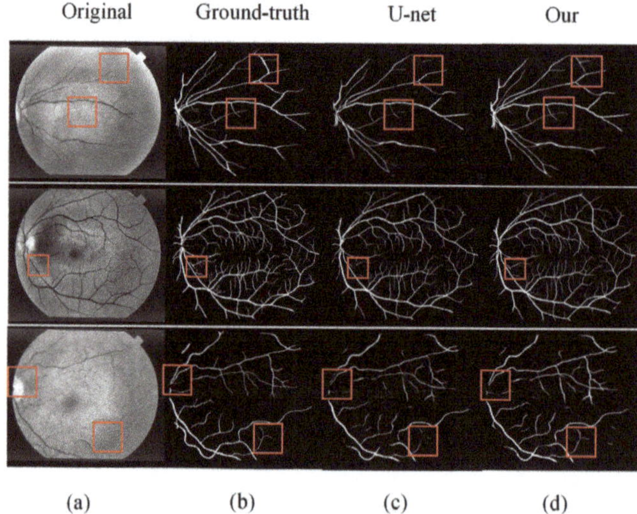

Figure 11. Comparison of STARE segmentation results: (**a**) original image, (**b**) ground-truth, (**c**) U-Net segmentation results, and (**d**) MRU-Net segmentation results.

In order to show the effect of our proposed method more intuitively, the ROC curves of the two data sets shown in Figure 12 are given. From the ROC curve, it can be intuitively seen that the algorithm in this paper has a higher true positive rate and a lower false positive rate and that the blood vessel segmentation error is smaller.

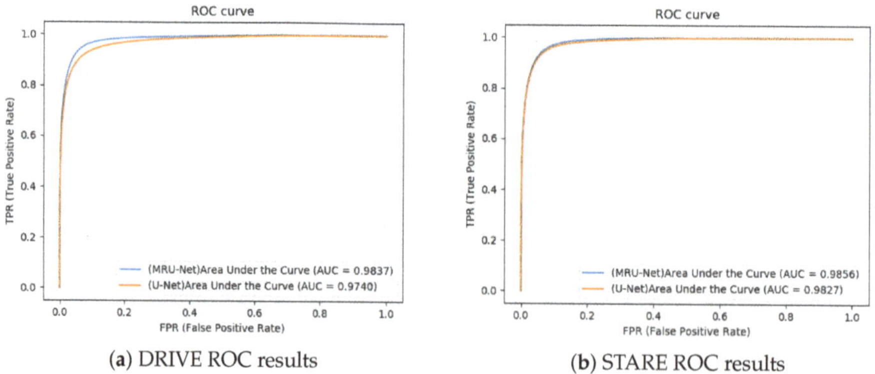

Figure 12. ROC curve: (**a**) DRIVE ROC results and (**b**) STARE ROC results.

3.2.2. Indicator Evalution Results

We conducted two comparative experiments, including (a) model structure comparison and (b) comparison of methods.

In order to explore the improvement effect of the model, the experimental results of different model structures were compared and analyzed, aiming to explore the influence of the improvement of each module on the experimental model. As shown in Figures 13 and 14, Non-FF indicates the experimental results without feature fusion, Non-RD refers to the experimental effect without the Res-Dilated module, and Non-R refers to the experimental results without the Resnet module for encoder and decoder. Through the comparison of experimental results, it can be found that adding feature fusion module can effectively improve the sensitivity (SE) of model segmentation and that the fundus image can be more detailed and effective segmentation; the addition of the Res-Dilated module and Resnet module can reduce the semantic gap between encoder and decoder and effectively improve the transmission of features and reduce the loss of features, so that the overall evaluation index of the model is also improved. On the other hand, although the Resnet and Res-Dilated methods can improve the model, the most significant improvement is the addition of the feature fusion method. Because the size of blood vessels in fundus retinal images is different, the motivation of multi-scale information fusion design is to use different expansion rates to capture multi-scale features, so it can effectively retain the information of microvessels, thus improving the detection accuracy of vascular edges and microvessels. From the data point of view, the value of sensitivity (SE) has been significantly improved, which also reflects that the feature fusion method designed by us improves the segmentation accuracy of more blood vessels.

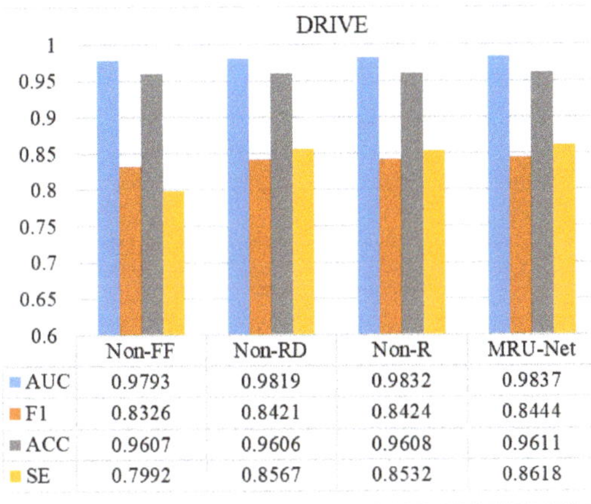

	Non-FF	Non-RD	Non-R	MRU-Net
AUC	0.9793	0.9819	0.9832	0.9837
F1	0.8326	0.8421	0.8424	0.8444
ACC	0.9607	0.9606	0.9608	0.9611
SE	0.7992	0.8567	0.8532	0.8618

Figure 13. Comparison of DRIVE results: comparison of the indexes of different model structures in our proposed method.

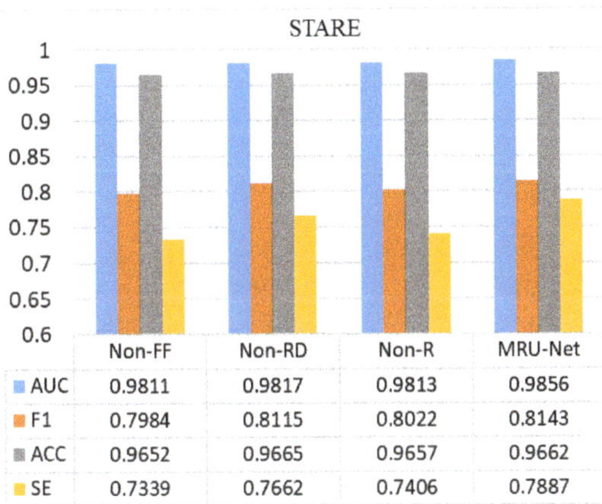

	Non-FF	Non-RD	Non-R	MRU-Net
AUC	0.9811	0.9817	0.9813	0.9856
F1	0.7984	0.8115	0.8022	0.8143
ACC	0.9652	0.9665	0.9657	0.9662
SE	0.7339	0.7662	0.7406	0.7887

Figure 14. Comparison of STARE results: comparison of the indexes of different model structures in our proposed method.

In addition, in order to evaluate the effectiveness of our method, the AUC value, F1 value, sensitivity, and accuracy of this method and other methods are compared. Tables 1 and 2 show the comparison results of this method and other methods. In the DRIVE dataset, the AUC, F1, sensitivity, and accuracy of the proposed method segmentation results reached 0.9837, 0.8444, 0.8618, and 0.9611, respectively. Among them, the AUC, F1, and SE achieved the best results. Compared with other methods, the accuracy rate also has a higher improvement effect, which is only slightly lower than that in [20]. In the STARE dataset, the AUC value, F1 value, sensitivity, and accuracy of the proposed method segmentation results reached 0.9856, 0.8143, 0.7887, and 0.9662, respectively. Among them, the AUC and F1 reached the optimums, ACC was slightly lower than Xiao et al. [20], SE was slightly lower than Lu et al. [21], and good segmentation results are obtained in general. In addition, the confidence ranges of the accuracy of DRIVE and STARE are [0.9604, 0.9625] and [0.9647, 0.9671], respectively.

Table 1. Evaluation of DRIVE Results.

Method	Year	AUC	F1	ACC	SE	Jaccard
U-Net	-	0.9740	0.8203	0.9517	0.7978	-
Gao et al.[22]	2017	0.9772	-	0.9636	0.7802	-
Hu et al. [23]	2018	0.9759	-	0.9533	0.7772	-
Xiao et al. [20]	2018	-	-	0.9655	0.7715	-
Feng et al. [24]	2019	0.9678	-	0.9528	0.7625	-
Jin et al. [25]	2019	-	0.8237	0.9566	0.7963	-
Guo et al. [26]	2019	0.9806	0.8249	0.9561	0.7891	-
Gu et al. [6]	2019	0.9779	-	0.9545	0.8309	-
Pan et al. [13]	2019	0.9811	-	0.9650	0.8310	-
Sekou et al. [27]	2019	0.9874	0.8252	0.9690	0.8398	0.7026
Cheng et al. [11]	2020	0.9793	-	0.9559	0.7672	-
Orujov et al. [28]	2020	-	0.5500	0.9390	-	0.3800
Our	2020	0.9837	0.8444	0.9611	0.8618	0.7291

Table 2. Evaluation of STARE Results.

Method	Year	AUC	F1	ACC	SE	Jaccard
U-Net	-	0.9827	0.7917	0.9648	0.7128	-
Dasgupta et al. [29]	2017	-	0.8133	-	0.7872	-
Hu et al. [23]	2018	0.9751	-	0.9632	0.7543	-
Xiao et al. [20]	2018	-	-	0.9693	0.7469	-
Lu et al. [21]	2018	0.9801	-	0.9628	0.8090	-
Feng et al. [24]	2019	0.9700	-	0.9633	0.7709	-
Jin et al. [25]	2019	0.9832	0.8143	0.9641	0.7595	-
Sekou et al. [27]	2019	0.9849	0.7929	0.9705	0.7695	0.6682
Tamim et al. [10]	2020	-	0.7717	0.9632	0.7806	-
Orujov et al. [28]	2020	-	0.5335	0.8650	-	0.3677
Our	2020	0.9856	0.8143	0.9662	0.7887	0.6864

We performed t-tests for each performance index, and the experimental results are shown in Table 3. From the results of the t-test, we can see that, compared with other methods, our method has significant differences in the performance of AUC, F1, and SE. In addition, we compare the operation time of the model, and the results are shown in Table 4. Our method takes about 15.6 min to train the model and 2.3 s to segment an image in the test set, which is totally acceptable in the medical field.

Table 3. *t*-test result of each index ($p < 0.05$ significant).

Data Set	AUC	F1	ACC	SE
DRIVE	0.0060	0.0002	0.0956	5.14×10^{-6}
STARE	0.0216	0.0454	0.1787	0.0173

Table 4. Computation time for processing one image.

Methods	Time	GPU
[30]	92 s	Tesla K20c
[31]	0.64 s	-
[32]	1.3 s	Tesla K40
[33]	0.5 s	-
[24]	0.063	GTX1070
[34]	70 s	-
Our	2.2 s	GTX 1080Ti

4. Discussion

In this paper, the application of a full convolution neural network in fundus blood vessel segmentation is studied. Traditional machine learning methods cannot effectively segment low-contrast microvessels, and there are problems of low segmentation accuracy and sensitivity. Based on this, an improved U-Net model is proposed, which not only realizes the information fusion of high-dimensional features and low-dimensional features but also improves the efficiency of feature use. In this study, experiments were conducted on DRIVE and STARE fundus image databases, and image blocks were randomly extracted to increase the amount of network training data to avoid network overfitting.

In terms of network structure design and parameter setting, based on the traditional U-shaped network, the network is defined as a structure block connected by upsampling or downsampling operations. According to the characteristics of small amount of experimental data, this study randomly extracts image blocks to reduce the dimension of image data, so as to reduce the complexity of the network model. Only three downsampling and three upsampling operations are used. In order to make full use of the feature information extracted from the network and to minimize the loss of feature information, the convolution operation with step size of 2 is used to replace the traditional pooling operation and each structural block is changed to Res-Block. In order to enhance the generalization ability of the network and to better identify the small vessel structure, dropout, batch normalization layer, and relu activation function were added before each convolution layer. Then, in order to better extract the context semantic information of the image and to effectively segment the vascular structure of different scales, we propose a multi-scale dilation convolution fusion method based on dilation convolution. In addition, in order to solve the semantic gap between the encoder and decoder, we propose a Res-Dilated block to replace the skip connection in the traditional U-Net model. As shown in Figures 10–13, the traditional U-shaped network can be improved to better segment the blood vessels with central line reflection and the microvessels with low contrast, so that the sensitivity, F1 value, accuracy, and area under ROC curve (AUC) index of this research algorithm are improved.

In order to more intuitively compare the improvement effect of the U-Net model and the segmentation performance of blood vessels, this study compared a variety of deep learning methods including U-Net model. In general, compared with the original U-Net model, the MRU-Net model has less noise points in the vessel probability map. From the details, the results of this study can better segment the small vessels and retain the integrity of the vessels, and the continuity of the vessels is better. In order to observe the influence of lesions on vascular segmentation more intuitively, by comparing with the experimental results of the original U-Net model, it can be seen from Figure 10 that the MRU-Net network proposed in this study can better segment blood vessels and is less affected by the lesion area. Compared with other methods, it can be seen from the detection indicators that our method achieves an optimal effect in three indicators of F1, SE, and Jaccard in the DRIVE dataset and that the AUC index is only slightly lower than that in the literature [27]. Although the ACC index is better than most methods, it needs to be improved. In the STARE dataset, our method achieves the best results in AUC, F1, and Jaccard; the ACC index is only slightly lower than in the literature [20,27]; and SE is slightly lower than in the literature [20].

Our main contribution is to propose a simple and effective multi-scale information fusion module, which uses the parallel convolution layer with different dilated rates to sample the feature map and to get the feature information of different scales, which improves the detection performance of vessel edges and microvasculature. In order to visually show the contribution of this paper, we compare and analyze the effect of feature fusion and Resnet. The results on the DRIVE dataset are shown in Table 5. It can be seen from Table 5 that feature fusion has a good effect on model improvement.

Table 5. Comparative analysis of feature fusion and Resnet.

DRIVE	FF	Resnet
AUC	0.9825	0.9773
F1	0.8420	0.8167
ACC	0.9618	0.9548
SE	0.8310	0.7905

It can be seen from the intuitive results and indicator evaluation results that our proposed method has excellent effects on retinal vessel image segmentation. For complex blood vessel images, standards close to manual segmentation by experts can be obtained. In addition, we invited five experts with medical image processing background to evaluate the subjective results. Among them, four experts think our method is effective for the segmentation of fundus images, and another expert points out that some of our segmentation results are excessive segmentations hoping to be improved in future research.

5. Conclusions

Using artificial intelligence to assist doctors in segmenting retinal blood vessel images is helpful to doctors quickly diagnosing the disease and has important practical significance. We propose an improved U-Net model structure for the current fundus vessel image segmentation with poor segmentation effect and difficult microvessel segmentation. The model structure adds a residual learning unit to the encoder structure and decoder structure of the traditional U-Net model, which is more conducive to the transmission and retention of information; a feature fusion module is added in the transition phase of the model's encoding structure and decoding structure, which aims to extract feature information of different granularities in the original image, which is more conducive to retaining fine-grained image features, so that tiny blood vessels can be better segmented; the Res-Dilated module is introduced in the skip connection, which can make the decoding structure network supplement the original information while reducing the semantic gap between the encoding and decoding stages. Through experimental verification analysis on two datasets of DRIVE and STARE, our proposed method has an excellent segmentation effect. This proposed method achieved 0.9611, 0.8618, and 0.9837 in DRIVE and 0.9662, 0.7887, and 0.9856 on STARE for respectable accuracy, sensitivity, and AUC performance metrics. Compared with the current methods, our method has better performance on the whole.

Although our method has good performance, it still needs further improvement. Our method can effectively segment microvessels, but some of them may still be discontinuous. In addition, based on the current experimental conditions, we only tested on DRIVE and STARE datasets, and more clinical data are needed for verification in the future. Our future work is to continue to optimize the effect of retinal microvascular segmentation. On the one hand, contrast between the blood vessel and the background is increased by optimizing the preprocessing steps; on the other hand, the segmentation effect is improved by continuously improving the model.

Author Contributions: Conceptualization, H.D. and X.C.; investigation, H.D., L.C., and K.Z.; writing—original draft, H.D.; writing—review and editing, X.C. and L.C. All authors have read and agreed to the published version of the manuscript.

Funding: This work has been supported by National Key Research and Development Program of China (NO.2018YFC1604000).

Conflicts of Interest: The authors declare no conflict of interest.

References

1. Dash, J.; Bhoi, N. Retinal Blood Vessel Extraction Using Morphological Operators and Kirsch's Template. In *Soft Computing and Signal Processing*; Springer: Berlin/Heidelberg, Germany, 2019; pp. 603–611.

2. Zhang, J.; Dashtbozorg, B.; Bekkers, E.; Pluim, J.P.; Duits, R.; ter Haar Romeny, B.M. Robust retinal vessel segmentation via locally adaptive derivative frames in orientation scores. *IEEE Trans. Med. Imaging* **2016**, *35*, 2631–2644. [CrossRef]

3. Ben Abdallah, M.; Malek, J.; Azar, A.T.; Montesinos, P.; Belmabrouk, H.; Esclarín Monreal, J.; Krissian, K. Automatic extraction of blood vessels in the retinal vascular tree using multiscale medialness. *Int. J. Biomed. Imaging* **2015**, *2015*. [CrossRef]

4. Orlando, J.I.; Prokofyeva, E.; Blaschko, M.B. A discriminatively trained fully connected conditional random field model for blood vessel segmentation in fundus images. *IEEE Trans. Biomed. Eng.* **2016**, *64*, 16–27. [CrossRef]

5. Wang, X.; Chen, D.; Luo, L. Retinal blood vessels segmentation based on multi-classifier fusion. In Proceedings of the 2018 Chinese Control And Decision Conference (CCDC), Shenyang, China, 9–11 June 2018; pp. 3542–3546.

6. Gu, Z.; Cheng, J.; Fu, H.; Zhou, K.; Hao, H.; Zhao, Y.; Zhang, T.; Gao, S.; Liu, J. CE-Net: Context encoder network for 2D medical image segmentation. *IEEE Trans. Med. Imaging* **2019**, *38*, 2281–2292. [CrossRef] [PubMed]

7. Zhou, Z.; Siddiquee, M.M.R.; Tajbakhsh, N.; Liang, J. Unet++: Redesigning skip connections to exploit multiscale features in image segmentation. *IEEE Trans. Med. Imaging* **2019**, *39*, 1856–1867. [CrossRef] [PubMed]

8. Soomro, T.A.; Afifi, A.J.; Shah, A.A.; Soomro, S.; Baloch, G.A.; Zheng, L.; Yin, M.; Gao, J. Impact of Image Enhancement Technique on CNN Model for Retinal Blood Vessels Segmentation. *IEEE Access* **2019**, *7*, 158183–158197. [CrossRef]

9. Kumawat, S.; Raman, S. Local phase U-Net for fundus image segmentation. In Proceedings of the ICASSP 2019—2019 IEEE International Conference on Acoustics, Speech and Signal Processing (ICASSP), Brighton, UK, 12–17 May 2019; pp. 1209–1213.

10. Tamim, N.; Elshrkawey, M.; Abdel Azim, G.; Nassar, H. Retinal Blood Vessel Segmentation Using Hybrid Features and Multi-Layer Perceptron Neural Networks. *Symmetry* **2020**, *12*, 894. [CrossRef]

11. Cheng, Y.; Ma, M.; Zhang, L.; Jin, C.; Ma, L.; Zhou, Y. Retinal blood vessel segmentation based on Densely Connected U-Net. *Math. Biosci. Eng.* **2020**, *17*, 3088. [CrossRef]

12. Francia, G.A.; Pedraza, C.; Aceves, M.; Tovar-Arriaga, S. Chaining a U-Net With a Residual U-Net for Retinal Blood Vessels Segmentation. *IEEE Access* **2020**, *8*, 38493–38500. [CrossRef]

13. Xiuqin, P.; Zhang, Q.; Zhang, H.; Li, S. A fundus retinal vessels segmentation scheme based on the improved deep learning U-Net model. *IEEE Access* **2019**, *7*, 122634–122643. [CrossRef]

14. Li, D.; Dharmawan, D.A.; Ng, B.P.; Rahardja, S. Residual U-Net for Retinal Vessel Segmentation. In Proceedings of the 2019 IEEE International Conference on Image Processing (ICIP), Taipei, Taiwan, 22–25 September 2019.

15. Staal, J.; Abràmoff, M.D.; Niemeijer, M.; Viergever, M.A.; Van Ginneken, B. Ridge-based vessel segmentation in color images of the retina. *IEEE Trans. Med. Imaging* **2004**, *23*, 501–509. [CrossRef] [PubMed]

16. Hoover, A.; Kouznetsova, V.; Goldbaum, M. Locating blood vessels in retinal images by piecewise threshold probing of a matched filter response. *IEEE Trans. Med. Imaging* **2000**, *19*, 203–210. [CrossRef] [PubMed]

17. Ronneberger, O.; Fischer, P.; Brox, T. U-Net: Convolutional Networks for Biomedical Image Segmentation. In *International Conference on Medical Image Computing and Computer-Assisted Intervention, Proceedings of the Medical Image Computing and Computer-Assisted Intervention—MICCAI 2015, Munich, Germany, 5–9 October 2015*; Springer: Berlin/Heidelberg, Germany, 2015.

18. Chen, L.C.; Papandreou, G.; Kokkinos, I.; Murphy, K.; Yuille, A.L. Deeplab: Semantic image segmentation with deep convolutional nets, atrous convolution, and fully connected crfs. *IEEE Trans. Pattern Anal. Mach. Intell.* **2017**, *40*, 834–848. [CrossRef]

19. Ibtehaz, N.; Rahman, M.S. MultiResUNet: Rethinking the U-Net architecture for multimodal biomedical image segmentation. *Neural Netw.* **2020**, *121*, 74–87. [CrossRef] [PubMed]

20. Xiao, X.; Lian, S.; Luo, Z.; Li, S. Weighted Res-UNet for high-quality retina vessel segmentation. In Proceedings of the 2018 9th International Conference on Information Technology in Medicine and Education (ITME), Hangzhou, China, 19–21 October 2018; pp. 327–331.

21. Lu, J.; Xu, Y.; Chen, M.; Luo, Y. A Coarse-to-Fine Fully Convolutional Neural Network for Fundus Vessel Segmentation. *Symmetry* **2018**, *10*, 607. [CrossRef]

22. Gao, X.; Cai, Y.; Qiu, C.; Cui, Y. Retinal blood vessel segmentation based on the Gaussian matched filter and U-net. In Proceedings of the 2017 10th International Congress on Image and Signal Processing, BioMedical Engineering and Informatics (CISP-BMEI), Shanghai, China, 14–16 October 2017; pp. 1–5.

23. Hu, K.; Zhang, Z.; Niu, X.; Zhang, Y.; Cao, C.; Xiao, F.; Gao, X. Retinal vessel segmentation of color fundus images using multiscale convolutional neural network with an improved cross-entropy loss function. *Neurocomputing* **2018**, *309*, 179–191. [CrossRef]

24. Feng, S.; Zhuo, Z.; Pan, D.; Tian, Q. CcNet: A cross-connected convolutional network for segmenting retinal vessels using multi-scale features. *Neurocomputing* **2020**, *392*, 268–276. [CrossRef]

25. Jin, Q.; Meng, Z.; Pham, T.D.; Chen, Q.; Wei, L.; Su, R. DUNet: A deformable network for retinal vessel segmentation. *Knowl. Based Syst.* **2019**, *178*, 149–162. [CrossRef]

26. Guo, S.; Wang, K.; Kang, H.; Zhang, Y.; Gao, Y.; Li, T. BTS-DSN: Deeply supervised neural network with short connections for retinal vessel segmentation. *Int. J. Med. Inform.* **2019**, *126*, 105–113. [CrossRef]

27. Sekou, T.B.; Hidane, M.; Olivier, J.; Cardot, H. From Patch to Image Segmentation using Fully Convolutional Networks—Application to Retinal Images. *arXiv* **2019**, arXiv:1904.03892.

28. Orujov, F.; Maskeliunas, R.; Damaševičius, R.; Wei, W. Fuzzy based image edge detection algorithm for blood vessel detection in retinal images. *Appl. Soft Comput.* **2020**, *94*, 106452. [CrossRef]

29. Dasgupta, A.; Singh, S. A fully convolutional neural network based structured prediction approach towards the retinal vessel segmentation. In Proceedings of the 2017 IEEE 14th International Symposium on Biomedical Imaging (ISBI 2017), Melbourne, Australia, 18–21 April 2017; pp. 248–251.

30. Liskowski, P.; Krawiec, K. Segmenting Retinal Blood Vessels With Deep Neural Networks. *IEEE Trans. Med. Imaging* **2016**, *35*, 2369–2380. [CrossRef] [PubMed]

31. Jebaseeli, T.J.; Durai, C.A.D.; Peter, J.D. Segmentation of retinal blood vessels from ophthalmologic Diabetic Retinopathy images. *Comput. Electr. Eng.* **2018**, *73*, 245–258. [CrossRef]

32. Fu, H.; Xu, Y.; Lin, S.; Wong, D.W.K.; Liu, J. DeepVessel: Retinal Vessel Segmentation via Deep Learning and Conditional Random Field. In Proceedings of the International Conference on Medical Image Computing and Computer-Assisted Intervention, Medical Image Computing and Computer-Assisted Intervention—MICCAI 2016, Athens, Greece, 17–21 October 2016; Springer: Berlin/Heidelberg, Germany, 2016.

33. Girard, F.; Kavalec, C.; Cheriet, F. Joint segmentation and classification of retinal arteries/veins from fundus images. *Artif. Intell. Med.* **2019**, *94*, 96–109. [CrossRef] [PubMed]

34. Li, Q.; Feng, B.; Xie, L.; Liang, P.; Zhang, H.; Wang, T. A cross-modality learning approach for vessel segmentation in retinal images. *IEEE Trans. Med. Imaging* **2015**, *35*, 109–118. [CrossRef]

Article

Intracranial Hemorrhage Detection in Head CT Using Double-Branch Convolutional Neural Network, Support Vector Machine, and Random Forest

Agata Sage * and Pawel Badura

Faculty of Biomedical Engineering, Silesian University of Technology, Roosevelta 40, 41-800 Zabrze, Poland; pawel.badura@polsl.pl
* Correspondence: agata.sage@polsl.pl

Received: 13 August 2020; Accepted: 24 October 2020; Published: 27 October 2020

Abstract: Brain hemorrhage is a severe threat to human life, and its timely and correct diagnosis and treatment are of great importance. Multiple types of brain hemorrhage are distinguished depending on the location and character of bleeding. The main division covers five subtypes: subdural, epidural, intraventricular, intraparenchymal, and subarachnoid hemorrhage. This paper presents an approach to detect these intracranial hemorrhage types in computed tomography images of the head. The model trained for each hemorrhage subtype is based on a double-branch convolutional neural network of ResNet-50 architecture. It extracts features from two chromatic representations of the input data: a concatenation of the image normalized in different intensity windows and a stack of three consecutive slices creating a 3D spatial context. The joint feature vector is passed to the classifier to produce the final decision. We tested two tools: the support vector machine and the random forest. The experiments involved 372,556 images from 11,454 CT series of 9997 patients, with each image annotated with labels related to the hemorrhage subtypes. We validated deep networks from both branches of our framework and the model with either of two classifiers under consideration. The obtained results justify the use of a combination of double-source features with the random forest classifier. The system outperforms state-of-the-art methods in terms of F1 score. The highest detection accuracy was obtained in intraventricular (96.7%) and intraparenchymal hemorrhages (93.3%).

Keywords: intracranial hemorrhage; computer-aided diagnosis; computed tomography; deep learning; random forest

---

1. Introduction

Intracranial hemorrhage (ICH) relates to bleeding occurring within the intracranial vault. Possible reasons include, i.a., vascular abnormalities, venous infarction, tumor, trauma effects, therapeutic anticoagulation, and cerebral aneurysm [1–4]. Regardless of the actual cause, a hemorrhage constitutes a major threat. Therefore, an accurate and rapid diagnosis is crucial for the treatment process and its success. ICH diagnosis relies on patient medical history, physical examination, and non-contrast computed tomography (CT) examination of the brain region. CT examination enables bleeding localization and can indicate the primary causes of ICH [5]. There are several challenges related to the ICH diagnosis and treatment: the urgency of the procedure, a complex and time-consuming decision-making process, an insufficient level of experience in the case of novice radiologists, and the fact that most emergencies occur at nighttime. Thus, there is a significant need for a computer-aided diagnosis tool to assist the specialist. Nevertheless, the accuracy of automated hemorrhage detection should be sufficiently high for medical purposes.

Depending on the brain's anatomic site of bleeding, several ICH subtypes can be distinguished (Figure 1). Subdural hemorrhage (SDH) refers to bleeding between the dura and the arachnoid, whereas the epidural subtype (EDH) involves bleeding between the dura and the bone. Both frequently result from traumatic injuries. Intraparenchymal hemorrhage (IPH) is bleeding within the area of brain parenchyma. A hemorrhage inside the ventricular system is known as intraventricular (IVH). Finally, blood within the subarachnoid space indicates subarachnoid hemorrhage (SAH) [1,5]. One of the leading causes of SAH is a ruptured cerebral aneurysm [3,4]. Hemorrhage detection and classification are challenging due to similarities between the various ICH subtypes (e.g., SDH vs. EDH) and subtle differences between healthy and bleeding tissues. These are barely noticeable to the inexperienced observer. Figure 1 presents examples of selected CT slices featuring the physiological brain appearance and ICH subtypes under consideration.

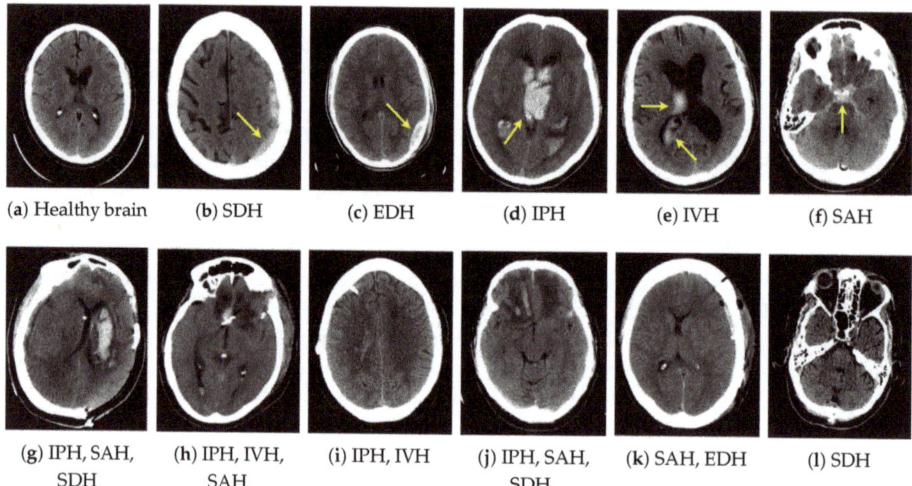

(a) Healthy brain **(b) SDH** **(c) EDH** **(d) IPH** **(e) IVH** **(f) SAH**

(g) IPH, SAH, SDH **(h) IPH, IVH, SAH** **(i) IPH, IVH** **(j) IPH, SAH, SDH** **(k) SAH, EDH** **(l) SDH**

Figure 1. Sample non-contrast computed tomography (CT) slices with various intracranial hemorrhage (ICH) subtypes in the top row (**a–f**). Yellow arrows indicate the areas of hemorrhage. Complex cases are shown in the bottom row: combinations of multiple ICH subtypes simultaneously (**g–k**) or barely visible ICH symptoms (**l**).

In recent years, we have observed increasing interest among researchers in deep learning methods for image classification and segmentation. Convolutional neural networks (CNNs) have gained popularity due to their reliability and efficiency, becoming significant factors in medical diagnosis support [6,7]. In general, CT scans are 3D structures composed of a stack of 2D slices. Thus, operating on image voxels is possible, but may require large computational complexity. A technique to avoid the latter is to either process slices individually or employ the 3D context in a less complex way. Several deep learning approaches to intracranial hemorrhage detection and classification have been proposed, most of them in the last three or four years. Different studies address either one-class detection related to a single class of ICH present in a CT scan [8–10], multi-class classification distinguishing ICH subtypes [11–15], or ICH pixel area detection within individual images [7,16,17]. The popularity of pre-trained CNN models can be observed, for example for VGG [8], MobileNet [15], AlexNet [18], and ResNet-18 [14].

Nguyen et al. [11] reported the results of a combination of a convolutional neural network and a long short-term memory (LSTM) for hemorrhage classification. The slice-wise pre-trained CNN (ResNet-50 and SE-ResNeXt-5 architectures) extracted features from every image, while the LSTM linked them across slices. Arbabshirani et al. [9] proposed a CNN architecture, employing two fully connected layers and 37,084 training images to achieve ICH presence detection. The achieved accuracy was equal to 95%. Danilov et al. [19] designed a ResNexT CNN model for classification between five ICH subtypes, using the Adam optimizer and a dataset of 674,258 CT slices. The accuracies were equal to 82.8%, 81.8%, 82.0%, 89.3%, and 83.5% for EDH, SDH, SAH, IVH, and IPH, respectively. Ye et al. [12] employed a joint 3D CNN and recurrent neural network (RNN) to detect ICH and recognize its five subtypes. The accuracy of detecting any bleeding exceeded 98%, yet in individual subtypes it varied between 75% and 96%. Ker et al. [20] proposed a 3D CNN for various IPH subtype classification: healthy brain, SAH, IPH, acute subdural hemorrhage (ASDH), and brain polytrauma hemorrhage (BPH). The dataset included 399 volumetric CT scans (approx. 12,000 images). Two types of classification were performed: two-class (normal vs. a specific ICH) and four-class (normal vs. all considered ICH subtypes). The authors employed the F1 score to assess their approaches and obtained 70.6–95.2% for the two-class cases and 68.4% for the multiclass analysis. Toğaçar et al. [18] proposed a combination of a CNN, autoencoder network, and a heat map method. The dataset was processed using an autoencoder network with heatmaps generated based on every image. The outcome subjected to the augmentation process constituted the input to the pre-trained CNN (AlexNet architecture). The classification employed a support vector machine (SVM). The authors reported 98.6% accuracy, 98.1% sensitivity, and 99.0% specificity for detecting any hemorrhage using a dataset consisting of 2101 images with augmentation. Dawud et al. [21] compared the results of the AlexNet CNN-SVM combination and two conventional CNN architectures (CNN and pre-trained AlexNet model). The first one produced evaluation metrics better than traditional architectures: 93% accuracy, 95% sensitivity, and 90% specificity for the ICH presence detection over a database containing 12,635 images. Burduja et al. [22] proposed an integration of two neural network architectures to perform ICH subtype classification: ResNeXt-101 and bidirectional long short-term memory (BiLSTM). They compared model performance with evaluation metrics calculated based on annotations collected from three radiologists. The obtained accuracy exceeded 96% in ICH subtypes with a weighted mean log loss equal to 0.04989 in the test dataset. However, reduced sensitivity over highly imbalanced data suggests that the F1 score would reflect the system's performance more realistically.

In this study, we propose a method for detecting various subtypes of intracranial hemorrhage (SDH, EDH, IPH, IVH, and SAH) in the brain CT scans based on a double-branch CNN for feature extraction and two different classifiers. According to the literature reports, binary detectors of individual subtypes frequently feature higher precision than multiclass approaches [8,10,12,15,23]. Moreover, a definitive decision from the multiclass framework is often partially wrong since a single slice can present more than one subtype of the ICH. Thus, we use individually trained instances of the proposed architecture to detect each ICH subtype separately. Before feature extraction and classification, preprocessing is applied, including a skull removal algorithm. Three intensity windows are employed to transform the original CT slice. We concatenate the resulting grayscale images to set up inputs for the double-branch CNN. In the first branch, the method considers different-window images to gain more information from a single input. The second branch analyzes the slice in a local 3D context using the neighboring slices. A ResNet-50 architecture is implemented and trained in both paths to extract features subjected to the classification automatically. The classification itself involves two tools: the SVM and the random forest (RF). We compare the results of both classifiers in the detection of each ICH subtype. We also verify the efficiency of individual ResNet-50 networks from both branches. For the training and assessment, we employ a public database containing 372,556 head CT slices with the corresponding ground-truth labels indicating the presence of either SDH, EDH, IPH, IVH, or SAH. Finally, we compare our approach to the state-of-the-art reference methods described in the introduction. The employment of two various representations of data and the use of an external classifier constitute the main contributions of our study. Both spatial and intensity-related features feed a classifier through a joint vector. Approaches incorporating spatial information can be found in the ICH detection domain; however, to the best of our knowledge, none of them have used a hybrid technique similar to the one proposed in this study.

The remainder of the paper is structured as follows: Section 2 describes the materials and methods, including preprocessing, double-branch feature extraction, and the classification process. In Section 3, we present the results of intracranial hemorrhage classification using the proposed tool with different settings and compare them to the state-of-the-art methods. Section 3 also discusses the results, whereas Section 4 concludes the paper.

2. Materials and Methods

2.1. Materials

The database employed in this study includes 372,556 head CT slices acquired from 9997 patients and 11,454 complete CT series. The data are a part of the public Radiological Society of North America (RSNA) database used for the intracranial hemorrhage detection competition [24,25]. The database contains images presenting various types of bleeding: subdural, epidural, intraventricular, intraparenchymal, and subarachnoid. Many CT slices are not assigned to any of those types; some contain more than one hemorrhage. A ground-truth label determining the presence of every ICH subtype is attached to each case. The class distribution within the database is given in Table 1. The number of slices per scan varies among patients, although most of the CT studies present the whole brain area, not only its middle section. We prepared the data for the analysis by sorting the patient ID, CT study, and series, based on standard DICOM (Digital Imaging and Communications in Medicine)attributes.

Table 1. Class distribution within the database.

Subdural (SDH)	Epidural (EDH)	Intraparen-Chymal (IPH)	Intraventri-Cular (IVH)	Subarachnoid (SAH)	None
24,912	1482	13,666	19,026	18,353	316,082

2.2. Methods

The proposed framework involves three stages of data processing (Figure 2): (1) preprocessing and preparing the input data for deep learning, (2) automated extraction of features using two CNN branches, and (3) two-class classification based on the joint feature vector. As shown in Figure 2, an individual deep learning and classification model is dedicated to each ICH subtype, so the preprocessing is the only step common for all five models.

2.2.1. Preprocessing

The data preprocessing starts with applying dedicated intensity windows to the CT image, followed by the skull removal algorithm. We use three intensity windows (in Hounsfield units—HU): L = 100, W = 200 (subdural window), L = 600, W = 2800 (bone window), and L = 40, W = 80 (brain window). Images after the intensity windowing are normalized to the 0–1 range. Then, we determine the region of interest (ROI) embracing the cranial area as a bounding box framing the largest binary object in the subdural image after binarization using the Otsu method [26]. The skull removal process is performed on the subdural-window image by extracting pixels with the highest intensities corresponding to the bone regions and replacing them with zero-valued pixels. Due to soft tissue outside the skull, the morphological opening is applied to extract the intracranial area.

After the skull removal, the data preparation splits into two branches related to the architecture of the following CNNs. In either branch, preprocessing concludes with preparing RGB images for the feature extraction part of the CNNs. In the left branch in Figure 2, the 3D image structure is composed of three windowed slices with skull pixels replaced with zeros. The right branch employs spatial information by combining three subsequent CT slices from the current scan, all in the subdural window and with a removed skull. The *i*-th slice under consideration is put into the green (G) channel, while the red (R) and blue (B) channels contain slices preceding and following the *i*-th one, respectively. Figure 3 shows the results of subsequent preprocessing stages for three sample CT slices.

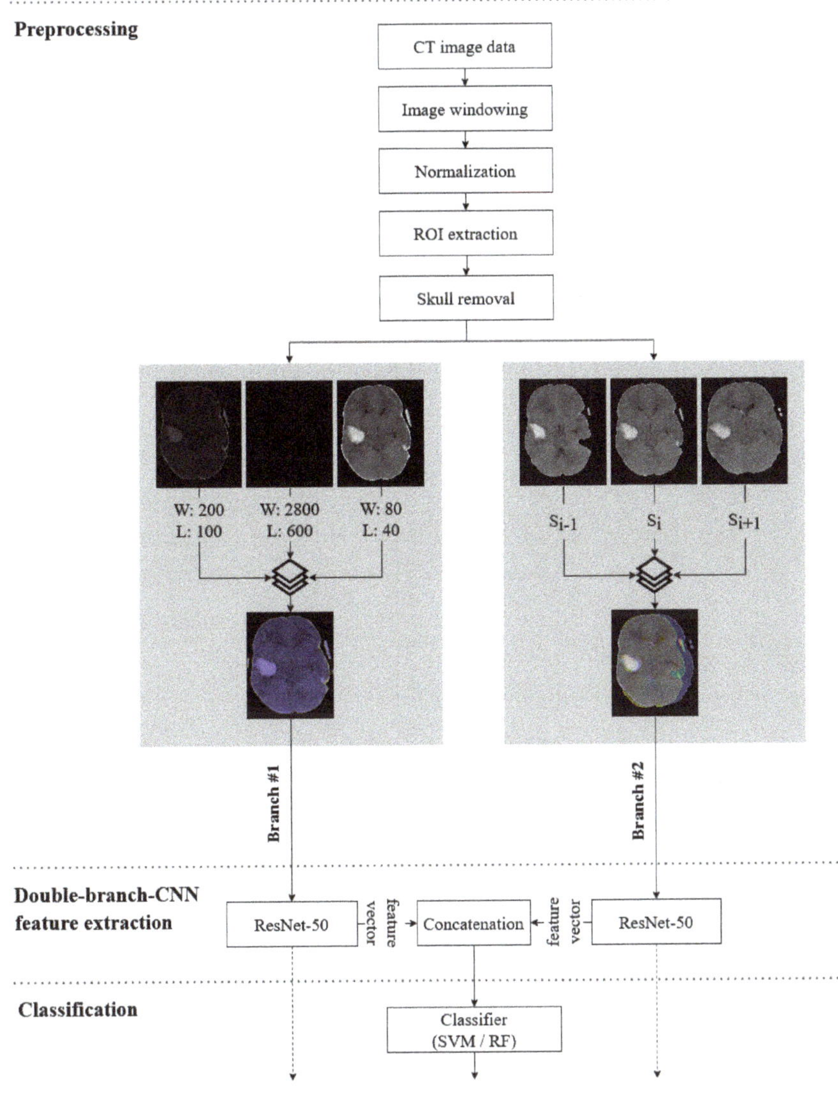

Figure 2. General scheme of a proposed hemorrhage detection model. The same architecture applies for each classification task (five ICH subtypes vs. healthy brain). The preprocessing is a common step for all five models.

2.2.2. Double-Branch-CNN Feature Extraction

To automatically extract features from the prepared RGB images, we use a double-branch convolutional neural network. Both paths employ similar pre-trained ResNet-50 architecture consisting of a set of convolutional and identity blocks (Figure 4) [27]. ResNet is a representative deep CNN yielding efficient results in the image classification domain. As a deeper network, it allows the extraction of more advanced features. However, a large number of layers might cause vanishing gradients and accuracy to decrease. The skip-connections in ResNet architecture minimize that risk [27]. The network accepts input data of a $224 \times 224 \times 3$ size, so the images prepared during preprocessing are resized to match the required format.

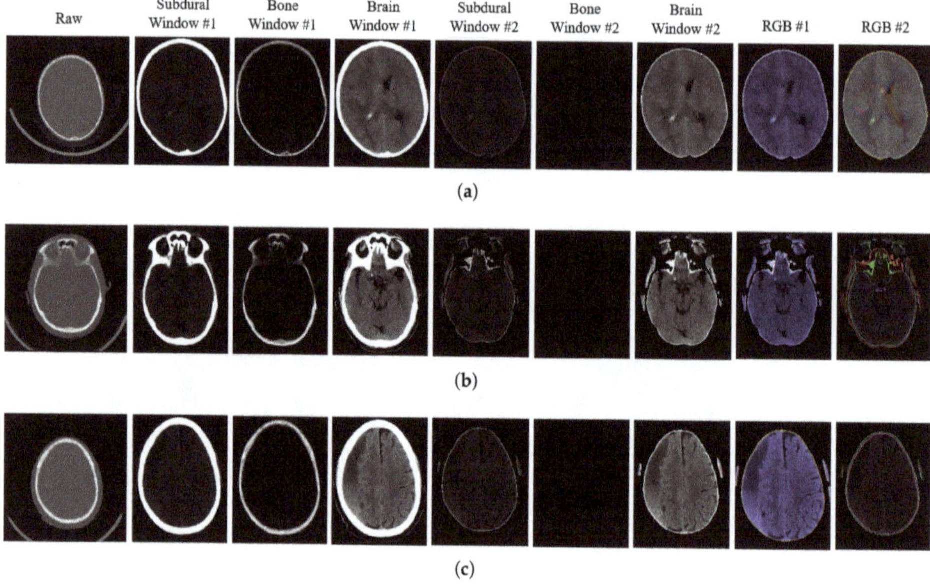

Figure 3. Illustration of the results of subsequent preprocessing stages in three sample CT slices (a–c). Columns (left to right): raw CT slice before region of interest (ROI) extraction, subdural-window image, brain-window image, bone-window image (first three before and next three after skull removal), a stack of three CT-windowed images (Branch #1), a stack of three neighboring slices (Branch #2).

We trained both ResNet networks separately using our training data. As mentioned in Section 1, we tested both ResNets to assess their performance (note Experiment #1 described in Section 3). However, in our model, we have taken the features extracted by either branch after the average pooling layer and concatenated them to set up a joint feature vector for classification.

The training of either network was performed using the adaptive moment estimation optimizer (Adam). We applied image augmentation, including data rotation ($\pm 15°$), scaling (0.8–1.25), shear ($0°$–$45°$ horizontally and vertically), and translation (± 25 pixels horizontally and vertically). Due to the dominance of non-hemorrhage slices within the database, we performed class distribution weighting in each ICH subtype, leading to different numbers of images for classification into categories (this concerned both the balancing of class sizes and the application of weighted cross-entropy loss during training). In each experiment, we randomly partitioned the database into the training, validation, and test subsets with a 80%:10%:10% ratio in a patient-wise manner. The patient-wise assigning of datasets secures the presence of data from each patient in a single subset only. That makes the framework robust against the impact of a subject's slice-to-slice similarity on the classification accuracy. The network parameters, including the mini-batch size (8) and the number of epochs (60), were chosen experimentally. The training was terminated automatically if the validation loss did not decrease for ten epochs.

After the training, features from the last block preceding the ResNet-50's fully connected layer were taken from either branch and concatenated. The joint feature vector containing 4096 elements was subjected to the classification process.

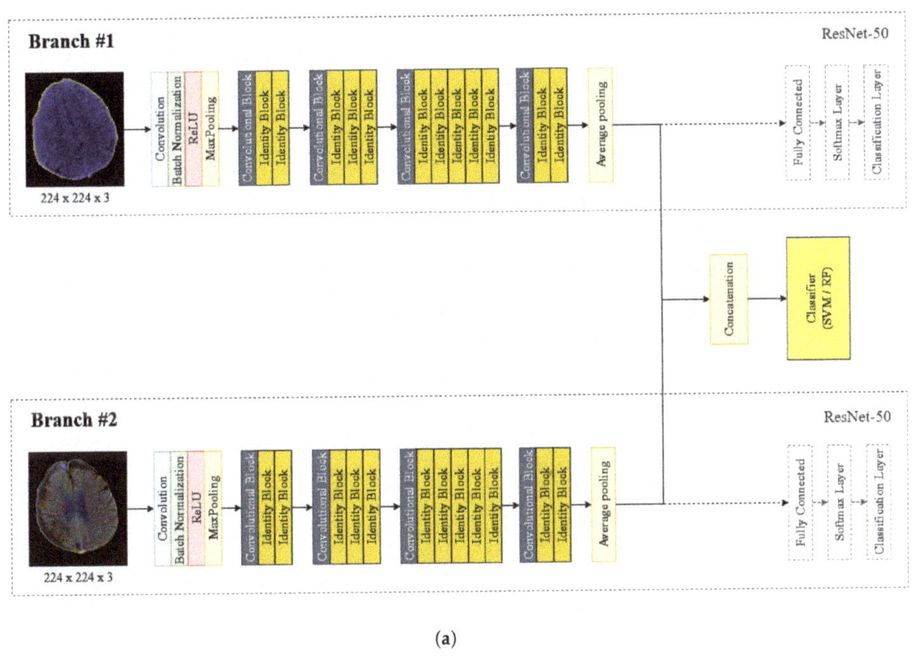

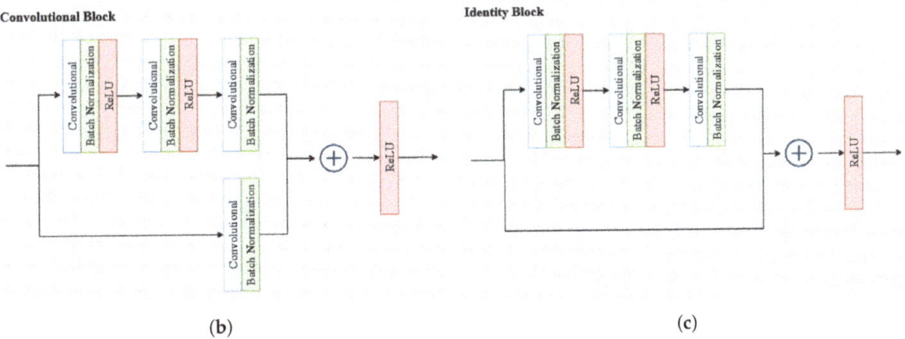

Figure 4. Illustration of the double-branch convolutional neural network (CNN) based on the ResNet-50 architecture (**a**). The feature extraction CNN is covered by a gray background. Dashed frames surround the full ResNet-50 structures. Rightmost blocks refer to the classification stage. Convolutional and identity blocks from ResNet-50 are presented in detail in (**b**,**c**), respectively

2.2.3. Classification

For classification purposes, we involved and compared two classifiers: the support vector machine and random forest.

SVM [28] is a well-recognized tool that is efficient in regression and classification, especially in two-class tasks. It aims to separate samples from different classes in the feature space with a surface that maximizes the margin between them using optimization methods. We used the SVM classifier with a linear kernel [29] due to its superior performance compared to other kernels under consideration, including Gaussian and RBF. As a result, the tool provides probabilities of a given ICH subtype's presence in the image. Hemorrhage is confirmed if the probability exceeds 50%.

RF [30] is an ensemble classifier that parallelly uses multiple models of decision trees (DTs). RF takes the decision of the majority of the trained DTs as its final choice. Single trees are trained in parallel using various subsets of a given dataset (bootstrapping) followed by an aggregation, together known as bagging. The training data are divided into bootstrap sample data and out-of-bag (OOB) data that allow RF for cross-validation. RF is considered more complex in interpretation than DTs, yet more convenient in respect of hyperparameter adjustment. We set the maximum number of splits to $n - 1$, where n is the number of observations in the training sample with the number of trees equal to 200.

3. Results and Discussion

3.1. Evaluation Metrics

We performed several experiments to train and evaluate our models. In each case, the classification performance was assessed using the following metrics:

- accuracy:

$$ACC = \frac{TP + TN}{TP + FP + TN + FN} \cdot 100\%, \tag{1}$$

- sensitivity (recall, true positive rate):

$$TPR = \frac{TP}{TP + FN} \cdot 100\%, \tag{2}$$

- specificity (true negative rate):

$$TNR = \frac{TN}{TN + FP} \cdot 100\%, \tag{3}$$

- F1 score (Dice index):

$$F1 = \frac{2 \cdot TP}{2 \cdot TP + FP + FN} \cdot 100\%, \tag{4}$$

where TP, TN, FP, and FN denote the number of true positive, true negative, false positive, and false negative classifications, respectively.

3.2. Experimental Results

In Experiment #1, we tested both full ResNet-50 architectures (corresponding to two branches of the feature extraction stage) separately. The concatenation of features and SVM or RF classification were omitted. The results of Experiment #1 are presented in Table 2. The best results were achieved for the intraventricular hemorrhage, whereas the worst performance was observed in the case of epidural hemorrhage. Neither branch outperformed the other in all cases.

In Experiment #2, the model described in Section 2.2 was prepared and validated with the concatenation of double-branch features and classification employing both the SVM and RF. Table 3 shows the obtained metrics for each ICH subtype and classifier. Similarly to Experiment #1, the best performance was observed in intraventricular hemorrhage, whereas the poorest outcomes were noted in epidural hemorrhage. RF produced higher quality measures in most categories except EDH. Moreover, in terms of accuracy and F1 score, the RF classifier outperformed any ResNet model from Experiment #1.

Table 2. Summary of evaluation metrics obtained in Experiment #1 for individual ICH subtypes and ResNet models.

ICH Subtype	Feature-Extraction Branch	ACC (%)	TPR (%)	TNR (%)	F1 (%)
Subdural (SDH)	Branch #1	88.7	87.6	89.9	88.9
	Branch #2	86.9	86.3	87.6	87.2
Epidural (EDH)	Branch #1	58.6	56.2	61.0	57.5
	Branch #2	66.4	47.7	85.1	58.6
Intraparenchymal (IPH)	Branch #1	92.4	91.7	93.1	92.2
	Branch #2	92.1	91.7	92.5	92.0
Intraventricular (IVH)	Branch #1	95.5	94.9	96.0	95.1
	Branch #2	95.7	96.0	95.4	95.6
Subarachnoid (SAH)	Branch #1	88.0	87.3	88.8	87.8
	Branch #2	88.2	91.0	85.4	88.7

Table 3. Summary of evaluation metrics obtained in Experiment #2 for individual ICH subtypes and classifiers.

ICH subtype	Classifier	ACC (%)	TPR (%)	TNR (%)	F1 (%)
Subdural (SDH)	SVM	87.1	83.7	90.7	87.0
	RF	89.1	86.7	91.7	89.1
Epidural (EDH)	SVM	76.9	73.4	80.1	75.3
	RF	74.3	68.6	79.9	72.7
Intraparenchymal (IPH)	SVM	91.9	93.1	90.7	91.7
	RF	93.3	93.9	92.6	93.1
Intraventricular (IVH)	SVM	96.0	97.1	94.9	95.9
	RF	96.7	96.7	96.7	96.6
Subarachnoid (SAH)	SVM	86.8	86.1	87.6	87.6
	RF	89.7	90.0	89.4	89.9

3.3. Discussion

The proposed classification tool works efficiently either as individual ResNet-50 networks or a double-branch deep learning architecture. There is a balance in the results produced by standalone single-branch ResNets with mostly slight differences. That suggests that both input data representations contribute almost equally to the proper hemorrhage classification. The IVH is the best-classified subtype featuring accuracy and F1 scores exceeding 95% in both branches. The worst performance can be observed in the EDH (accuracy below 70%, F1 score below 60%).

In a double-branch CNN architecture supported by SVM and RF classifiers, IVH and EDH classification performs most and least efficiently, respectively. The random forest has a slight advantage over the support vector machine except for the EDH. However, keep in mind that the EDH produces classification results significantly lower than the other subtypes. The subtype-classification-assessment ranking reflects the order yielded by individual ResNets. Nonetheless, double-branch architecture improves classification in each subtype. The gains in accuracy and F1 score differ between subtypes—from ca. 1–2% in SDH, IPH, IVH, and SAH to over 10% in EDH, where the room for growth is the largest. The consistency of improvement throughout various ICHs leads us to conclude that the employment of an external machine learning classifier taking advantage of the double-source deep features is profitable in ICH detection.

Improved performance of the proposed double-branch architecture can be associated with the use of multi-source features for classification. The fusion of features extracted by networks addressing specific image data representations provides a broader range of information. A slice-to-slice neighborhood context appears to be particularly relevant. Considering the entire CT study arranged in the correct order, hemorrhage should be observed in subsequent slices. Thus, such spatial information improves the cognitive capabilities of the automated system.

Slice-wise brain hemorrhage detection frameworks generally operate on the entire CT slices or, like our approach, perform some primary ROI extraction to prepare the data for the analysis. The central part of processing is left to the deep learning tool with a limited impact of the operator on the feature extraction. Some targeted localization of selected portions of the image could likely improve the detection and classification scores. Differences between various hemorrhage subtypes are associated with their location and shape. Taking the above into account in more specifically dedicated procedures might benefit classification accuracy improvement. Hemorrhages are frequently caused by mechanical head injuries. Thus, the skull's external area could be examined as well, apart from the deep learning analysis.

The lowest metrics obtained in EDH classification raise a question on the imbalance between classes in a single classification task. The number of epidural cases is an order smaller than the other subtypes. That is probably the case in this particular bleeding, and the class weighing employed during training is not sufficient to resolve this problem. However, there is no such correlation throughout all subtypes. The IVH with the top classification accuracy (correctly detecting 29 out of each 30 cases) is not the best-represented hemorrhage; second-best IPH (14 out of 15 correct) is the second-worst-represented subtype of ICH. Thus, the character of a particular hemorrhage appearance in CT has a crucial influence on the detection capabilities. Similarities between different bleeding symptoms and features are also possible (recall that the dataset of negative samples in each subtype classification also contains images with all other hemorrhages). Some complex cases represent a combination of multiple subtypes simultaneously. Therefore, correct classification is undoubtedly a challenging task.

To put our study in context, we gathered the F1 score results from corresponding state-of-the-art ICH detection methods [12,16,19,22] along with our double-branch RF model (DB-RF) in Table 4. The F1 score is a reliable measure in highly imbalanced datasets since it ignores the true negatives. The selected methods addressed the detection of at least a half of our hemorrhage subtype set (Chang et al. [16] operate on a joint SDH+EDH class and omit the IVH) and were published from 2018 to 2020. Most importantly, similarly to our approach, they detected hemorrhages in CT slices, not in the entire 3D series. Danilov et al. [19] and Burduja et al. [22] used the RSNA competition database. We used provided data and assessment results to determine the F1 score wherever it was not given explicitly.

Table 4. Comparison of classification F1 scores obtained using state-of-the-art methods [12,16,19,22] and our double-branch random forest (DB-RF) model. All values in (%).

Method	Dataset Size (No. of Slices)	SDH	EDH	IPH	IVH	SAH
Chang et al., 2018 [16]	536,266	{86.3}		93.1	—	77.2
Ye et al., 2019 [12]	76,621	84.0	72.0	93.0	87.0	78.0
Danilov et al., 2020 [19]	401 series × 32–64 slices	54.1	62.3	81.3	82.2	61.8
Burduja et al., 2020 [22]	870,301 slices	71.8	48.9	83.8	88.4	73.2
Our approach (DB-RF)	372,556	89.1	72.7	93.1	96.6	89.9

4. Conclusions

In this paper we addressed the detection of various intracranial hemorrhage types by employing double-branch deep feature extraction and machine learning classifiers. The results justify the idea of searching for relevant features in different representations of the data (normalized image in multiple intensity windows and 3D spatial context) and their concatenation for classification using random forest. The system offers the highest detection efficiency in intraventricular and intraparenchymal hemorrhage.

Author Contributions: Conceptualization, A.S. and P.B.; methodology, A.S.; software, A.S.; validation, A.S. and P.B.; formal analysis, P.B.; investigation, A.S.; resources, A.S.; data curation, A.S.; writing—original draft preparation, A.S. and P.B.; writing—review and editing, P.B.; visualization, A.S.; supervision, P.B.; project administration, P.B.; funding acquisition, A.S. All authors have read and agreed to the published version of the manuscript.

Funding: This research was supported by the Silesian University of Technology, Poland, grant number 12/FSW18/0003-03/2019.

Conflicts of Interest: The authors declare no conflicts of interest. The funders had no role in the design of the study; in the collection, analyses, or interpretation of data; in the writing of the manuscript, or in the decision to publish the results.

References

1. Naidech, A.M. Intracranial Hemorrhage. *Am. J. Respir. Crit. Care Med.* **2011**, *184*, 998–1006. [CrossRef] [PubMed]
2. Marcolini, E.; Stretz, C.; DeWitt, K.M. Intracranial Hemorrhage and Intracranial Hypertension. *Emerg. Med. Clin. N. Am.* **2019**, *37*, 529–544. [CrossRef] [PubMed]
3. Tominari, S.; Morita, A.; Ishibashi, T.; Yamazaki, T.; Takao, H.; Murayama, Y.; Sonobe, M.; Yonekura, M.; Saito, N.; Shiokawa, Y.; et al. Prediction model for 3-year rupture risk of unruptured cerebral aneurysms in Japanese patients. *Ann. Neurol.* **2015**, *77*, 1050–1059. [CrossRef] [PubMed]
4. Yasugi, M.; Hossain, B.; Nii, M.; Kobashi, S. Relationship Between Cerebral Aneurysm Development and Cerebral Artery Shape. *J. Adv. Comput. Intell. Intell. Inform.* **2018**, *22*, 249–255. [CrossRef]
5. Freeman, W.D.; Aguilar, M.I.. Intracranial Hemorrhage: Diagnosis and Management. *Neurol. Clin.* **2012**, *30*, 211–240. [CrossRef] [PubMed]
6. Litjens, G.; Kooi, T.; Bejnordi, B.E.; Arindra, A.; Setio, A.; Ciompi, F.; Ghafoorian, M.; van der Laak, J.A.W.M.; van Ginneken, B.; Sánchez, C.I. A survey on deep learning in medical image analysis. *Med. Image Anal.* **2017**, *42*, 60–88. [CrossRef]
7. Hssayeni, M.D.; Croock, M.S.; Salman, A.D.; Al-khafaji, H.F.; Yahya, Z.A.; Ghoraani, B. Intracranial Hemorrhage Segmentation Using Deep Convolutional Model. *Data* **2020**, *5*, 14. [CrossRef]
8. Castro, J.; Chabert, S.; Saavedra, C.; Salas, R. Convolutional neural networks for detection intracranial hemorrhage in CT images. In Proceedings of the 5th Congress on Robotics and Neuroscience, Valparaíso, Chile, 27–29 February 2020; Volume 2564.
9. Arbabshirani, M.; Fornwalt, B.; Mongelluzzo, G.; Suever, J.; Geise, B.; Patel, A.; Moore, G. Advanced machine learning in action: identification of intracranial hemorrhage on computed tomography scans of the head with clinical workflow integration. *npj Dig. Med.* **2018**, *1*, 9. [CrossRef]
10. Li, Y.; Wu, J.; Li, H.; Li, D.; Du, X.; Chen, Z.; Jia, F.; Hu, Q. Automatic detection of the existence of subarachnoid hemorrhage from clinical CT images. *J. Med. Syst.* **2012**, *36*, 1259–1270. [CrossRef]
11. Nguyen, N.; Tran, D.; Nguyen, N.; Nguyen, H. A CNN-LSTM Architecture for Detection of Intracranial Hemorrhage on CT scans. *Med. Imaging Deep Learn.* **2020**. [CrossRef]
12. Ye, H.; Gao, F.; Yin, Y.; Guo, D.; Zhao, P.; Lu, Y.; Wang, X.; Bai, J.; Cao, K.; Song, Q.; et al. Precise diagnosis of intracranial hemorrhage and subtypes using a three-dimensional joint convolutional and recurrent neural network. *Eur. Radiol.* **2019**, *29*, 6191–6201. [CrossRef] [PubMed]
13. Kuo, W.; Häne, C.; Mukherjee, P.; Malik, J.; Yuh, E.L. Expert-level detection of acute intracranial hemorrhage on head computed tomography using deep learning. *Proc. Natl. Acad. Sci. USA* **2019**, *116*, 22737–22745. [CrossRef]
14. Chilamkurthy, S.; Ghosh, R.; Tanamala, S.; Biviji, M.; Campeau, N.; Venugopal, V.; Mahajan, V.; Rao, P.; Warier, P. Deep learning algorithms for detection of critical findings in head CT scans: A retrospective study. *Lancet* **2018**, *392*, 2388–2396. [CrossRef]
15. Anaya, E.; Beckinghausen, M. A Deep Learning Approach to Classifying Intracranial Hemorrhages. Available online: http://cs230.stanford.edu/projects_fall_2019/reports/26260039.pdf (accessed on 13 August 2020).
16. Chang, P.; Kuoy, E.; Grinband, J.; Weinberg, B.; Thompson, M.; Homo, R.; Chen, J.; Abcede, H.; Shafie, M.; Sugrue, L.; et al. Hybrid 3D/2D Convolutional Neural Network for Hemorrhage Evaluation on Head CT. *Am. J. Neuroradiol.* **2018**, *39*, 1609–1616. [CrossRef] [PubMed]

17. Majumdar, A.; Brattain, L.; Telfer, B.; Farris, C.; Scalera, J. Detecting Intracranial Hemorrhage with Deep Learning. In Proceedings of the 2018 40th Annual International Conference of the IEEE Engineering in Medicine and Biology Society (EMBC), Honolulu, HI, USA, 18–21 July 2018; Volume 2018, pp. 583–587. [CrossRef]

18. Toğaçar, M.; Cömert, Z.; Ergen, B.; Budak, Ü. Brain Hemorrhage Detection based on Heat Maps, Autoencoder and CNN Architecture. In Proceedings of the 2019 1st International Informatics and Software Engineering Conference (UBMYK), Ankara, Turkey, 6–7 November 2019; pp. 1–5. [CrossRef]

19. Danilov, G.; Kotik, K.; Negreeva, A.; Tsukanova, T.; Shifrin, M.; Zakharova, N.; Batalov, A.; Pronin, I.; Potapov, A. Classification of Intracranial Hemorrhage Subtypes Using Deep Learning on CT Scans. *Stud. Health Technol. Inf.* **2020**, *272*, 370–373. [CrossRef]

20. Ker, J.; Singh, S.; Bai, Y.; Rao, J.; Lim, T.; Wang, L. Image Thresholding Improves 3-Dimensional Convolutional Neural Network Diagnosis of Different Acute Brain Hemorrhages on Computed Tomography Scans. *Sensors* **2019**, *19*, 2167. [CrossRef] [PubMed]

21. Dawud, A.M.; Yurtkan, K.; Oztoprak, H. Application of Deep Learning in Neuroradiology: Brain Haemorrhage Classification Using Transfer Learning. *Comput. Intell. Neurosci.* **2019**, *2019*, 4629859. [CrossRef] [PubMed]

22. Burduja, M.; Ionescu, R.T.; Verga, N. Accurate and Efficient Intracranial Hemorrhage Detection and Subtype Classification in 3D CT Scans with Convolutional and Long Short-Term Memory Neural Networks. *Sensors* **2020**, *20*, 5611. [CrossRef]

23. Cho, J.; Park, K.S.; Karki, M.; Lee, E.; Ko, S.; Kim, J.K.; Lee, D.; Choe, J.; Son, J.; Kim, M.; et al. Improving Sensitivity on Identification and Delineation of Intracranial Hemorrhage Lesion Using Cascaded Deep Learning Models. *J. Digit. Imaging* **2019**, *32*, 450–461. [CrossRef]

24. RSNA Intracranial Hemorrhage Detection. Available online: https://www.kaggle.com/c/rsna-intracranial-hemorrhage-detection/data (accessed on 5 August 2020).

25. Flanders, A.; Prevedello, L.; Shih, G.; Halabi, S.; Kalpathy-Cramer, J.; Ball, R.; Mongan, J.; Stein, A.; Kitamura, F.; Lungren, M.; et al. Construction of a Machine Learning Dataset through Collaboration: The RSNA 2019 Brain CT Hemorrhage Challenge. *Radiol. Artif. Intell.* **2020**, *2*, e190211. [CrossRef]

26. Otsu, N. A Threshold Selection Method from Gray-Level Histograms. *IEEE Trans. Syst. Man Cybern.* **1979**, *9*, 62–66. [CrossRef]

27. He, K.; Zhang, X.; Ren, S.; Sun, J. Deep Residual Learning for Image Recognition. In Proceedings of the 2016 IEEE Conference on Computer Vision and Pattern Recognition (CVPR), Las Vegas, NV, USA, 27–30 June 2016; pp. 770–778. [CrossRef]

28. Cortes, C.; Vapnik, V. Support Vector Networks. *Mach. Learn.* **1995**, *20*, 273–297. [CrossRef]

29. Cervantes, J.; García-Lamont, F.; Rodríguez, L.; Lopez-Chau, A. A comprehensive survey on support vector machine classification: Applications, challenges and trends. *Neurocomputing* **2020**. [CrossRef]

30. Breiman, L. Random Forests. *Mach. Learn.* **2001**, *45*, 5–32. [CrossRef]

Publisher's Note: MDPI stays neutral with regard to jurisdictional claims in published maps and institutional affiliations.

Article

Comorbidity Pattern Analysis for Predicting Amyotrophic Lateral Sclerosis

Chia-Hui Huang [1], Bak-Sau Yip [2], David Taniar [3], Chi-Shin Hwang [4] and Tun-Wen Pai [5,*

[1] Department of Computer Science and Engineering, National Taiwan Ocean University, Keelung 20224, Taiwan; 10757030@email.ntou.edu.tw

[2] Department of Neurology, National Taiwan University Hospital Hsin Chu Branch, Hsin Chu 30059, Taiwan; neuro@hch.gov.tw

[3] Department of Software Systems and Cybersecurity, Monash University, Victoria 3800, Australia; David.Taniar@monash.edu

[4] Department of Neurology, ShuTien Hospital, Taipei 10662, Taiwan; DAH88@tpech.gov.tw

[5] Department of Computer Science and Information Engineering, National Taipei University of Technology, Taipei 10608, Taiwan

* Correspondence: twp@csie.ntut.edu.tw; Tel.: +886-2-27712171 (ext. 4222)

Abstract: Electronic Medical Records (EMRs) can be used to create alerts for clinicians to identify patients at risk and to provide useful information for clinical decision-making support. In this study, we proposed a novel approach for predicting Amyotrophic Lateral Sclerosis (ALS) based on comorbidities and associated indicators using EMRs. The medical histories of ALS patients were analyzed and compared with those of subjects without ALS, and the associated comorbidities were selected as features for constructing the machine learning and prediction model. We proposed a novel weighted Jaccard index (WJI) that incorporates four different machine learning techniques to construct prediction systems. Alternative prediction models were constructed based on two different levels of comorbidity: single disease codes and clustered disease codes. With an accuracy of 83.7%, sensitivity of 78.8%, specificity of 85.7%, and area under the receiver operating characteristic curve (AUC) value of 0.907 for the single disease code level, the proposed WJI outperformed the traditional Jaccard index (JI) and scoring methods. Incorporating the proposed WJI into EMRs enabled the construction of a prediction system for analyzing the risk of suffering a specific disease based on comorbidity combinatorial patterns, which could provide a fast, low-cost, and noninvasive evaluation approach for early diagnosis of a specific disease.

Keywords: Electronic Medical Record (EMR); disease prediction; Amyotrophic Lateral Sclerosis (ALS); weighted Jaccard index (WJI)

Citation: Huang, C.-H.; Yip, B.-S.; Taniar, D.; Hwang, C.-S.; Pai, T.-W. Comorbidity Pattern Analysis for Predicting Amyotrophic Lateral Sclerosis. *Appl. Sci.* **2021**, *11*, 1289. https://doi.org/10.3390/app11031289

Academic Editors: Michał Strzelecki and Pawel Badura

Received: 9 October 2020

Accepted: 28 January 2021

Published: 31 January 2021

1. Introduction

Amyotrophic Lateral Sclerosis (ALS) is a multi-syndrome and fatal neurodegenerative disorder that results in progressive loss of bulbar and limb function [1]. The prevalence of ALS is uniformly distributed across most countries [2]. According to the ALS Association, the incidence rate of ALS is approximately between 1.8 and 2 per 100,000 person-years. Death from ALS occurs from respiratory failure, generally within 2–3 years from the onset of the bulbar symptom and 3–5 years after the limb onset [3]. For the majority of patients with ALS, it usually takes 9 to 15 months from symptom onset to definitive diagnosis by two or three specialists [1]. The diagnostic processes for ALS are very complicated compared to other diseases. Several steps are required to ensure that all medical evaluation items are completely performed. These processes include neurological examination and a series of diagnostic tests such as electromyography (EMG), magnetic resonance imaging (MRI), blood and urine tests, etc. However, there is yet no definitive diagnostic testing standard for ALS. Different medical evaluations are always performed after disease progression in different combinations of suggestive clinical signs of pathologies. Currently, the etiology

147

of sporadic ALS is unclear, and there is no cure for ALS. Treatment is mainly aimed at delaying the progression of the disease and not at eliminating the symptoms [4,5]. Hence, if at-risk subjects of ALS can be identified during the early stages, they can be advised to focus on strengthening their immune system and on improving the environmental conditions in which they live. Early prognosis may slow the progressive deterioration of ALS symptoms. Since there is a near two-year lead-time interval prior to definitive diagnosis of ALS, the symptoms and comorbidities could provide important clues for ALS prediction. Hence, we consider Electronic Medical Records (EMRs) as useful for discriminating ALS and non-ALS patients in the early stages, and the developed classifier could serve as an alternative diagnostic approach or an early warning signal for doctors and ALS patients.

This study was aimed at identifying high-risk Amyotrophic Lateral Sclerosis (ALS) subjects in the early stage using a novel detection mechanism solely based on EMRs. The novel prediction index used in this study is a modified version of the Jaccard index (JI). The traditional JI is a statistical value for comparing the similarity and diversity between two different sample sets. It measures the similarity between two limited sample groups. The value is directly proportional to the similarity between the two groups [6]. Studies have demonstrated that it is more effective to consider the weighted coefficients on similarity analysis for specific problems [7].

To enhance the traditional JI indicator and the evidence of different populations with an identical disease possessing high comorbidity similarities, we proposed a novel weighted JI (WJI) that can effectively reflect the comorbidity distribution of EMRs, instead of using the binary status of comorbidity; further, compared with traditional approaches, we formulated indices to provide accurate prediction results. The details and experimental results are described in the following sections.

Electronic Medical Records

EMRs are medical history of patients systematically collected by hospitals and/or insurance institutions; they contain general information on clinical practices, such as medical diagnoses and associated treatments. Wide usage of EMRs can reduce healthcare expenses and medical errors and can improve patients' health [8]. An EMR system has the potential to improve healthcare delivery and to reduce medical costs by enhancing data management capabilities and by mining valuable information from the comprehensive clinical practice database [9]. According to research reports, EMRs play an important role in medical services, such as clinical decision-making support, medical quality monitoring, disease prediction model construction, clinical trial analysis, and treatment personalization.

There are many reports on the different applications of EMRs. Several studies have focused on expanding the clinical contexts of genomic diversities based on EMR-linked genetic data. Denny et al. revealed the associations among rare diseases and genetic effects in relation to prognosis, treatments, drug responses, and comorbidity risk [10]. Based on EMRs, other studies have identified patterns among multiple comorbidities and defined non-random associations between diseases; using different methodologies for feature analysis and similarity detection among these patterns, they found various comorbidity modes [11]. Freund et al. proposed predictive models based on historical insurance claim data to identify and explore patterns of multimorbidity in new customers to detect the high risk of future hospitalizations for constructing intervention management for primary care [12]. Kirchberger et al. identified patterns of comorbidity and multimorbidity by analyzing prevalence figures, logistic regression models, and exploratory factor analysis [13]. The objective of this study was to explore patterns of comorbidity and multimorbidity in the elderly population using different analytical approaches. The report confirmed the existence of certain co-occurring diseases in elderly persons that were not caused by chance.

In some studies, a prediction model featuring patient similarities based on a large amount of medical data was constructed. Different approaches are used for evaluating patient similarities and clustering patient groups, such as analyzing the similarities or distinguishing characteristics among a variety of feature components from patient data. The commonly used Euclidean vector adopts the cosine angle, also known as the cosine similarity measure, between two patients' feature vectors to define the associated patient similarity metric [14]. For example, Tashkandi et al. and Lee et al. evaluated the patient similarities in the ICU dataset (MIMIC-III) using the cosine-similarity-based patient similarity measure (PSM) [15,16].

Taiwan has an internationally well-known National Health Insurance Research Database (NHIRD) [17] that maintains general information of clinical practices in all hospital clinics and integrates the medical records of all Taiwanese citizens. Using this database, researchers can integrate, extract, and convert long-term historical longitudinal medical records from multi-aspect and multi-function databases into a required specification format according to a patient's diagnosis records, medication, and hospitalization information for specific medical applications. In this study, the similarities between the disease records of the control and experimental groups were calculated to identify those with high-risk factors through EMRs. In the following sections, we describe the construction of the prediction model using the proposed WJI and verify the effectiveness of the novel proposed indicator.

2. Materials and Methods

2.1. Source of Materials

The data used in this study were anonymous medical data authorized by the NHIRD (Taiwan), consisting of one million insured people collected between 1996 and 2013 (IRB: 104-5430B). The disease classification code followed the international disease classification standard ICD-9-CM, which contains 17 chapters and 2 supplementary categories. The 17 major chapters can be further labeled and classified into 143 mid-level classification and 999 individual disease classes. The information used in this study was mainly based on an analysis of the disease codes at the individual- and the mid-levels. According to the definition of the disease group, the individual-level classification was represented by a three-digit code for a single disease, and each mid-level classification represented a disease group. Hence, 999 disease categories were defined based on individual-level classification; these individual disease categories were grouped into 143 disease groups, which were defined as mid-level classification groups.

2.2. Medical Histories and Feature Extraction Analysis

Firstly, positive and negative data groups were defined. The positive and negative data groups were composed of subjects with and without ALS, respectively. The positive subjects were obtained directly from the one million NHIRD medical database, and the relatively larger number of negative subjects were retrieved from the same database based on matching gender and age attributes according to the positive subjects.

The experimental data obtained from the NHIRD database underwent data cleaning and data integration to yield two complete sets of medical diagnosis records for both groups. Furthermore, the subjects for the different disease codes for each group were counted. Statistical analysis was performed on the different disease codes to select associated disease comorbidities for both the experimental and control groups for constructing the prediction systems.

Figure 1a,b are simple schematic diagrams of the disease set corresponding to the experimental and control groups. These two diagrams were designed for illustration; they do not represent actual medical data contents. The data represent the comprehensive disease records of the positive subjects (diagnosed with the ALS disease) within two years before diagnosis. The total number of subjects suffering from a specific disease in the experimental or control group was noted, and the comorbidity codes were preliminarily screened by evaluating a minimum support threshold setting. A minimum support threshold value represents the minimum number of subjects or the minimum percentage of total subjects within the experimental group. In other words, comorbidities which were present in less than a certain percentage of subjects within the experimental group were discarded from the feature set in this study.

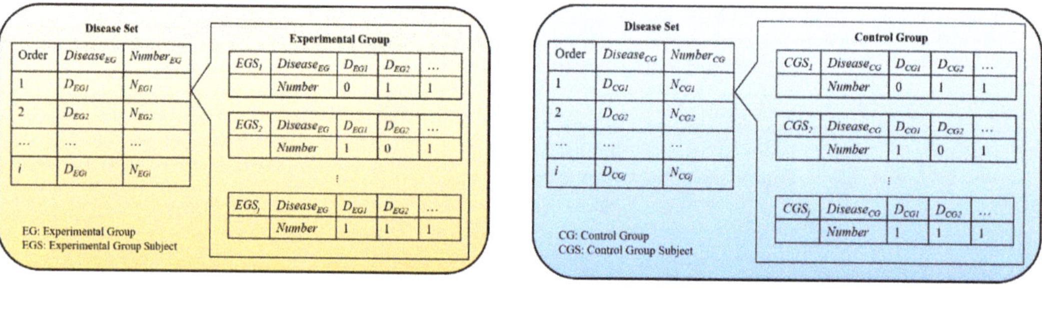

(**a**)   (**b**)

Figure 1. Illustration of the disease set from (**a**) the experimental group and (**b**) the control group: on the right part of the two tables are disease records of each subject, and each disease is displayed within the defined interval binary representation of 0 and 1. The left side is the disease set of the experimental/control group and the total number of subjects suffering various diseases. $Disease_{EG}$ and $Disease_{CG}$ refer to disease classification codes in the experimental and control groups, respectively. $Number_{EG}$ and $Number_{CG}$ refer to the total number of subjects in the experimental and control groups suffering from a certain disease. D_{EGi} and D_{CGi} refer to the ith disease classification code in the experimental and control groups, respectively. N_{EGi} and N_{CGi} refer to the total number of subjects in the experimental/control group suffering from the ith disease under a classification code. EGS_j and CGS_j refer to the jth subject in the experimental and control groups, respectively.

Figure 2 is a schematic diagram of the comorbidity feature set. For the disease code sets from both the experimental and control groups, the same lead-time interval and an identical minimum support threshold setting were applied to define the associated disease records for comorbidity analysis. To identify highly associated comorbidities with ALS, the classification codes of all the comorbidities were further analyzed by evaluating the corresponding odds ratios (ORs) for ALS. In other words, the OR of each comorbidity was calculated to evaluate the chances of a comorbidity associated with ALS. A set of comorbidity codes with ORs > 2 was constructed and defined as an associated comorbidity feature set for ALS. Since the associated comorbidity set of ALS was retrieved from the disease set of the original experimental group, after the OR analysis, the selected comorbidities must be a subset from the disease set of the experimental group. Therefore, the number of subjects for the filtered associated diseases in the associated comorbidity feature set should be the same as that of the subjects within the original experimental group. This process may remove background noise (unassociated comorbidities) from the constructed prediction model. For example, the disease code sets of both the experimental and control groups may simultaneously contain a seasonal cold; hence, a small OR could be obtained and no specific correlation was detected between seasonal cold and ALS. Thus, following the statistical significance analysis of all comorbidities, the system defined a comorbidity feature set with strong association with ALS, which can explain the strong relationship between the comorbidity codes and ALS.

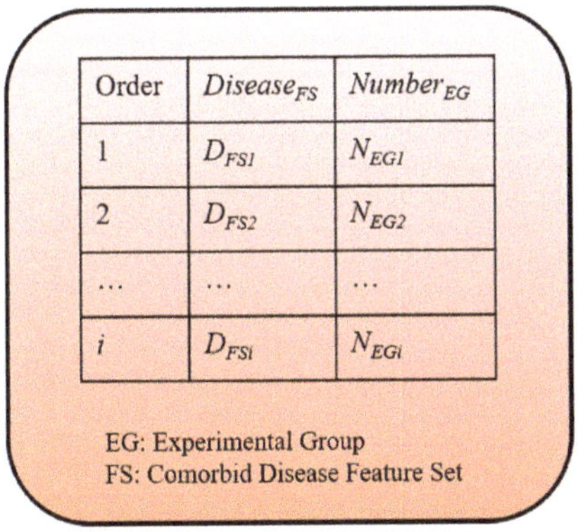

Order	$Disease_{FS}$	$Number_{EG}$
1	D_{FS1}	N_{EG1}
2	D_{FS2}	N_{EG2}
...	...	...
i	D_{FSi}	N_{EGi}

EG: Experimental Group
FS: Comorbid Disease Feature Set

Figure 2. Illustration of the comorbidity feature set: $Disease_{FS}$ refers to comorbidity classification codes with odds ratios > 2. $Number_{EG}$ refers to the total number of subjects in the experimental group suffering from a specific disease. The number of subjects suffering from a specific disease in the comorbidity feature set is the same as the number of subjects in the disease set in the experimental group ($Number_{EG}$).

2.3. Proposed Patient Prediction Module

The proposed diagnostic model is mainly based on the similarity measurement of comorbidity patterns between testing subjects and known ALS patients. From the experimental and control data groups, we initially calculated the odds ratios for each comorbidity (within a two-year interval) for measuring the association between comorbidities and ALS disease. Based on odds ratio analysis, comorbidities with strong associations were selected and constructed as an important comorbidity feature set, and the proposed WJI similarity measurements were further calculated for the experimental group vs. the comorbidity feature set and for the control group vs. the comorbidity feature set. A binary outcome prediction model for supporting clinical decision making was trained on the calculated WJI indicators of the experimental (ALS) and control (non-ALS) groups versus the ALS-associated comorbidity feature set. Once the WJI indicators of the experimental and control data groups were trained and constructed as a prediction model, only the selected associated comorbidities of testing subjects were considered and evaluated according to the WJI similarity measurement and compared to the previously trained thresholding value. The ALS prediction module was constructed and described as the following. Obtaining the comorbidities for each subject in the experimental or control groups, and its corresponding WJI similarity was calculated with respect to the defined associated comorbidity feature set. Subsequently, a Z-score standardization procedure was performed for similarity measurement and for training and construction of the probability prediction model. Excluding the subjects in the experimental and control data groups, for which the data were utilized as the training dataset (80% of samples), the data of the remaining subjects were applied as the testing data (20% of samples).

In this study, four different machine learning models were utilized for data training: Logistic Regression (LR) [18], Support Vector Classifier (SVC) [19,20], random forest [21], and eXtreme Gradient Boosting (XGBoost) [22]; five-fold cross validation was adopted to verify the accuracy and prediction stability of the trained models. All four machine learning techniques are discrimination supervised learning models that learn from the training dataset and build a model to make predictions for unseen data in different classification applications. Both the LR and SVC are prediction models commonly used in traditional medical applications. The random forest and XGBoost, which are the most popular machine learning models in recent years, are both ensemble learning approaches designed by integrating a variety of learning algorithms to achieve better prediction capabilities.

2.4. Jaccard Index

The traditional Jaccard index (JI) defines the similarity between two different sample sets, A and B. The index is shown in Equation (1):

$$\text{Jaccard}(A,\ B) = \frac{|A \cap B|}{|A \cup B|} = \frac{|A \cap B|}{|A| + |B| - |A \cap B|} \tag{1}$$

where $|A \cap B|$ represents the number of overlapped items and $|A \cup B|$ represents the number of union items. Regardless of the occurrence frequency of each item (such as disease classification code), "1" and "0" denote existing and nonexisting conditions, respectively. This similarity indicator confers equal importance of co-occurring items.

However, to predict a target disease through the comorbidity analysis, the co-occurrence frequencies of different comorbidities should possess different weights for disease evaluation. For example, pregnant women with high blood pressure-related diseases would have a relatively higher probability of premature delivery than those who suffer from skin diseases. However, the traditional JI can only mark different comorbidity codes as "1" or "0," ignoring the incidence of certain important comorbidities.

To strengthen the occurrence frequency of various comorbidities from the EMRs of known patients, we proposed an improved WJI to replace the traditional JI for better evaluation of the similarities between two comorbidity sets. This novel index utilizes the proportional incident rate for corresponding weight calculation and enhances the accuracy of the similarity measurement between positive and negative training datasets. The WJI-related terminologies are defined in Table 1; several examples are illustrated in the supplementary document I.

Table 1. Number of patients and corresponding weights for $AG*$ and $BG*$.

	$Disease_A$	d_{A1}	d_{A2}	$\cdots$	d_{Ai-1}	d_{Ai}
$AG*$	$Number_A$	N_{A1}	N_{A2}	$\cdots$	N_{Ai-1}	N_{Ai}
	$Weight_A$	$\frac{N_{A1}}{\sum N_{Ai}}$	$\frac{N_{A2}}{\sum N_{Ai}}$	$\cdots$	$\frac{N_{Ai-1}}{\sum N_{Ai-1}}$	$\frac{N_{Ai}}{\sum N_{Ai}}$
	$Disease_B$	d_{B1}	d_{B2}	$\cdots$	d_{Bj-1}	d_{Bj}
$BG*$	$Number_B$	N_{B1}	N_{B2}	$\cdots$	N_{Bj-1}	N_{Bj}
	$Weight_B$	$\frac{N_{B1}}{\sum N_{Bj}}$	$\frac{N_{B2}}{\sum N_{Bj}}$	$\cdots$	$\frac{N_{Bj-1}}{\sum N_{Bj-1}}$	$\frac{N_{Bj}}{\sum N_{Bj}}$

AG: set of comorbidities for patient group A within the defined interval before the patient is diagnosed with specific target disease. $AG*$: top i frequently co-occurring diseases content set of AG. $Disease_A$: comorbidities in patient group A. d_{Ai}: i^{th} comorbidity in group A. $Number_A$: number of patients with a specific comorbidity in group A. N_{Ai}: number of patients with the ith comorbidity in group A. $Weight_A$: corresponding weights of specific comorbidities in group A. $\frac{N_{Ai}}{\sum NA}$: the corresponding weight of the ith specific comorbidity in group A. BG: the comorbidity records for group B within the same interval as AG. $BG*$: top j frequently co-occurring diseases content set of BG. $Disease_B$: comorbidities in patient group B. d_{Bj}: j^{th} comorbidity in group B. $Number_B$: number of patients with a specific comorbidity in group B. N_{Bj}: number of patients with the jth comorbidity in group B. $Weight_B$: corresponding weights of specific comorbidities in group B. $\frac{N_{Bj}}{\sum NB}$: corresponding weight of the jth specific comorbidity in group B.

There are two different types of comorbidities, and different methods are used to calculate the corresponding WJIs. For the first type, the comorbidity classification codes appeared exclusively in either set $AG*$ or $BG*$. When a comorbidity occurred only in $AG*$, the corresponding weighting coefficient for the specific disease code was obtained by dividing the number of patients with such a comorbidity within $AG*$ by the total number of patients with each type of comorbidity in $AG*$. Similarly, the weighting coefficient algorithm can be applied to any single comorbidity occurring in $BG*$ exclusively.

In contrast, the second type of comorbidity code occurred within $AG*$ and $BG*$ simultaneously. In this case, the corresponding weights of the co-occurring comorbidity codes were calculated by taking proportional measurements in each patient group as the first type and then by taking an average from both $AG*$ or $BG*$ as the final corresponding weights for the common comorbidities. The formula is denoted as Equation (2):

$$Weight = \begin{cases} \frac{N_{Ai}}{\sum N_{Ai}} \ or \ \frac{N_{Bi}}{\sum N_{Bi}} \ , if \ d_{Ai} \neq d_{Bj} \\ \frac{\frac{N_{Ai}}{\sum N_{Ai}} + \frac{N_{Bj}}{\sum N_{Bj}}}{2} \ , if \ d_{Ai} = d_{Bj} \end{cases} \tag{2}$$

The WJI was then obtained by summing up all the weighting coefficients corresponding to $AG*$ and $BG*$ of the commonly occurring comorbidity codes and then by dividing by the sum of all the weights of the disease classification codes within $AG*$ and $BG*$. A few examples are given in the Supplementary Materials.

2.5. Scoring Methods

To evaluate the proposed WJI similarity-based ALS prediction system, two different approaches including the traditional JI similarity (without weighting concern) and the scoring mechanism (score) were constructed for comparison. The scoring method is similar to the symptom indexing chart used in the past for the initial diagnosis of a specific disease. Different approaches were used for estimating the risk of suffering a target disease. A common approach was to design a score sheet for evaluating disease symptoms. Another common approach was to administer questionnaires. For example, the International Prostate Symptom Score, the Edinburgh Postnatal Depression Scale, and kidney disease include scores sheet to evaluate associated symptoms [23–25]

In the following experimental analysis, the scoring index was defined by taking the comorbidity classification codes as the symptom feature set; subsequently, a related weighted proportional score was given according to the historical number of patients. Finally, the weighted scores of the comorbidity records of the test subject within a specific duration were summed up, and the score for the test subject was obtained. The score acts as a preliminary prediction of the possibility of the subject suffering the disease in the near future. The simple illustration in Table 2 shows the conception of the intuitive scoring mechanism defined in this study.

Table 2. Simple illustration of scoring mechanism.

$Disease_{FS}$	$Number_{FS}$	Weight	Patient Score
d_{A1}	4	$\frac{4}{20} = 0.20$	V
d_{A2}	6	$\frac{6}{20} = 0.30$	
d_{A3}	10	$\frac{10}{20} = 0.50$	V
Total	20	1	$0.20 + 0.50 = 0.70$

The left side is the disease classification codes and the total number of subjects in the comorbidity feature set; the related weighted proportional score was given according to the historical number of patients. The right side is the final score of the test subject, which is the sum of the weighted scores of comorbidity records of a subject within a specific period. $Disease_{FS}$ is the disease classification codes in the comorbidity feature set. $Number_{FS}$ is the total number of subjects in the comorbidity feature set suffering from a certain disease. *Weight* is the corresponding weights of specific comorbidities. *Patient Score* is the final score of a test subject.

2.6. Amyotrophic Lateral Sclerosis (ALS)

In this study, a total of 162 ALS patients were identified, and their corresponding EMRs were retrieved from the selected governmental medical database composed of one million data (IRB: 104-5430B). To confirm that the initial subjects were indeed ALS patients (excluding suspected cases), we only considered patients who were hospitalized due to the disease. Hence, only 71 subjects were selected for this study (male–female ratio = 45:26).

Among them, 41 were hospitalized directly upon diagnosis; the remaining 30 subjects were hospitalized after some time following the diagnosis. The average period of hospitalization for ALS following the first diagnosis was 2 years.

The 71 confirmed subjects were considered the experimental group, and their historical disease records within 2 years of diagnosis were defined as *ALS_EG*. The total number of subjects suffering from each disease was recorded as the disease set of the experimental group, and the comorbidity classification codes were preliminarily screened and defined as *ALS_EG_FEA*. To further analyze and compare the different parameter settings, the comorbidity classification code of *ALS_EG_FEA* that was found among more than 10% of the total number of patients (71) was extracted and defined as $ALS\_EG\_FEA_{0.1}$.

The subjects of the control group were retrieved from the database based on the age and ratio of each gender. A total number of 399 subjects without the ALS disease were selected as the control group (male–female ratio = 249:150). According to the governmental database in Taiwan, the average onset age of the retrieved ALS subjects is 51 years old. Therefore, the historical medical records between 49 and 51 years old were retrieved for each control group subject and defined as *ALS_CG*. All the various disease records for the control group were enumerated and counted, and the corresponding comorbidity codes were preliminarily screened and defined as *ALS_CG_FEA*. To analyze and compare the different parameter settings, a comorbidity code occurring among more than 10% of the 399 subjects was additionally extracted and defined as $ALS\_CG\_FEA_{0.1}$.

Finally, we calculated the odds for each comorbidity code within both *ALS_EG_FEA* and *ALS_CG_FEA* and divided them to obtain an OR for explaining the strength of specific comorbidities associated with ALS. In other words, the OR identifying a specific comorbidity was calculated. When OR > 2, the associated disease code was selected and defined as an important feature for subsequent identification; the comorbidity set collected was defined as *ALS_SELECT_FEA*. These associated comorbidities were defined as the feature set of the ALS comorbidities that were used to explain the important correlation between certain comorbidities and ALS diseases. Further, to select a feature disease set with stronger representation, a screening condition based on a proportion greater than 10% of the total subjects was established and defined as $ALS\_EG\_FEA_{0.1}$ and $ALS\_CG\_FEA_{0.1}$. Similar procedures for calculating the ORs for each disease code from $ALS\_EG\_FEA_{0.1}$ and for selecting comorbidities with ORs > 2 were performed for subsequent identification, and the comorbidity set was defined as $ALS\_SELECT\_FEA_{0.1}$.

Because there were two different comorbidity feature sets, it was necessary to construct two different ALS prediction models for at-risk patient recognition. The first prediction system focused on each subject within *ALS_EG* and *ALS_CG* and adopted the WJI similarity analysis, with respect to *ALS_SELECT_FEA*. The similarity between each subject within the comorbidity feature set was calculated and normalized, following which a probability prediction model was constructed. Subsequently, the model was used to predict the probability of a subject suffering from ALS. Finally, five-fold cross validation was adopted to verify the accuracy and prediction stability of the trained model. In contrast, the second prediction system used a disease occurrence rate >10%, that is, $ALS\_SELECT\_FEA_{0.1}$, which was obtained using $ALS\_EG\_FEA_{0.1}$ and $ALS\_CG\_FEA_{0.1}$ for significance analysis. The final prediction results were obtained by analyzing each subject of *ALS_EG* and *ALS_CG* using $ALS\_SELECT\_FEA_{0.1}$.

2.7. Model Evaluation Index

The model score, precision, sensitivity, specificity, accuracy, F1 score, and area under the receiver operating characteristic curve (AUC) were utilized as the evaluation indicators for comparing different predictive models [26]. A true positive (TP) is the number of subjects in the positive class correctly predicted as belonging to the positive class; a true negative (TN) is the number of subjects in the negative class correctly predicted as belonging to the negative class; a false positive (FP) is the number of subjects in the negative class incorrectly predicted as belonging to the positive class; and a false negative (FN) is the number of subjects in the positive class incorrectly predicted as belonging to the negative class.

Precision is defined as TP/(TP+FP).

Sensitivity, also referred to as the recall rate, is defined as TP/(TP+FN).

Specificity is defined as TN/(FP+TN);

Accuracy was the accuracy rate of prediction based on the testing data; it is defined as (TP+TN)/(TP+FP+TN+FN); the model score is the accuracy rate of prediction obtained from the training data itself.

F1 score refers to the harmonic average of the precision and the recall rate.

AUC refers to the area under the receiver operating characteristic curve, which represents a statistical value of the predictive ability of the classifier; the larger its value, the better the performance of the prediction model.

3. Results and Discussion

We analyzed comorbidities of ALS disease with a strong association through odds ratio analysis. The selected comorbidities were categorized into the individual level and mid-level. Both selected comorbidities with strong associations in different levels are shown in the Supplementary Materials. We also compared pair similarities among the disease set of the experimental group vs. control group and thee comorbidity feature set vs. control group. The value of JI similarity between the disease set of the experimental group and control group is always larger than the JI similarity between the comorbidity feature set and control group. It is due to the comorbidity feature set removing a lot of unassociated diseases by statistical analytics. Hence, several common diseases such as the seasonal flu and common cold were removed and the similarity indicator decreased. Similarly, the similarity measurement for the WJI indicator possesses the same decreasing trend. From Table 3, it can be observed that the differentiation value of JI indicator (0.22) is less than the WJI indicator (0.43). It is mainly due to each disease code in the feature set weighted in the WJI indicator according to the disease population, and the difference was enlarged due to unevenly distributed subjects. Therefore, the WJI indicator provided a better differentiating factor than the JI indicator in general. It is an important property to provide a better prediction performance for discriminating ALS patients and health subjects. The similarities in the individual disease and mid-level clustered disease code levels among *ALS_EG_FEA*, *ALS_CG_FEA*, and *ALS_SELECT_FEA* are shown in Tables 3 and 4. For different disease code levels (either individual or mid-level codes), the differentiation values also preserved an identical trend between the WJI and JI indicators.

Table 3. Jaccard index and weighted Jaccard index similarities pairwise *ALS_EG_FEA*, *ALS_CG_FEA*, and *ALS_SELECT_FEA* (individual-level classification).

Individual-Level Classification	Jaccard Index	Weighted Jaccard Index
	Disease Set from Control Group [2]	Disease Set from Control Group [2]
Disease Set from Experimental Group [1]	0.55	0.87
Comorbidity Feature Set [3]	0.33	0.44
Differentiation	0.22	0.43

[1] Disease set from the experimental group: *ALS_EG_FEA*; [2] Disease set from the control group: *ALS_CG_FEA*; [3] Comorbidity feature set: *ALS_SELECT_FEA*.

Table 4. Jaccard index and weighted Jaccard index similarities pairwise *ALS_EG_FEA*, *ALS_CG_FEA*, and *ALS_SELECT_FEA* (mid-level classification).

Mid-Level Classification	Jaccard Index	Weighted Jaccard Index
	Disease Set from Control Group [2]	Disease Set from Control Group [2]
Disease Set from Experimental Group [1]	0.75	0.98
Comorbidity Feature Set [3]	0.43	0.53
Differentiation	0.32	0.45

[1] Disease set from experimental group: *ALS_EG_FEA*; [2] Disease set from control group: *ALS_CG_FEA*; [3] Comorbidity feature set: *ALS_SELECT_FEA*.

Tables 5 and 6 show the pair similarities of the individual disease code level and middle-clustered disease code level among constrained disease feature groups ($ALS\_EG\_FEA_{0.1}$ vs. $ALS\_CG\_FEA_{0.1}$, and $ALS\_SELECT\_FEA_{0.1}$ vs. $ALS\_CG\_FEA_{0.1}$). From Tables 3–6, the differences in the WJI between the disease set of the experimental and control groups are higher than those of the JI. Again, the main reason for this is that the WJI involves weight coefficient calculations and possesses more detailed associations than the JI. When the number of subjects suffering from an identical disease between the two sets was large, the relative weight coefficient for the specific comorbidity and the degree of influence became associated. However, when the number of subjects suffering from diseases under various classification codes was large but the number of subjects was small, the relative weight coefficient for the specific comorbidity was reduced and the corresponding impact decreased. The traditional JI similarity is based solely on the presence or absence of a disease classification code; it does not consider the difference in the prevalence of the associated comorbidity codes. The binary counting of 0 and 1 was used to analyze the comorbidity relationship. Therefore, the value of the WJI provides better distinguishability than that using JI.

Table 5. Jaccard index and weighted Jaccard index similarities pairwise $ALS\_EG\_FEA_{0.1}$, $ALS\_CG\_FEA_{0.1}$, and $ALS\_SELECT\_FEA_{0.1}$ (individual-level classification).

Individual-Level Classification	Jaccard Index	Weighted Jaccard Index
	Disease Set from Control Group [2]	Disease Set from Control Group [2]
Disease Set from Experimental Group [1]	0.48	0.65
Comorbidity Feature Set [3]	0.19	0.23
Differentiation	0.29	0.42

[1] Disease set from the experimental group: $ALS\_EG\_FEA_{0.1}$; [2] Disease set from control group: $ALS\_CG\_FEA_{0.1}$; [3] Comorbidity feature set: $ALS\_SELECT\_FEA_{0.1}$.

A high similarity between the comorbidity set of the experimental and control groups implies a small difference between the two sets. This might be due to too many common comorbidity codes within the similarity analysis. For example, seasonal flu is a common disease in both groups; therefore, this disease code has no association in distinguishing between both two groups. Hence, we used statistical verification to eliminate disease codes that failed to satisfy the significance analysis. A new feature set with the associated disease codes was constructed for comparison, which showed a larger difference between the associated feature set and the original disease code set of the control group. The tables show that the difference in the similarities according to the WJI was larger than that using the JI. Therefore, through our proposed method, two original feature sets with no discriminative attributes could be improved by constructing the associated feature set to realize a better prediction model. The similarity measurement according to the middle-clustered classification codes was generally higher than that of the individual-level classification; this is mainly because the individual-level classification representing each disease code was a specific disease, and the mid-level clustered classification implied that each code represents a group of similar diseases. A disease group contains many individual disease codes, which increase the similarity measurement in relation to the middle-clustered classification models.

Table 6. Jaccard index and weighted Jaccard index similarities pairwise $ALS\_EG\_FEA_{0.1}$, $ALS\_CG\_FEA_{0.1}$, and $ALS\_SELECT\_FEA_{0.1}$ (mid-level classification).

Mid-Level Classification	Jaccard Index	Weighted Jaccard Index
	Disease Set from Control Group [2]	Disease Set from Control Group [2]
Disease Set from Experimental Group [1]	0.61	0.73
Comorbidity Feature Set [3]	0.27	0.34
Differentiation	0.34	0.39

[1] Disease set from the experimental group: $ALS\_EG\_FEA_{0.1}$; [2] Disease set from the control group: $ALS\_CG\_FEA_{0.1}$; [3] Comorbidity feature set: $ALS\_SELECT\_FEA_{0.1}$.

According to the individual-level and middle-clustered classification of disease codes and different parameter settings, to construct the training data and prediction system, each subject of *ALS_EG* and *ALS_CG* was analyzed based on the similarities between two different comorbidities. The verification of the prediction results is shown in Tables 7–10.

Tables 7 and 8 show the prediction results obtained by training *ALS_EG*, *ALS_CG*, and *ALS_SELECT_FEA*. In addition to the XGBoost, compared with the JI and traditional scoring analysis, the WJI improved the performance of the training models and prediction results. Table 7 shows the prediction results obtained using individual disease code-level classification for model training. Excluding the XGBoost, the prediction results of the other three machine learning models based on the WJI was higher than that of the JI and traditional scoring analysis in terms of sensitivity, accuracy, F1 score, and AUC value. The results obtained by applying the middle-clustered classification for model training are shown in Table 8. In addition to the XGBoost, all the training and prediction results of the other three machine learning models based on the WJI outperformed the JI and traditional scoring in terms of model score, precision, sensitivity, specificity, accuracy, F1 score, and AUC value.

According to the WJI training verification results shown in Tables 7 and 8, the performance of the XGBoost based on the individual disease code-level classification was generally worse than that of the mid-level clustered classification, unlike the other three machine learning models. The comparisons between the individual- and mid-level clustered disease codes for the three machine learning models were analyzed as follows. In the mid-level clustered classification, the models trained using the SVC and random forest approach could achieve a high accuracy of 0.808, sensitivity of 0.729, specificity of 0.886, and AUC value of 0.844. However, the individual-level classification outperformed the mid-level clustered classification. The SVC and random forest models achieved an accuracy of 0.837, sensitivity of 0.788, specificity of 0.857, and AUC values of 0.907.

Table 7. Training and verification results of *ALS_EG*, *ALS_CG* and *ALS_SELECT_FEA* (individual level classification).

Individual Level Classification (Mean of 5-fold Values)	Logistic Regression			Support Vector Classification			Random Forest			XGBoost		
	JI [1]	WJI [2]	Score	JI	WJI	Score	JI	WJI	Score	JI	WJI	Score
Model Score	0.785	0.840	0.753	0.662	0.852	0.715	0.805	0.866	0.776	0.949	0.977	0.933
Precision	0.883	0.854	0.824	1.0	0.87	0.823	0.84	0.883	0.784	0.835	0.79	0.643
Sensitivity	0.659	0.83	0.659	0.322	0.816	0.491	0.688	0.788	0.688	0.816	0.745	0.562
Specificity	0.915	0.843	0.859	1.0	0.857	0.9	0.872	0.886	0.816	0.814	0.787	0.704
Accuracy	0.787	**0.836**	0.759	0.661	**0.837**	0.695	0.78	**0.837**	0.752	**0.815**	0.766	0.633
F1 Score	0.751	0.836	0.725	0.481	0.835	0.613	0.752	0.829	0.729	0.817	0.755	0.596
AUC	0.872	0.907	0.816	0.411	0.907	0.816	0.87	0.915	0.807	0.815	0.766	0.633

[1] JI represents Jaccard index; [2] WJI represents weighted Jaccard index.

Table 8. Training and verification results of *ALS_EG*, *ALS_CG* and *ALS_SELECT_FEA* (mid-level classification).

Mid-Level Classification (Mean of 5-fold Values)	Logistic Regression			Support Vector Classification			Random Forest			XGBoost		
	JI	WJI	Score	JI	WJI	Score	JI	WJI	Score	JI	WJI	Score
Model Score	0.739	0.792	0.743	0.755	0.812	0.743	0.774	0.828	0.751	0.924	0.977	0.921
Precision	0.762	0.787	0.733	0.77	0.866	0.699	0.794	0.865	0.694	0.827	0.784	0.65
Sensitivity	0.716	0.801	0.728	0.659	0.731	0.728	0.659	0.729	0.703	0.772	0.788	0.659
Specificity	0.761	0.774	0.733	0.789	0.886	0.689	0.801	0.887	0.689	0.818	0.774	0.635
Accuracy	0.739	**0.788**	0.731	0.724	**0.808**	0.709	0.73	**0.808**	0.696	**0.795**	0.781	0.647
F1 Score	0.731	0.788	0.723	0.703	0.791	0.707	0.71	0.787	0.695	0.789	0.784	0.649
AUC	0.837	0.876	0.812	0.816	0.844	0.794	0.833	0.866	0.788	0.795	0.781	0.647

Table 9. Training and verification results of *ALS_EG*, *ALS_CG* and *ALS_SELECT_FEA$_{0.1}$* (individual level classification).

Individual Level Classification (Mean of 5-fold Values)	Logistic Regression			Support Vector Classification			Random Forest			XGBoost		
	JI	WJI	Score	JI	WJI	Score	JI	WJI	Score	JI	WJI	Score
Model Score	0.753	0.785	0.746	0.75	0.776	0.736	0.759	0.794	0.759	0.912	0.963	0.919
Precision	0.76	0.789	0.754	0.76	0.78	0.739	0.752	0.795	0.794	0.698	0.765	0.733
Sensitivity	0.744	0.773	0.715	0.744	0.719	0.715	0.702	0.759	0.688	0.673	0.773	0.676
Specificity	0.761	0.788	0.747	0.761	0.787	0.732	0.76	0.802	0.817	0.718	0.761	0.731
Accuracy	0.752	**0.781**	0.731	0.752	**0.753**	0.724	0.731	**0.781**	0.752	0.696	**0.767**	0.704
F1 Score	0.747	0.778	0.724	0.747	0.742	0.718	0.721	0.775	0.731	0.679	0.768	0.698
AUC	0.821	0.832	0.8	0.796	0.809	0.773	0.811	0.841	0.797	0.696	0.767	0.704

Table 10. Training and verification results of *ALS_EG*, *ALS_CG* and *ALS_SELECT_FEA$_{0.1}$* (mid-level classification).

Mid-Level Classification (Mean of 5-fold Values)	Logistic Regression			Support Vector Classification			Random Forest			XGBoost		
	JI	WJI	Score	JI	WJI	Score	JI	WJI	Score	JI	WJI	Score
Model Score	0.738	0.75	0.734	0.746	0.767	0.743	0.767	0.781	0.746	0.901	0.977	0.896
Precision	0.744	0.735	0.733	0.762	0.812	0.684	0.769	0.786	0.683	0.769	0.713	0.55
Sensitivity	0.728	0.729	0.728	0.687	0.702	0.714	0.617	0.689	0.674	0.733	0.731	0.503
Specificity	0.747	0.746	0.733	0.788	0.829	0.676	0.815	0.8	0.689	0.774	0.704	0.605
Accuracy	**0.738**	0.737	0.731	0.737	**0.765**	0.695	0.716	**0.744**	0.682	**0.754**	0.718	0.554
F1 Score	0.729	0.724	0.723	0.716	0.746	0.69	0.68	0.726	0.675	0.748	0.72	0.523
AUC	0.83	0.846	0.806	0.799	0.807	0.79	0.831	0.843	0.783	0.754	0.718	0.554

Tables 9 and 10 show the various verification results for *ALS_EG*, *ALS_CG*, and *ALS_SELECT_FEA$_{0.1}$*. It can be observed that the four machine learning models based on the WJI generally outperformed the JI and traditional scoring analysis. Table 9 is the prediction result obtained from using the individual disease codes for model training. For model training, the four different machine learning models incorporating the WJI yielded scores better in terms of precision, accuracy, and AUC values compared with the JI and traditional scoring analysis. Table 10 shows the results obtained using the middle-clustered-level classification for model training. The SVC and random forest learning models based on the WJI, rather than the JI and traditional scoring indicator, achieved better performance.

The comparison of the performance of the WJI-based training models for the individual disease code and the middle clustered-level classification for the four different machine learning models are shown in Tables 9 and 10. For the mid-level clustered classification, the SVC learning model achieved a high accuracy of 0.765, sensitivity of 0.702, specificity of 0.829, and AUC value of 0.807. However, the individual-level classification was more effective than the middle-clustered-level classification. Among the four different learning models, the LR and random forest achieved an accuracy of 0.781, sensitivity of 0.759, specificity of 0.788, and AUC value of 0.832.

Tables 7–10 show that the learning models based on the WJI generally outperformed the ones based on the JI and traditional scoring analysis. Specifically, Tables 7 and 8 show that the prediction results for each subject in *ALS_EG* and *ALS_CG* against *ALS_SELECT_FEA* were better than that in *ALS_SELECT_FEA$_{0.1}$*. In addition, the results obtained at the individual disease code level (shown in Table 7) were better than those of the mid-level clustered classification (Table 8). Because each mid-level clustered code represented a group of similar diseases, the corresponding weighted coefficients might be diluted during feature training, resulting in reduced effectiveness in distinguishing the various comorbidity features. This leads to poor performance during model training and cross-verification prediction.

In this study, ALS was applied as the target disease for validating the proposed method. As a matter of fact, different multivariate models were considered initially, but the performance was not good and not reported in this study. It is believed that the many comorbidities and few ALS subjects were the main reasons causing unsatisfactory results. Therefore, we proposed the novel similarity-based approach instead of the statistical modeling approach. To validate the proposed efficient and effective prediction systems, four different machine learning models were adopted. To further analyze the prediction models with different parameter settings, two different levels of comorbidity, namely individual disease codes and mid-level clustered disease codes, were applied for constructing the prediction model and for comparing the system performance. To enhance the effectiveness of the proposed model, in addition to the proposed WJI similarity, the traditional JI similarity and a scoring mechanism were applied.

According to our survey results, our proposed machine learning method using WJI indicators is the first approach based on Electronic Medical Records for ALS prediction. Previous ALS diagnosis mainly focused on clinical biomarkers, biological biomarkers, genetic/proteomics biomarkers, and neuroimaging indicators [27]. Several machine learning techniques have been extensively applied to assist ALS diagnosis using the mentioned biomarkers. Among all different biomarker usages, using a neuroimaging approach achieved the best performance of higher than 80% sensitivity in general [28,29]. However, these diagnostic approaches often occurred after serious symptom onset and led to diagnostic delay. How to apply noninvasive approaches with outstanding sensitivities and specificities at an early stage has become an important issue for ALS patients, and such a diagnostic model could help at-risk ALS patients to be recruited into earlier clinical trials. One advantage of our proposed method is that using individual EMRs as features to compare with comorbidity patterns of ALS patients is a noninvasive, cost-free, and efficient approach for early detection. However, no single test could provide a definitive diagnosis of ALS so far. Several diseases such as multiple sclerosis and Parkinson's disease hold similar symptoms to ALS, and these symptoms may be initially neglected by non-ALS-trained physicians. Hence, the proposed approach only provides an early alert for physicians. Complete neurologic examinations for muscle weakness, spasticity, and atrophy are still required for precise diagnosis of ALS.

Through historical disease records, each subject in the experimental and the control groups was analyzed, and training was performed using two different sets of comorbidity features for constructing the prediction model. We proposed the novel WJI with four different machine learning techniques to construct an ALS prediction system. Since there is no diagnostic model adopting EMRs for ALS prediction, we applied both the traditional JI indicator and scoring mechanism for comparison. Furthermore, two different levels of comorbidity, individual-level disease codes and mid-level clustered disease codes, were applied for model construction and comparison of system performance.

For the ALS prediction model, the new WJI indicator-based system performed better than the traditional JI indicator and scoring mechanism-based methods. Predicting ALS disease using the WJI indicator and individual-level disease codes could yield a high accuracy of 83.7%, sensitivity of 78.8%, specificity of 85.7%, and AUC value of 0.907. When the mid-level clustered disease codes were used, the performance was slightly degraded, showing an accuracy of 80.8%, sensitivity of 72.9%, specificity of 88.6%, and AUC value of 0.844.

From these results, it can be observed that the learning models were more effective when the individual-level disease codes rather than the mid-level clustered classification were used. Thus, the WJI is more effective for model construction based on individual-level disease codes. This is because the mid-level clustered disease codes can dilute the weighted relationship and can reduce the effectiveness of the comorbidity features. The difference between the WJI and JI is that the former enhances the uneven relationship of comorbidities. When a large proportion of subjects in both the experimental and control groups suffer from an identical disease, the relative weighting coefficients and degrees of influence increase. In contrast, when the number of disease codes for various diseases is large but only a small number of subjects suffer from diseases under the same codes, the relative weighting coefficient and corresponding impact are reduced. The traditional JI similarity indicator is based on either the presence or absence of a disease code. It does not consider the different proportions and prevalence of the selected comorbidities. As only the binary conditions of 0 and 1 can be applied to evaluate comorbidities, the value of WJI could yield better distinguishability than the JI indicator.

4. Conclusions

In this study, the original feature sets without discriminative attributes can be improved by the novel proposed indicator and the newly modified feature sets can be trained effectively to realize a good prediction system. Although the data size is small in this study, the prediction performance with accuracy rates higher than 80% was comparable to traditional neuroimaging-based approaches. The proposed ALS prediction model is a time-saving and convenient noninvasive way to detect and evaluate at-risk ALS subjects. However, many mimicking diseases holding similar symptoms are likely to cause similar historical EMRs, and this would increase the false positive rate in general. Nevertheless, early alerts for physicians to identify at-risk ALS patients is one of the main goals of this study, and the proposed WJI indicators can be applied to construct a prediction model for a defined disease with specific comorbidities. Based on the prediction results of the personal disease records, detecting and treating potential patients in the early stages can be achieved. Therefore, the model can strengthen the prevention of specific diseases, can prolong survival years, and can improve the quality of life. It can be used to obtain an efficient in silico analytical tool to effectively aid medical applications for detecting or evaluating difficult diseases. By exploring a large number of medical records to improve the preventive medical applications, we hope that the proposed method can enable doctors to make good decisions in terms of medical treatment and risk assessment for precise diagnosis in the future.

Supplementary Materials: The following are available online at https://www.mdpi.com/2076-3417/11/3/1289/s1. Five different examples are illustrated to show the conception of the novel proposed weighted Jaccard index.

Author Contributions: Conceptualization, C.-H.H. and T.-W.P.; methodology, C.-H.H. and T.-W.P.; software, C.-H.H.; validation, B.-S.Y., D.T. and C.-S.H.; data curation, B.-S.Y. and C.-S.H.; writing—original draft preparation, C.-H.H.; writing—review and editing, T.-W.P.; supervision, B.-S.Y., D.T. and C.-S.H. All authors have read and agreed to the published version of the manuscript.

Funding: The work was supported by Ministry of Science and Technology, Taiwan (MOST 104-2321-B-019-009 to Tun-Wen Pai) and National Taipei University of Technology International Joint Research Project (NTUT-IJRP-109-08).

Institutional Review Board Statement: The study was conducted according to the guidelines of the Declaration of Helsinki, and approved by the Institutional Review Board of Chang Gung Memorial Hospital, Taiwan (protocol code 104-5430B and date of approval 2015/08/20). The IRB approved a request to waive the documentation of informed consent.

Data Availability Statement: Data sharing not applicable.

Conflicts of Interest: The authors declare that they have no competing interests.

References

1. Hardiman, O.; Berg, L.H.V.D.; Kiernan, M.C. Clinical diagnosis and management of Amyotrophic Lateral Sclerosis. *Nat. Rev. Neurol.* **2011**, *7*, 639–649. [CrossRef] [PubMed]
2. Kiernan, M.C.; Vucic, S.; Cheah, B.C.; Turner, M.R.; Eisen, A.; Hardiman, O.; Burrell, J.R.; Zoing, M.C. Amyotrophic Lateral Sclerosis. *Lancet* **2011**, *377*, 942–955. [CrossRef]
3. Fang, F.; Ingre, C.; Roos, P.M.; Kamel, F.; Piehl, F. Risk factors for Amyotrophic Lateral Sclerosis. *Clin. Epidemiol.* **2015**, *7*, 181–193. [CrossRef] [PubMed]
4. Wijesekera, L.C.; Leigh, P.N. Amyotrophic Lateral Sclerosis. *Orphanet J. Rare Dis.* **2009**, *4*, 3. [CrossRef] [PubMed]
5. Knibb, J.A.; Keren, N.; Kulka, A.; Leigh, P.N.; Martin, S.; Shaw, C.E.; Tsuda, M.; Al-Chalabi, A. A clinical tool for predicting survival in ALS. *J. Neurol. Neurosurg. Psychiatry* **2016**, *87*, 1361–1367. [CrossRef] [PubMed]
6. Gower, J.C. A General Coefficient of Similarity and Some of Its Properties. *Biometrics* **1971**, *27*, 857. [CrossRef]
7. Candillier, L.; Meyer, F.; Fessant, F. Designing Specific Weighted Similarity Measures to Improve Collaborative Filtering Systems. In Proceedings of the Mining Data for Financial Applications, Leipzig, Germany, 16–18 July 2008.
8. Hillestad, R.; Bigelow, J.; Bower, A.; Girosi, F.; Meili, R.; Scoville, R.; Taylor, R. Can Electronic Medical Record Systems Transform Health Care? Potential Health Benefits, Savings, And Costs. *Health Aff.* **2005**, *24*, 1103–1117. [CrossRef]
9. Myers, D.L.; Culp, K.S. Culp, Electronic Medical Record Using Text Database. U.S. Patent 5,832,450, 3 November 1998.
10. Denny, J.E.; Bastarache, L.; Ritchie, M.D.; Carroll, R.J.; Zink, R.; Mosley, J.D.; Field, J.R.; Pulley, J.M.; Ramirez, A.H.; Bowton, E.; et al. Systematic comparison of phenome-wide association study of Electronic Medical Record data and genome-wide association study data. *Nat. Biotechnol.* **2013**, *31*, 1102–1111. [CrossRef]
11. Prados-Torres, A.; Calderón-Larrañaga, A.; Hancco-Saavedra, J.; Poblador-Plou, B.; Akker, M.V.D. Multimorbidity patterns: A systematic review. *J. Clin. Epidemiol.* **2014**, *67*, 254–266. [CrossRef]
12. Freund, T.; Kunz, C.U.; Ose, D.; Szecsenyi, J.; Peters-Klimm, F. Patterns of Multimorbidity in Primary Care Patients at High Risk of Future Hospitalization. *Popul. Health Manag.* **2012**, *15*, 119–124. [CrossRef]
13. Kirchberger, I.; Meisinger, C.; Heier, M.; Zimmermann, A.-K.; Thorand, B.; Autenrieth, C.S.; Peters, A.; Ladwig, K.-H.; Döring, A. Patterns of Multimorbidity in the Aged Population. Results from the KORA-Age Study. *PLoS ONE* **2012**, *7*, e30556. [CrossRef] [PubMed]
14. Brown, S.-A. Patient Similarity: Emerging Concepts in Systems and Precision Medicine. *Front. Physiol.* **2016**, *7*, 561. [CrossRef] [PubMed]
15. Tashkandi, A.; Wiese, I.; Wiese, L. Efficient In-Database Patient Similarity Analysis for Personalized Medical Decision Support Systems. *Big Data Res.* **2018**, *13*, 52–64. [CrossRef]
16. Lee, J.; Maslove, D.M.; Dubin, J.A. Personalized Mortality Prediction Driven by Electronic Medical Data and a Patient Similarity Metric. *PLoS ONE* **2015**, *10*, e0127428. [CrossRef] [PubMed]
17. Wu, T.-Y.; Majeed, A.; Kuo, K.N. An overview of the healthcare system in Taiwan. *Lond. J. Prim. Care* **2010**, *3*, 115–119. [CrossRef] [PubMed]
18. Peng, C.-Y.J.; Lee, K.L.; Ingersoll, G.M. An Introduction to Logistic Regression Analysis and Reporting. *J. Educ. Res.* **2002**, *96*, 3–14. [CrossRef]

19. Westreich, D.; Lessler, J.; Funk, M.J. Propensity score estimation: Neural networks, support vector machines, decision trees (CART), and meta-classifiers as alternatives to logistic regression. *J. Clin. Epidemiol.* **2010**, *63*, 826–833. [CrossRef]
20. Gunn, S.R. Support vector machines for classification and regression. *ISIS Tech. Rep.* **1998**, *14*, 5–16.
21. Liaw, A.; Wiener, M. Classification and regression by randomForest. *R News* **2002**, *2*, 18–22.
22. Chen, T.; Guestrin, C. Xgboost: A scalable tree boosting system. In Proceedings of the 22nd Acm Sigkdd International Conference on Knowledge Discovery and Data Mining, San Francisco, CA, USA, 13–17 August 2016.
23. Bosch, J.; Hop, W.; Kirkels, W.; Schröder, F. The International Prostate Symptom Score in a community-based sample of men between 55 and 74 years of age: Prevalence and correlation of symptoms with age, prostate volume, flow rate and residual urine volume. *BJU Int.* **1995**, *75*, 622–630. [CrossRef]
24. Cox, J.L.; Holden, J.M.; Sagovsky, R. Detection of postnatal depression: Development of the 10-item Edinburgh Postnatal Depression Scale. *Br. J. Psychiatry* **1987**, *150*, 782–786. [CrossRef] [PubMed]
25. Mehran, R.; Aymong, E.D.; Nikolsky, E.; Lasic, Z.; Iakovou, I.; Fahy, M.; Mintz, G.S.; Lansky, A.J.; Moses, J.W.; Stone, G.W. A simple risk score for prediction of contrast-induced nephropathy after percutaneous coronary interventionDevelopment and initial validation. *J. Am. Coll. Cardiol.* **2004**, *44*, 1393–1399. [CrossRef]
26. Powers, D.M. Evaluation: From precision, recall and F-measure to ROC, informedness, markedness and correlation. *arXiv* **2020**, arXiv:2010.16061.
27. Grollemund, V.; Pradat, P.-F.; Querin, G.; Delbot, F.; Le Chat, G.; Pradat-Peyre, J.-F.; Bede, P. Machine Learning in Amyotrophic Lateral Sclerosis: Achievements, Pitfalls, and Future Directions. *Front. Neurosci.* **2019**, *13*, 135. [CrossRef] [PubMed]
28. Müller, H.-P.; Turner, M.R.; Grosskreutz, J.; Abrahams, S.; Bede, P.; Govind, V.; Prudlo, J.; Ludolph, A.C.; Filippi, M.; Kassubek, J. A large-scale multicentre cerebral diffusion tensor imaging study in Amyotrophic Lateral Sclerosis. *J. Neurol. Neurosurg. Psychiatry* **2016**, *87*, 570–579. [CrossRef] [PubMed]
29. Bede, P.; Iyer, P.M.; Finegan, E.; Omer, T.; Hardiman, O. Virtual brain biopsies in Amyotrophic Lateral Sclerosis: Diagnostic classification based on in vivo pathological patterns. *NeuroImage Clin.* **2017**, *15*, 653–658. [CrossRef]

applied sciences

Article

Computer Tools to Analyze Lung CT Changes after Radiotherapy

Marek Konkol [1,2], Konrad Śniatała [3], Paweł Śniatała [3,*], Szymon Wilk [3], Beata Baczyńska [3] and Piotr Milecki [1,4]

[1] Department of Electroradiology, Poznan University of Medical Sciences, 61-701 Poznań, Poland; marek.konkol@wco.pl (M.K.); piotr.milecki@wco.pl (P.M.)
[2] Radiation Oncology Department IV, Greater Poland Cancer Centre, 61-866 Poznań, Poland
[3] Institute of Computing Science, Poznan University of Technology, 60-965 Poznań, Poland; konrad.sniatala@put.poznan.pl (K.Ś.); szymon.wilk@put.poznan.pl (S.W.); beata.baczynska@student.put.poznan.pl (B.B.)
[4] Radiation Oncology Department I, Greater Poland Cancer Centre, 61-866 Poznań, Poland
* Correspondence: pawel.sniatala@put.poznan.pl

Abstract: The paper describes a computer tool dedicated to the comprehensive analysis of lung changes in computed tomography (CT) images. The correlation between the dose delivered during radiotherapy and pulmonary fibrosis is offered as an example analysis. The input data, in DICOM (Digital Imaging and Communications in Medicine) format, is provided from CT images and dose distribution models of patients. The CT images are processed using convolution neural networks, and next, the selected slices go through the segmentation and registration algorithms. The results of the analysis are visualized in graphical format and also in numerical parameters calculated based on the images analysis.

Keywords: lung cancer; CT images; CNN; pulmonary fibrosis; radiotherapy

Citation: Konkol, M.; Śniatała, K.; Śniatała, P.; Wilk S.; Baczyńska B.; Milecki, P. Computer Tools to Analyze Lung CT Changes after Radiotherapy. *Appl. Sci.* **2021**, *11*, 1582. https://doi.org/10.3390/app11041582

Academic Editors: Francesco Bianconi and Michał Strzelecki
Received: 4 January 2021
Accepted: 30 January 2021
Published: 10 February 2021

1. Introduction

Lung cancer remains one of the most critical oncology challenges, with more than 2.2 million new cases diagnosed and almost 1.8 million deaths in 2020. As the five year survival rate hardly reaches 20% even in well-developed countries, screening, diagnostic, and treatment improvement is needed. The use of IT tools in lung cancer diagnostics has been an important research topic for many years [1]. With growing computational power and artificial intelligence (AI) solutions, developing large-scale screening programs using computed tomography (CT) is possible. Early detection of lung cancer increases patients' chances to obtain an effective, radical treatment like surgery, systemic chemo- and immuno-therapy, or radiotherapy (RT). Radiation oncology itself uses sophisticated IT technologies to plan and deliver the treatment most safely and assess the early and late toxicity.

Radical radiotherapy is an essential part of lung cancer treatment in all patients not eligible for radical surgery because of disease extent or medical comorbidities. It is also a necessary post-surgical adjuvant in some cases (positive surgical margins or lymph nodes) and plays a crucial role in palliative treatment. Despite the clinical scenario, the aim is to deliver a high radiation dose to the tumor cells, sparing normal tissue nearby (so-called OARs (organs at risk). Modern delivery techniques like IMRT (intensity modulated radiation therapy), VMAT (volumetric arc technique), or SBRT (stereotactic body radiotherapy), sophisticated particle modalities (protons or heavy-ions), and positioning accuracy (cone beam-CT imaging, respiratory motion gating) let us precisely conform the radiation beam and meet the OARs' dose constraints' criteria. However, there is still an issue of late and acute toxicity, especially concerning the lung tissue. Ten to 30% of patients develop a

Appl. Sci. **2021**, *11*, 1582. https://doi.org/10.3390/app11041582

subacute radiation-induced pneumonitis (RIP) observed within six months post-treatment or radiation-induced lung fibrosis (RILF) as late toxicity (6–12 months after conventional RT) [2,3]. In clinical practice, the assessment of RILF is based on various grading scales including the Common Terminology Criteria for Adverse Events (CTCAE), Radiation Therapy Oncology Group (RTOG) criteria, or LENT-SOMA(EORTC) scoring. They mainly focus on clinical presentation, partially supported by imaging finding [4]. Some other semi-quantitative fibrosis scoring methods like the Warrick score [5] are commonly used in connective tissue diseases. Nevertheless, in oligosymptomatic patients, the presented scales do not meet radiation oncologists needs, and it seems necessary to use precise, numeric assessment of CT changes based on density values. This approach confirms the time evolution and the impact of mid- and high-radiation doses on RILF [6]. However, in the era of modern dynamic RT delivery methods, it is important to focus on the impact of the lowest doses (0–5 Gy), which is not clear yet.

This study aims to develop a software tool that unifies the workflow of state-of-the-art solutions for an automatic, fast, large-scale radiomic comparison of lung cancer patients' CT images after radiotherapy. The future clinical application is to observe subtle tissue density changes (represented by Hounsfield unit values) in anatomically corresponding parts of lungs in time. Our elaborated system supports the analysis of the association between the tissue changes and the dose delivered to the patient and other dosimetric and clinical factors. The analysis of many real patient cases can help define more precise dose and dose-volume constraints for future RT to prevent affecting lung morphology and function. Initially, the proposed tool was dedicated to analyzing the RILF only. However, currently, during the COVID-19 pandemic, the system can be used to measure post-COVID-19 pulmonary fibrosis.

The paper is organized in the following way. After the Introduction, the next section presents the general structure of the proposed system. Section 3 describes the preprocessing module, which utilizes a CNN to select CT slices, which contain lungs and can be used for further processing. Segmentation and registration modules are presented in Section 4. Next, implementation remarks are provided in Section 5. The last sections presents the conclusion and the possible application in COVID related problems.

2. Structure of the System

Analyzing CT slices provides information about the tissue density. It is expressed in different shades of grey in relation to its X-ray absorption. The full scale is in the range of (-1000, $+3095$). The example values of the Hounsfield scale for different matter are presented in Table 1.

Table 1. Hounsfield scale.

Matter	Hounsfield Scale Range
Bone	$+400 \rightarrow +1000$
Soft tissue	$+40 \rightarrow +80$
Water	0
Fat	$-60 \rightarrow -100$
Lung	$-400 \rightarrow -600$
Air	-1000

Having information about the Hounsfield value of a given pixel, we can decide about the item located at this point according to the scale presented in Table 1, and the classification of pulmonary fibrosis can be quantified.

The proposed system consist of several main blocks:

- The import of DICOM (Digital Imaging and Communications in Medicine) files (including radiation doses),
- The election of CT slices containing lungs,
- The reduction of the number of slices in the moving set (after radiation series),

- Lung segmentation,
- Affine lung registration,
- Elastic lung registration,
- Calculations on preprocessed images,
- The export of results to a CSV file.

The following sections present the most important modules, and finally, some implementation details will be described. The data flow of the proposed system is presented in Figure 1. The input is a set of CT examinations of a patient in the DICOM format. First, this set of CT images is fed into the preprocessing CNN module in order to select slices with lung images. Next, the selected images are transferred for further processing, i.e., lung segmentation and registration. Finally, such prepared lung images are used in the calculation module, and the results are presented in a chosen form.

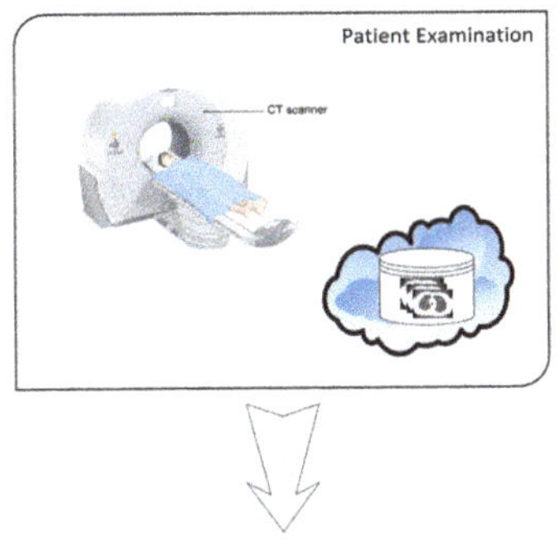
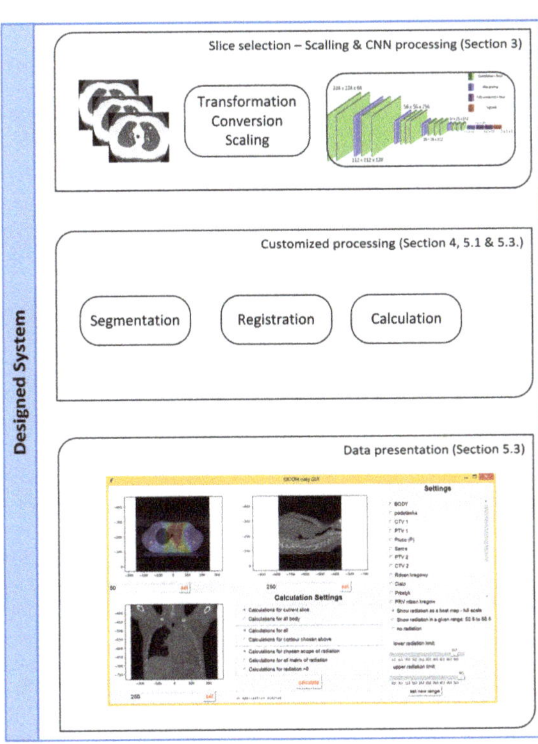

Figure 1. The general dataflow of the proposed system.

3. CT Images' Preprocessing Using CNN

The application of AI and deep learning (DL) in the field of diagnosis from CT scans was proposed in several research works. Mahmud et al. in [7] presented a comprehensive survey on the application of DL, reinforcement learning (RL), and deep RL techniques in mining biomedical data, including medical images. Song et al. [8] analyzed three types of artificial neural networks, i.e., non-convolutional deep neural networks (DNNs), CNNs, and sparse autoencoders (SAEs), and the CNN model was chosen as the most accurate in this type application. The CNN was also successfully used by Gonzales et al. [9] to analyze CT scans of smokers to identify those having chronic obstructive pulmonary disease. CT-based examination was proven to be an important diagnostic procedure for COVID-19 and has been applied as a major diagnostic modality in confirming positive COVID-19 cases [10–13].

An important task to be completed before starting the lung images' analysis is the selection of the CT slices, which include lungs, out of all the set of slices obtained from the multi-slice CT. In our system, for this purpose, we decided to use two of the neural network tested in [14], i.e., VGG16 and VGG19.The structure of the VGG16 network is presented in Figure 2.

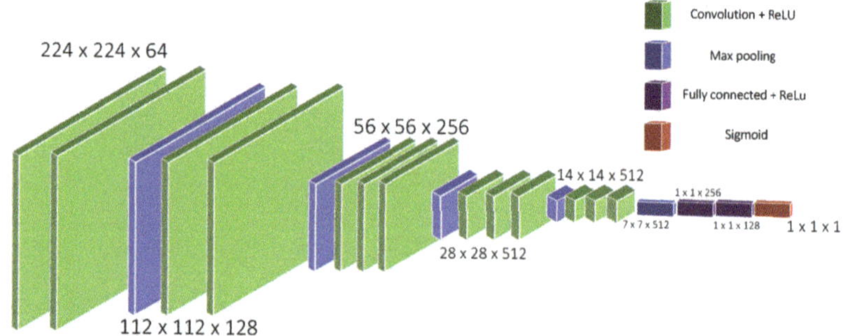

Figure 2. The structure of the VGG network used in our system.

The data flow of the implemented procedure is presented in Figure 3.

The input data required for CNN training, validation, and testing were downloaded from the Cancer Imaging Archive website [15]. The selection filter was set to the following values:

- Collections: NSCLC-radiomics
- Image modality: CT
- Anatomical site: lung

The structure of the training, validation and testing data is described in Table 2.

After reading the slices given in DICOM, the necessary scaling was applied. First the read slices were converted to the Hounsfield scale. The minimum threshold was set to -1024 (-1000 is the air level). This operation reset the *PixelPaddingValue* parameter, which was present in some DICOM files. The next operation moved the scale to include only positive values (simply a minimum value in each slice was added). Next, the maximum threshold was set. All values above X were changed to X ($X = 2000$, 2500). The resulting range was rescaled to real numbers in the range <0.0, 1.0>. Since the VGG network was designed for RGB input images, we tripled the one input channel. It was also necessary to change the slice image size to 224×224. We used a sigmoid activation function for the last layer in the VGG networks and the ReLU (rectified linear unit) function for the other layers. The *padding* parameter was set to *same* value, and the training process was set for 32 epochs.

As mentioned above, we tested two sets of input data with different maximum thresholds X ($X = 2000, 2500$) and two CNN configurations: VGG16 and VGG19. We also set the threshold imposed on the CNN's output to 0.5 in order to obtain two responses: *not lung* and *lung*. The confusion matrices for the validation set are presented in Table 3. In order to be able to compare the performance of the tested networks, sensitivity and specificity were calculated on the same data; they are presented in Table 4.

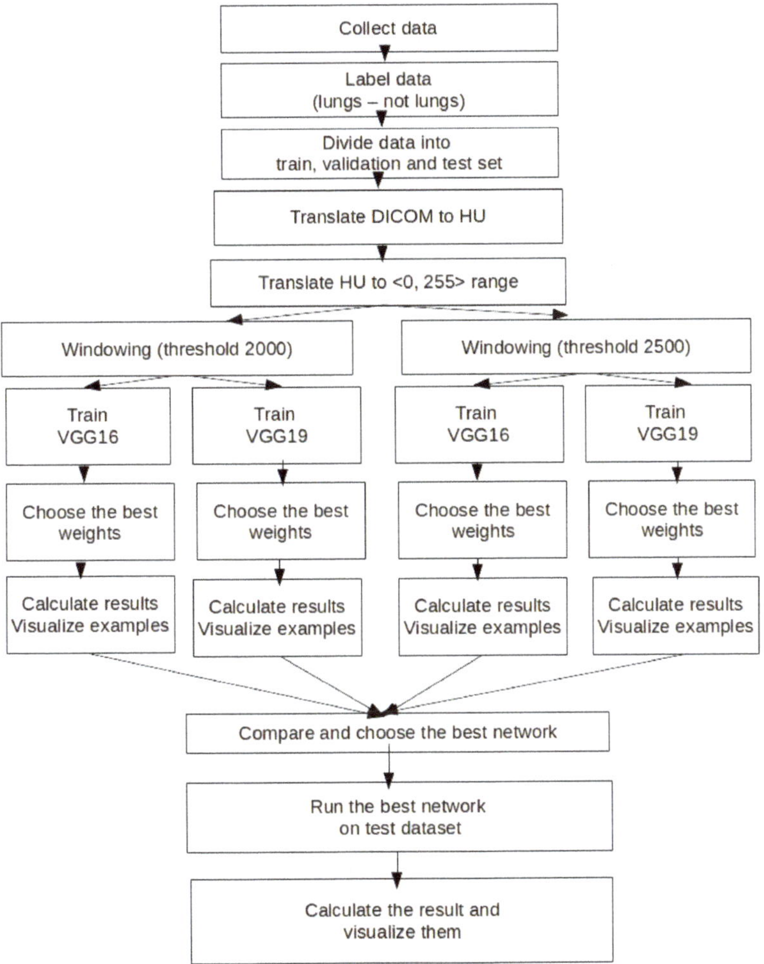

Figure 3. CNN selection procedure. HU, Hounsfield unit.

Table 2. Number of CT slices used as the input data for the CNN.

	Number of CT Slices		
	Not Lungs	**Lungs**	**Total**
Training data	2409 (50 patients)	3293 (45 patients)	5702
Validation data	567 (10 patients)	887 (12 patients)	1454
Testing data	655 (13 patients)	946 (13 patients)	1601

Table 3. Confusion matrices for the validation set.

	Notation	
Real: Lung Real: Not lung	False negative (FN) True negative (TN) Predicted: Not lung	True positive (TP) False positive (FP) Predicted: Lung
	VGG16-2000	
Real: Lung Real: Not lung	23 558 Predicted: Not lung	864 9 Predicted: Lung
	VGG16-2500	
Real: Lung Real: Not lung	11 544 Predicted: Not lung	876 23 Predicted: Lung
	VGG19-2000	
Real: Lung Real: Not lung	27 555 Predicted: Not lung	860 12 Predicted: Lung
	VGG19-2500	
Real: Lung Real: Not lung	25 554 Predicted: Not lung	862 13 Predicted: Lung

Table 4. Accuracy, sensitivity, and specificity for the validation set.

	VGG19-2000	**VGG19-2500**	**VGG16-2000**	**VGG16-2500**
Accuracy = (TP + TN)/All	0.9732	0.9739	0.9780	0.9766
Sensitivity = TP/(TP + FN)	0.9696	0.9718	0.9741	0.9876
Specificity = TN/(TN + FP)	0.9788	0.9771	0.9841	0.9594

VGG16-2500 was discarded because of *not lung* images being classified as *lung*. These mistakes did not appear for the other networks, which in fact were similar in accuracy. We noticed a difference for 1–2 images; however, in these cases, the expert's decisions for the validation sets were also not clear. Figure 4 illustrates the set of images that were classified by all tested networks as FN, which means the image was classified as *not lung*, whereas it contained a lung.

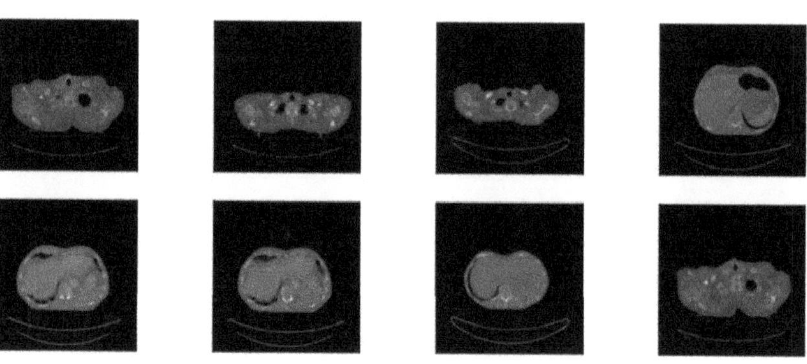

Figure 4. Example of images classified as *false negative* by all tested networks.

Figure 5 illustrates the set of images that were classified by all tested networks as FP, which means the image was classified as *lung*, whereas during labeling, a clinician decided there was no lung. It can be seen that in that case, the clinician made a mistake, because we can find a small part of a lung in each image.

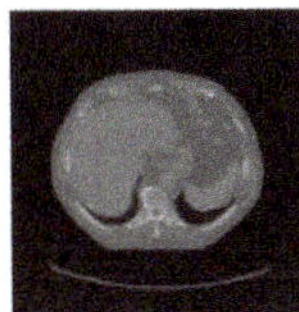 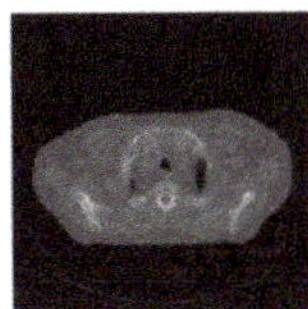 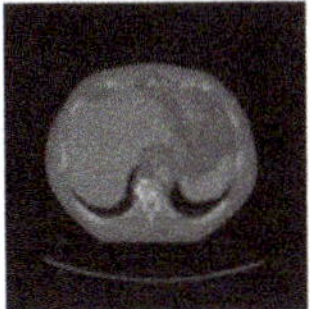

Figure 5. Example of images classified as *false positive* by all tested networks.

Finally, VGG16-2000 was chosen because of the slightly better results. The achieved results on the testing set are presented in Table 5.

Table 5. Results for VGG16-2000 applied to the testing data.

Real: Lung	33	913	Accuracy	0.9763
Real: Not lung	650	5	Sensitivity	0.9651
	Predicted: Not lung	Predicted: Lung	Specificity	0.9924

4. Lung Images' Segmentation and Registration

4.1. Segmentation Process

In order to extract the examined region of interest (ROI) areas from whole images, the segmentation process is necessary. According to [16], lung segmentation may be challenging, because there are differences in pulmonary inflation, which can lead to large variability in volumes and margins when trying to implement automatic algorithms. There is still no universal segmentation method that will properly work for all lung pathological conditions. The amount of unique lung disease cases makes accurate segmentation without human verification very difficult. We can group traditional segmentation methods into fourmain categories: thresholding-based, region-based, shape-based, and neighboring anatomy-guided [16]. We will focus on the first two approaches, i.e., thresholding-based and region-based, which were used in our system. It is important to note that usually to achieve a properly segmented image, multiple techniques are combined.

4.1.1. Thresholding-Based Segmentation

The most basic and easy to implement is a thresholding-based method. We can set a certain global threshold value for the whole gray-scale image. The picture will be transformed into a binary region map, with ones where pixel values are above or equal to the threshold level and zeroes where they are below. Although the algorithm is simple, the most difficult step is to automatically find the threshold level for a specific image. Having lung CT images taken by different machines, the values of pixels representing lungs may differ. Generating a histogram from an image can help to choose the threshold values.

Unfortunately, even choosing a proper threshold does not guarantee that the desired object will be acceptably segmented. Sometimes, images have different lightning in different areas, and applying a simple threshold would give unacceptable results. That is why an adaptive threshold algorithm was introduced. It calculates the threshold "adaptively" for each region of an image. Thanks to this solution, the output is more accurate. It takes into consideration the different brightness of the pixels of the same object, which may be caused by lightning or a faulty camera [17]. The main advantages of this method are the calculation speed and simplicity of implementation. However, it works efficiently only when there is a large contrast between separated objects (which usually is not true in the case of lung CT images). It is very difficult to obtain a satisfying result if there is no significant value difference in the gray levels [18,19].

4.1.2. Region-Based Segmentation

Another approach to lung segmentation includes region-based methods. The most popular from this group is a region-growing algorithm. In general, it compares each pixel to its neighbors, and if a certain condition is met, it is added to the chosen region. One of the main algorithms representing this technique is region-growing. It uses the technique of pixel connectivity as described in [19]. This method segments the whole image into smaller regions, based on the growing (manually or automatically) of the chosen initial pixels (seeds) [17]. The algorithm begins at the initial pixel and checks whether its neighbors are within the threshold range given as an input parameter. If yes and the pixel does not belong to any other region, it is marked as visited, and all pixels from the neighborhood are recurrently checked. When a region is labeled (there are no pixels within the threshold in the neighborhood), a new starting point is chosen, and the algorithm begins to search for another connected region [20]. The main idea is to have pixels with similar properties located together within a region. Compared to simple thresholding, the computational cost of this solution is greater [18]. However, due to taking into consideration the spacial information and region criteria, these methods can conduct lung segmentation much more efficiently and with higher accuracy than the thresholding-based ones.

Another region-based algorithm is split-and-merge. At the beginning, the whole image is treated as a single region. Next, it is iteratively split into smaller sub-regions, until no further splitting is necessary. After that, similar regions are merged together, and a new one is created [17]. The stopping condition of the algorithm is reaching the expected number of regions given as the input or region uniformity. The split-and-merge algorithms are often implemented employing a quadtree data structure [21]. This approach overcomes the need for choosing initial seed points. Unfortunately, it is computationally expensive, because it requires building a pyramidal grid structure of the image [22].

The solution that was a starting point in the approach used in our system is called the watershed transform. The whole image can be treated as a surface. High insensitivity pixels correspond to peaks and low to craters. An intuitive example with a description can be found in [23]. The main goal is to identify the center of each crater—the local minimum called the marker. They give us an approximate idea of where different objects can be possibly located. Each marker is at the bottom of a unique basin, and the algorithm starts filling these basins with different colors, until reaching the boundary (the watershed line) of the adjacent marker [17]. This algorithm was implemented and tested in our system for lung segmentation.

A possible problem of this approach is a lung oversegmentation. This occurs because each regional minimum forms its own small basin, which will be later filled while applying the transform. To overcome this problem, in [24], an extended version of the watershed transform was proposed. Region minima are decreased and next bounded within the region of interest in order to prevent oversegmentation. Choosing internal markers is the key step of this approach. In the described solution, these markers are connected components of pixels that have similar intensity values (in Hounsfield units (HUs)) and whose external boundary pixel values are above a certain gray level. According to [25], the lung region is in the range from -600 HU to -400 HU. That is why to specify internal markers, only pixels with a value lower than -400 are chosen. After eliminating the background, applying morphological transforms, imposing regional minima, applying the watershed transform, and filling the cavities, the segmented lung regions are obtained (Figure 6).

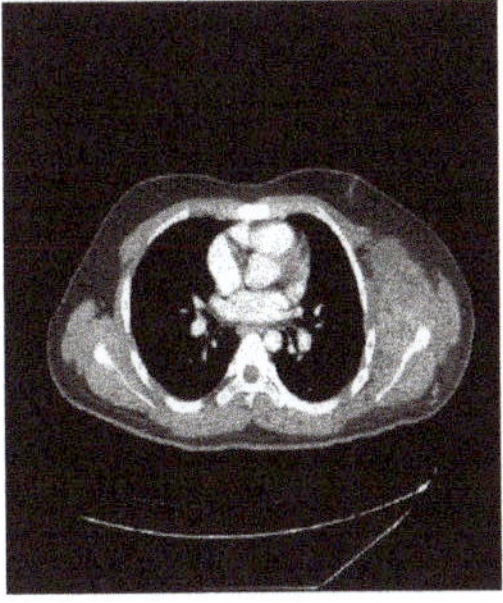 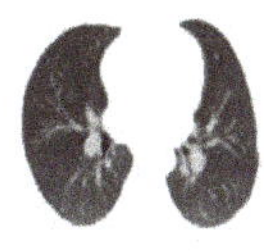

(**a**) Original CT image.　　　(**b**) Lung segmentation result.

Figure 6. Lung segmentation results.

4.2. Registration Process

Image registration is used to overlay multiple images of the same object or scene taken at different times with the use of the same or different devices [26]. It is used in different fields such as medicine, remote sensing (e.g., environmental monitoring, weather forecasting), cartography, and computer vision. Usually, some preprocessing steps like noise removal, normalizing image sizes, or smoothing filters are applied. As described in [27], medical image registration is an integration process applied in order to bring images acquired from different modalities into spatial alignment. An example of registration usage is radiotherapy treatment planning. Doctors can merge images from different devices and/or different times and prepare a proper dosing plan.

A survey of pulmonary image registration methods was given in [28], and it discussed the following approaches:

- Intensity based—registration relies on statistical criteria for matching different intensities between fixed and moving images.
- Feature based—registration is based on different geometrical structures, i.e., points or curves, very useful while registering pathological lungs.
- Segmentation based—in rigid registration, the same structures from both images are extracted; after that, they become the input parameters for the main registration method; unfortunately, registration accuracy is connected with the segmentation quality.
- Mass-preserving—registration relies on detecting density changes on CT images, which are related to different inhalation volumes (air volume in lungs).

5. System Implementation

Python was chosen as the programming language for this project. An important factor for this choice was the availability of many well-documented image processing libraries (e.g., scikit-image, SimpleITK, OpenCV). The syntax and simplicity of writing the code were additional arguments.

Our input data were two CT images series. The first one was taken before radiotherapy (RT), and the second one was acquired during the follow-up. Different CT devices were used during the examinations, and the obtained CT series had different size parameters. Images that were taken before RT had a slice thickness equal to 3 mm, whereas the later images had this value at the level of 1.25 mm. For further processing steps (segmentation, registration, calculations), we had to equalize the number of slices in both series, trying to choose the closest corresponding slices. The first tested approach was based on the slice thickness. We assumed that the first slices from both series were corresponding ones. This could be done because we selected the corresponding beginning and ending slices in both series. Next, we prepared a table where each slice (from both series) had its relative distance to the beginning one. Finally, we iterated over each slice from the "before" series, taking its relative distance and finding corresponding slice from the "after" series, where the difference between those relative distances was minimal. Having selected slices with lung images (this was done using the CNN network described in Section 3), the next implemented block was responsible for the segmentation.

5.1. Segmentation

5.1.1. Watershed Segmentation

The watershed lung segmentation algorithm was based on the solution described in [24], but in order to achieve a satisfying result, many additional operations were added. Finally, our watershed transform segmentation had the following structure:

1. The first step was aimed at finding internal labels. The threshold filter at the level of -360 HU was applied in order to distinguish lungs. Obviously, as presented in Figure 7a, after thresholding, we obtained not only lungs, but also other elements, e.g., the background, pixel intensity of which was at the level of -1000 HU. In order to remove the background, a `clear_border` function from the `skiimage.segmentation` module was used. Thanks to that, only lung pixels were left, as shown in Figure 7b.

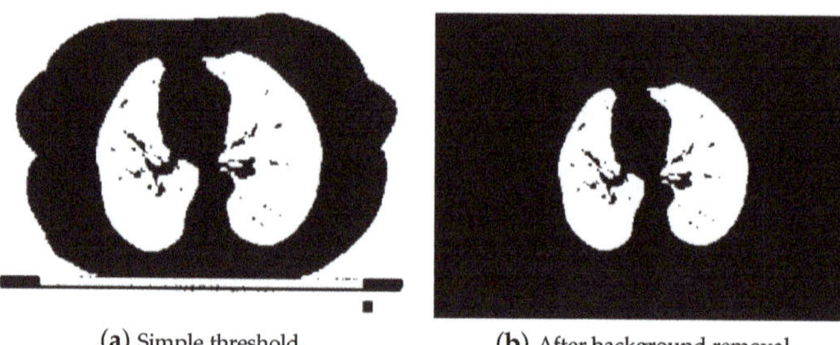

(**a**) Simple threshold.   (**b**) After background removal.

Figure 7. First steps of segmentation.

Afterwards a labeling function from the `scikit.measure` package was used, in order to mark pixels that belonged to the same groups with unique labels. Unfortunately, this solution worked well only for the slices from the middle part of the series. The first ones contained organs like trachea or other significant airways, which were also detected using the threshold (due to the air inside them; a similar Hounsfield value), as shown in Figure 8. In further steps, these areas could be mistakenly treated as internal markers.

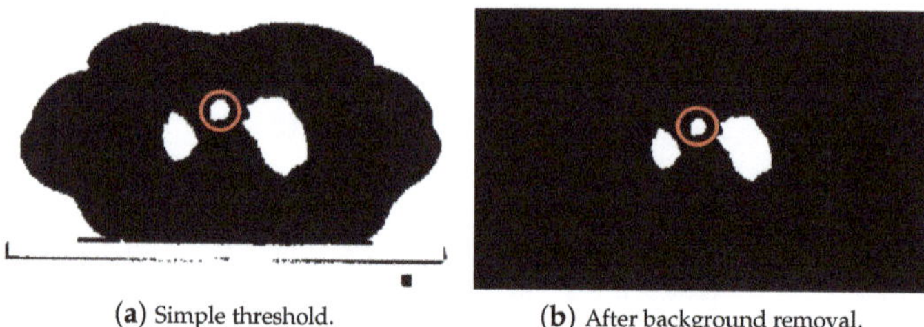

(**a**) Simple threshold.   (**b**) After background removal.

Figure 8. First steps of segmentation with trachea.

To solve this problem for the first 9% of the slices (which very likely contained trachea), additional steps were implemented in order to properly detect only lungs.

*   All regions with areas smaller than 0.00012 of the whole image size were removed. This was necessary in order to get rid of noise and very small insignificant objects, but at the same time, this value cannot be too high, because it would reject regions

with small sections of lungs (beginning slices). As a result of this step, we were able obtain two or three segmented regions.

- If only two regions were left, that meant that one of them was lung, and the other was trachea. We could assume that due to the fact that in all images containing lungs, lungs were bigger than trachea. We found the minimum y value of pixels in both areas (located at the top of each area) and removed this region, which had the smaller value (trachea is above lungs on the first slices). This analysis is illustrated in Figure 9.

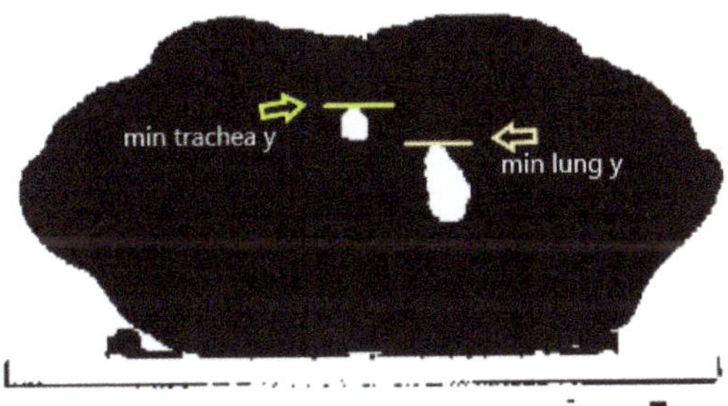

Figure 9. Minimum y values of trachea and one lung.

- If in one of the first slices, three regions were detected, that meant that two of them were lungs, and one was trachea. Unfortunately, we could not simply preserve the two largest ones, because we were not able to ensure that lungs were bigger than trachea (usually, it is the opposite). That is why for this purpose, we detected the centroids of each region (they had x and y coordinates). Next, we made a list of all possible pairs of centroids and calculated the Euclidean distance between each pair. As a result, we obtained a list of three distances, sorted them, and kept the two regions between which the distance was largest. A simple visualization is shown in Figure 10.

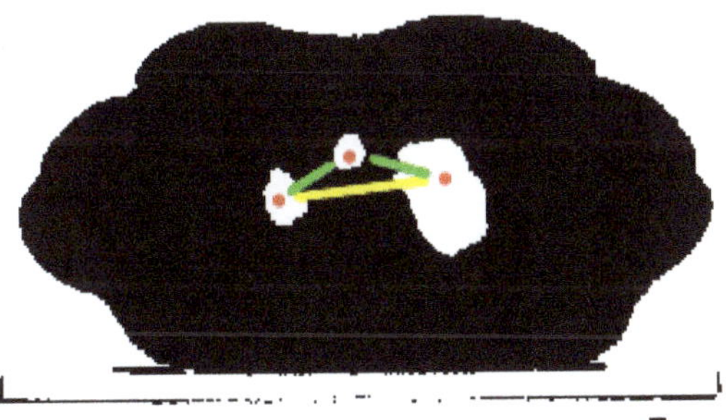

Figure 10. Distances between centroids calculations.

2. The next step was finding external labels. This was done by dilating the internal marker, creating two temporary matrices, and taking regions where they were different from each other (XOR operation). By changing the dilation iterations, we were able to control the distance between those two markers. The external marker looked like a wide border surrounding the internal one. The border had to be wide enough to cover all minima, which could be located in the neighborhood. The final watershed marker containing internal and external ones is presented in Figure 11.
3. Afterwards, the Sobel filter along both axes (x and y) was applied in order to detect edges on the input image (Figure 11).

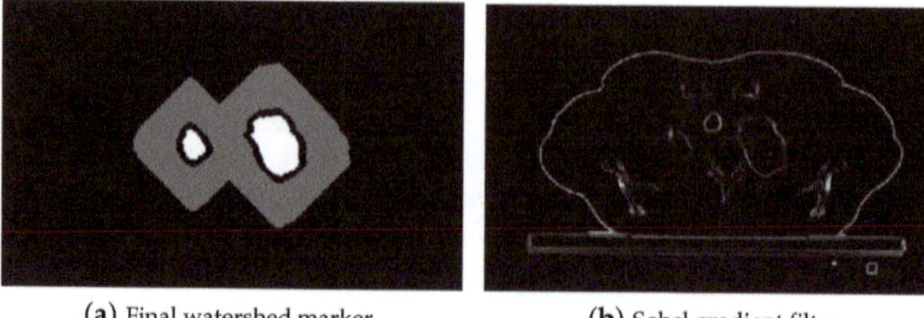

(**a**) Final watershed marker.　　　　　　　　(**b**) Sobel gradient filter.

Figure 11. Sobel filter application in the watershed algorithm.

4. The watershed algorithm was executed with the `sobel_gradient` and `watershed_markers` as inputs. In order to find the border of the result of the watershed algorithm, `morphological_gradient` was applied (with the difference between the dilation and erosion of the watershed as the input). In order to re-include significant nodules near the border, a black-hat operation was applied. Thanks to that, the areas that might carry significant information about the treatment were still in the segmented lungs. We decided that it was better to include an area that normally would not be treated as lung in the segmentation process. If we wanted to omit it during the calculations, we would just contour it with an ROI and exclude it from the calculation area. Removing it during segmentation would lead to a loss of the significant information that it carried.

5. Finally, binary opening, closing, and then, small erosion operations were applied in order to remove noise and fill the holes, which were inside lungs after thresholding. Examples of segmented lungs are presented in Figure 12.

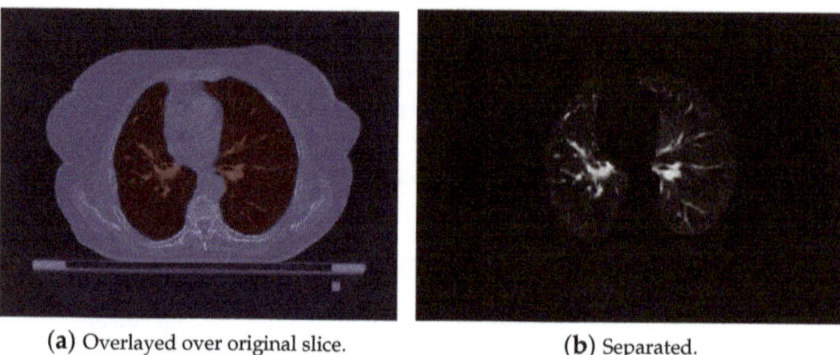

(**a**) Overlayed over original slice.　　　　　　　　(**b**) Separated.

Figure 12. Example of segmented lungs.

The total time of segmenting both series was 50.7 s.

Obviously, in order to make this algorithm more universal and able to segment a larger amount of different lung series, we would probably omit the step of detecting trachea on the first slices and simply remove objects that have a larger ratio. Unfortunately, small lung areas on the beginning slices would not be detected, but if we conducted the calculations on a big amount of patients (100–150), then this would become less significant.

5.1.2. Segmentation Using the SITK Library

The watershed algorithm segmented lungs quite properly, although sometimes, especially for the first or last slices, the resulting masks were too sharp. Lung borders should be smooth in order to properly reflect the reality. At the same time, such segmented lungs would be easier to register. The second approach was implemented using the SimpleITK (SITK) library and based on the example described in [29]. Below, we give the steps of the SITK segmentation algorithm:

1. In order to denoise and smooth the image, we used a `CurvatureFlow` image filter. Smoothed images are better for further processing.

2. In order to separate lungs from the background and other organs, a threshold value at -280 HU was set. It does not strictly correspond to the theoretical Hounsfield range of lungs (form -400 to -600 HU), but the acquired images came from different modalities, so we wanted to cover all lung areas, with pixels in a close neighborhood.

3. The next challenge was to remove the background. We decided to use a `NeighborhoodConneted` function. It labeled pixels connected to an initially selected seed and checked if all pixel neighbors were within the threshold range. The initial seed was chosen as point (10, 10), which was located near the upper left image corner. Pixels with a value equal to zero in the thresholded image (background) were set to one. In the thresholded image, pixels containing body parts other than lungs were also set to one. Adding a thresholded image to the neighborhood connected one gave us a result with segmented lungs and some noise. In Figure 13, the first segmentation steps are depicted.

4. After detecting lungs, a `ConnectedComponents` filter was applied to label the object on the input binary image. Pixels equal to zero were labeled as background, whereas those equal to one were treated as objects. Different groups of connected pixels were labeled with unique labels.

5. In the next step, the area of detected objects was calculated, and the largest one or two, whose size was larger than 0.002 of the whole image, were chosen. Due to this condition, right lung in the third slice was not detected by this algorithm, but it was in the watershed. In order to fix this issue, additional restrictions for the extreme slices should be added.

6. On such a segmented lung binary image, opening and closing were applied. Binary opening was responsible for removing small structures (smaller than the radius) from the image. It consisted of two functions executed on the output of each other: `Dilatation(Erosion(image))`. On the other hand, binary closing removed small holes present in the image.

7. Sometimes, after applying those filters, still, some large holes were left inside the images. Usually, they should stay there, because the area did not match the conditions. However, in our lung segmentation approach, there should not be any holes left inside the lung area. In order to satisfy this requirement, a `VotingBinaryHoleFilling` filter was applied. It also filled in holes and other cavities. The voting operation on each pixel was applied. If the result was positive, it was set to one. We chose a large radius for this operation (16) in order to ensure that no holes would be left inside.

8. Finally, we applied a small erosion filter with the radius set to one. We excluded the lung border from the segmented image. Tissue located in that area was irrelevant for the analysis and could even falsify the results.

Although this approach gave slightly better results, when it came to the final lung segmented shapes, still, some additional filters or methods should be added in order to make this algorithm more universal. Some of the last slices (Figure 14) were segmented more accurately, but the time of calculations—218.6 s for both series—were much longer than for the watershed algorithm—50.7 s.

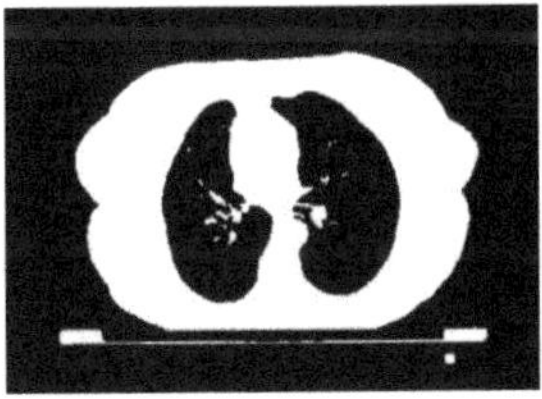

(**a**) Thresholded image.

(**b**) Neighborhood connected background.

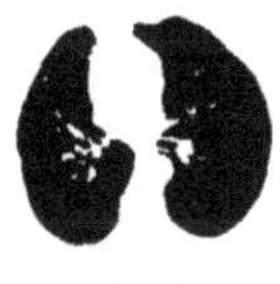

(**c**) Sum of those two images—initially segmented lungs.

Figure 13. Example of the segmentation steps using the SimpleITK library.

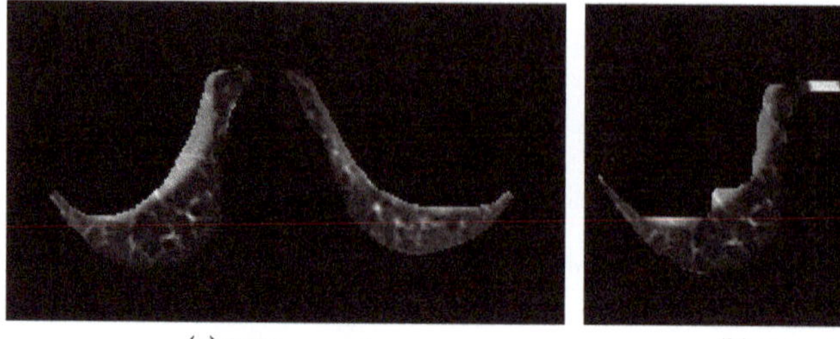

(**a**) SITK approach. (**b**) Watershed algorithm.

Figure 14. Segmented lungs in Slice No. 66 (end of the series).

5.2. Registration

The registration process started with an affine registration where transformations like rotation, scaling, and translation were used. The results of the affine registration were supposed to give a better initial step for further elastic registration.

5.2.1. Affine Slice Registration

Firstly, we analyzed the whole previously segmented images not as solid 3D figures, but as individual slices. Each pair of corresponding slices was registered, where the image from the after series was treated as `moving_image` and the one from the before series as `fixed_image`. For this task, we also used the SITK library [30]. Before executing the registration algorithm, many parameters had to be specified:

1. Initial transformation: In this step, we used `CenteredTransformInitializer` in order to align the centers of both slices. At the same time, the `AffineTransform` transformation type was chosen. Transformations like `Euler2DTransform` would not be proper in this case, because both series were acquired by different devices, which means they had various scales. Euler transformation is specified as rotation and translation, but without scaling. As seen in Figure 15, lungs on the "before" series were much smaller those on the "after" series, so scaling was one of the most significant needed transformations.

2. Measure metric: We used `MattesMutualInformation`. The metric sampling strategy was set to `REGULAR`, and the metric sampling percent was set to 100%. That means that to calculate the current measure, all pixels were taken into consideration. Due to the fact that lung segmentation was performed before registration, we could not compute the metric using a small percent of random points, as was proposed in [31]. If only black points (outside the lungs) had been randomly chosen from both images, then the cost function would have been very low, but it would not mean that the lungs were properly registered.

3. Interpolator: We set it as `Linear`. In most cases, linear interpolation is the default setting. It gives a weighted average of surrounding voxels with the usage of distances as weights.

4. Optimizer: The gradient descent optimizer was chosen for affine registration. It has many parameters to be set, and we used the following:

   - `learningRate = 1,`
   - `numberOfIterations = 255.`

5. Multi-resolution framework: The last set parameters were associated with the multi-resolution framework, which the SITK library provides. Due to this strategy, we were able to improve the robustness and the capture range of the registration process. The input image was firstly significantly smoothed and had a low resolution. Then, during the next iterations, it became less smoothed, and the resolution increased. This additional multi-resolution utility was realized with two functions: `SetShrinkFactorsPerLevel` (shrink factors applied at the change of level)

and `SetSmoothingSigmasPerLevel` (smoothing sigmas applied also at the change of level—in voxels or physical units). In affine registration, where the images differed significantly from each other, we decided to set four multi-resolution levels. The chosen parameters were as follows:

- `shrinkFactors = [6,4,2,1]`,
- `smoothingSigmas = [3,2,1,0]`.

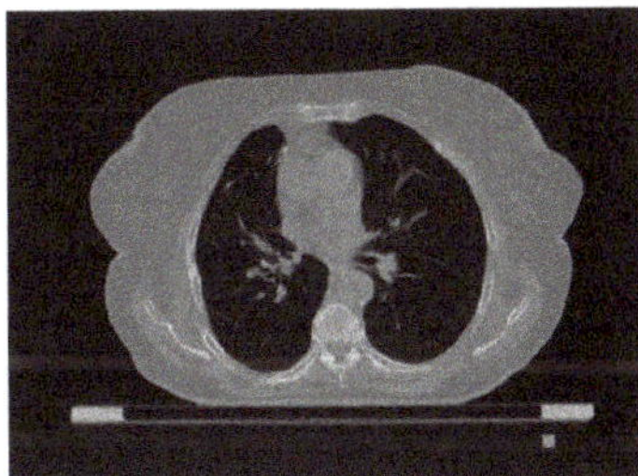

(**a**) Original before series slice.

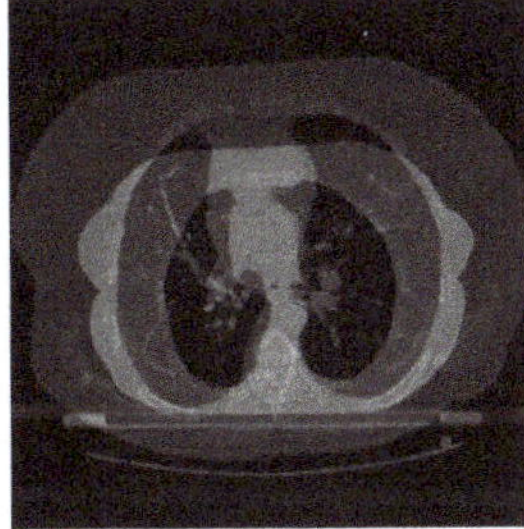

(**b**) Before and after overlayed slices.

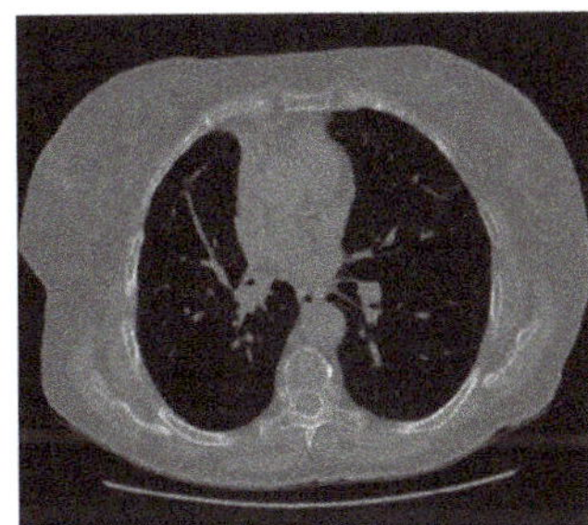

(**c**) Original after series slice.

Figure 15. Illustration of the size difference in before and after slices.

We executed this algorithm for each pair of corresponding slices. An example of an image before and after affine transformation is shown in Figure 16. The time of affine registration for one slice was 2.5 s, and the total time of affine registration for the whole series was 168.1 s.

Table 6 presents the SimpleITK "one slice" affine registration results.

Table 6. Result of one slice affine registration using SimpleITK.

Parameter	Value
Transformation	"Affine"
Final metric value	-0.4329
Total registration time	2.5 s

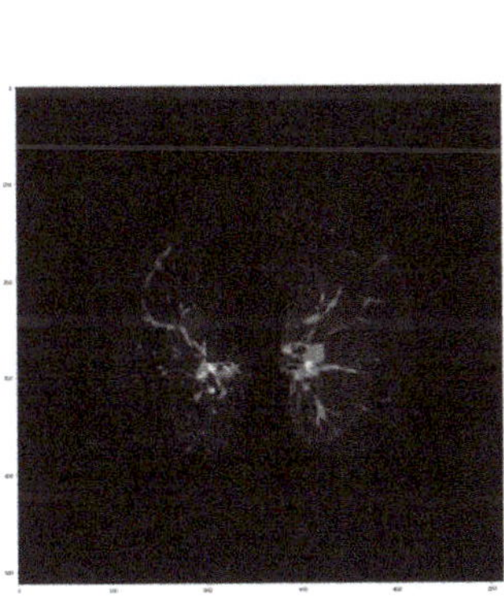

(**a**) Overlayed images before SITK registration.

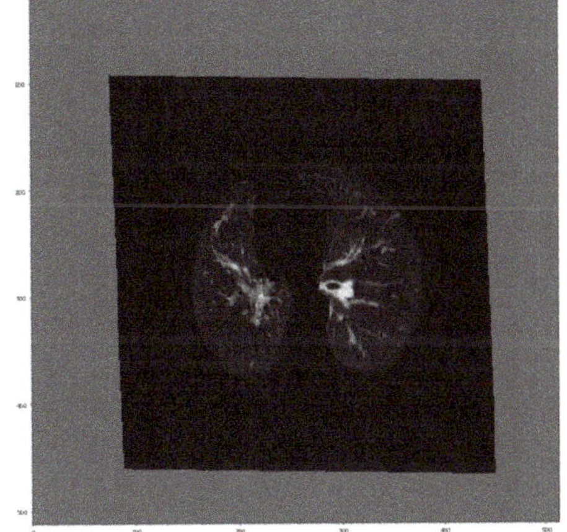

(**b**) Overlayed images after SITK registration.

Figure 16. Illustration of the registration results using the SimpleITK library.

5.2.2. SimpleITK 3D Affine Registration

Affine registration of the whole segmented and raw 3D images was also tested. We wanted to compare how the registration results and times of calculations depended on the input 3D image and chosen samplers (for the segmented image, we had to sample all points, not only those randomly chosen). In order to test how the SimpleITK library handles 3D image registration, we implemented two short programs. The first one was prepared for registering raw CT images, and the second one was responsible for registering lungs segmented from those images. As expected, it took longer to register 3D images with previously segmented lungs than the raw ones. Table 7 compares the results of the 3D affine registration using SimpleITK. In Figure 17, overlayed raw slices, before (a) and after (b) affine registration, are depicted. In order to visualize the differences in registration between raw and segmented images, Figure 18 presents overlayed segmented slices also before (a) and after (b) affine registration. As seen when it comes to 3D registration, the slices, from the middle of the series, are transformed properly.

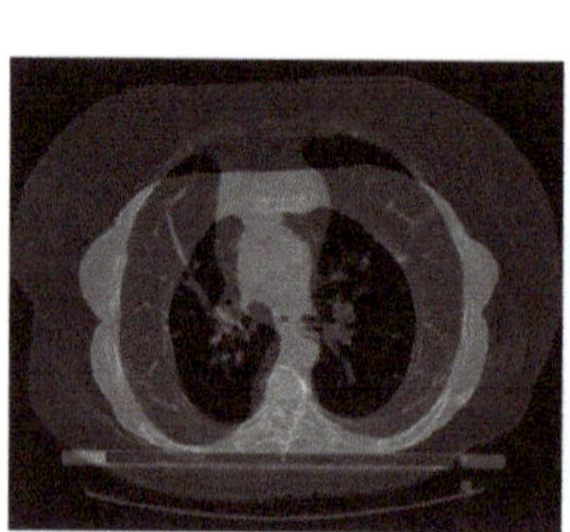
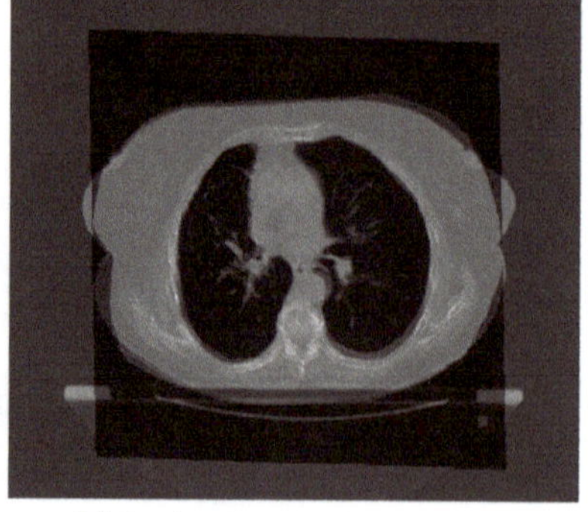

(**a**) Overlayed images before registration.   (**b**) Overlayed images after registration.

Figure 17. Illustration of the affine 3D registration with raw input slices.

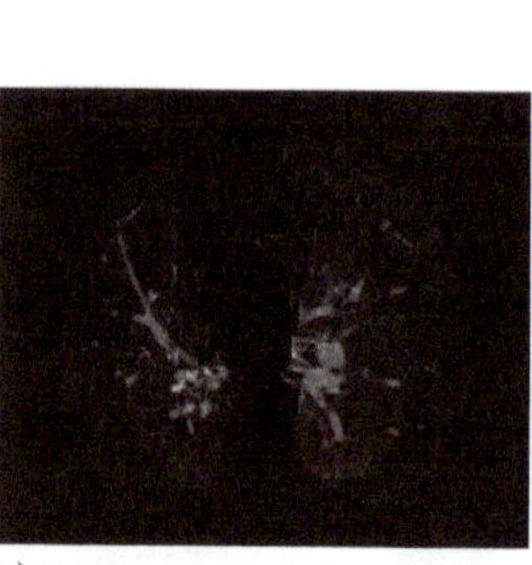
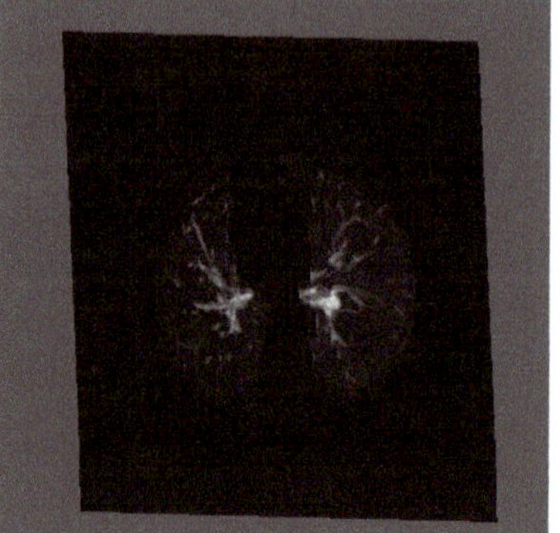

(**a**) Overlayed images before registration.   (**b**) Overlayed images after registration.

Figure 18. Illustration of the affine 3D registration with segmented input slices.

Table 7. Result of 3D affine registration using SimpleITK.

	Raw Images	Segmented Images
Transformation	"Affine"	"Affine"
Sampler	'Random'	"Regular"
Final metric value	−0.72	−0.201
Total registration time	59.5 s	157.4 s

5.2.3. Elastic Lung Registration

Another approach to the registration, implemented in our system, was the elastic registration. This algorithm was implemented using two libraries: SimpleITK and SimpleElastix. It needs to be noted that elastic registration is much slower and more sophisticated than affine. As seen in Figure 16, the first registration step gives quite satisfying results, so the elastic registration will not need to start the process from scratch. Many parameters are similar to those from the affine registration process, so below, we present only the chosen ones, which are different:

1. Initial transformation: In this step, we chose the `BSplineTransformation` type.
2. Measure metric: It is analogous to the affine one.
3. Interpolator: It was also set as `Linear`.
4. Optimizer: The gradient descent optimizer was also chosen for elastic registration; however, its parameters slightly changed:

   - `learningRate = 1,`
   - `numberOfIterations = 255.`

5. Multi-resolution framework: The parameters of the multi-resolution framework were also changed:

   - `shrinkFactors = [2,1],`
   - `smoothingSigmas = [1,0].`

The elastic registration algorithm was also executed for each pair of corresponding slices, where the moving slice was the one after affine registration, and the fixed one was invariably the segmented slice from the "before" series. An example of an image before and after elastic transformation is shown in Figure 19. The time of elastic registration for the whole series was 2103.4 s. Table 8 presents the SimpleITK "one slice" elastic registration results.

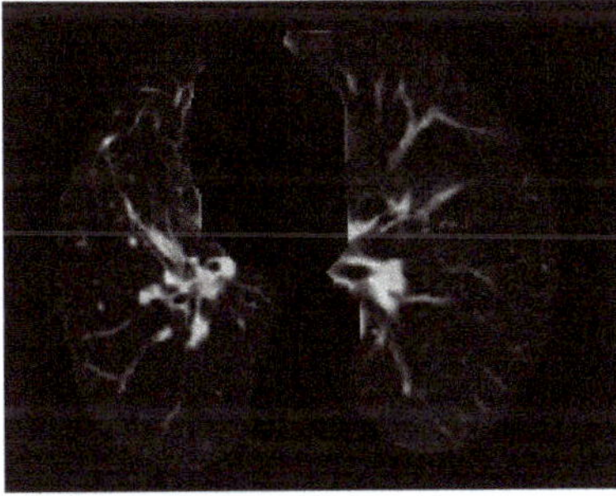

(**a**) Segmented slice from before series.

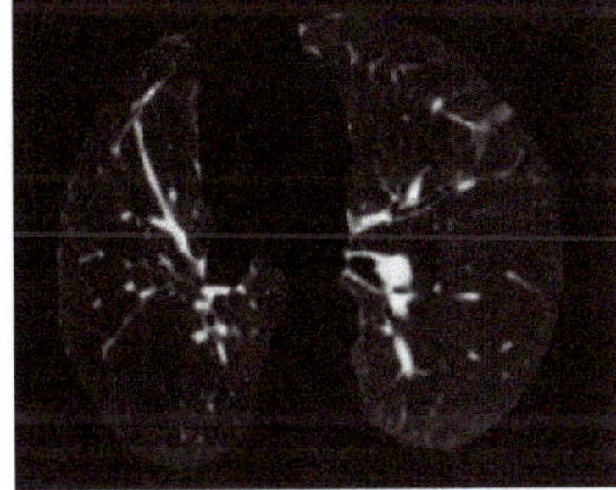

(**b**) Corresponding segmented slice from after series, after applying elastic registration.

Figure 19. Illustration of the elastic registration using the SimpleITK library.

Table 8. Result of one slice elastic registration using SimpleITK.

Parameter	Value
Transformation	"Bspline"
Final metric value	−0.3012
Total registration time	21.0 s

In order to compare different implementations, the registration using the SimpleElastix library was also applied. It was also run for a pair of corresponding slices, where the moving slice was the one from the "after" series and the fixed one was the segmented slice from the "before" series. An example of an image before and after registration is shown in Figure 20.

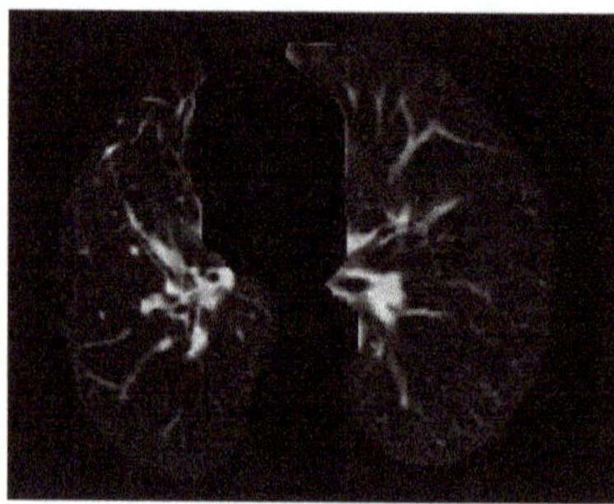 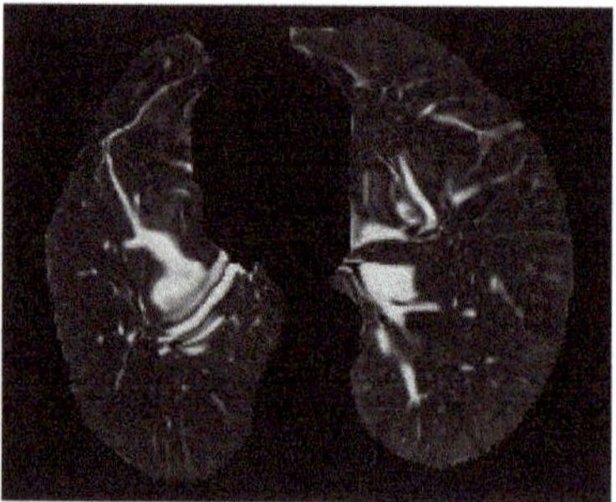

(**a**) Segmented slice from before series. (**b**) Corresponding segmented slice from after series, after applying elastic registration.

Figure 20. Illustration of the elastic registration using the SimpleElastix library.

Additionally, in order to evaluate the registration process, the SimpleElastix library gives the possibility to display the final metric (the `AdvancedMattesMutualInformation` metric was used) value. Table 9 presents the SimpleElastix "one slice" registration results.

Table 9. Result of one slice registration using SimpleElastix.

Parameter	Value
Transformation 1	"Affine"
Transformation 2	"Bspline"
Final metric value	−0.5657
Total registration time	144.9 s

Comparing these results with those presented in Figure 19, we can observe that these solutions differed from each other. Registering using SimpleITK did not change the lung structure significantly; however, at the same time, it did not make the images fully equivalent in terms of shape. The SimpleElastix approach preserved the shape better. Although SimpleElastix had a very good metric value (Table 9), we are not sure if the internal lung structure should be changed to such a large extent. Obviously, the parameters of the SimpleITK methods can also be changed in order to achieve results similar to those produced by SimpleElastix, but the calculation time increases dramatically.

5.3. Data Presentation Module

All aforementioned processing steps such as slice choosing, segmentation, and registration were conducted in order to finally obtain paired images of lungs with a similar shape so that "before" and "after" series can be compared slice-by-slice. Thanks to that, various statistics could be calculated on the corresponding images from the before and after series, including: the maximum, minimum, median, and mean of the pixels (in HU).

This is necessary to compare subtle density changes that appear in different parts of lungs after radiotherapy and could be correlated with the delivered radiation dose and to assess the impact on healthy lung tissue. Some visualization functions were developed. The calculation methods were implemented based partly on the solutions presented in [29].

The main window of the system interface, presented in Figure 21, consists of three areas:

1. Current CT slice (displayed for three directions) (red area—1): The slice can be selected by setting its number or by using a mouse scroll wheel. By moving the mouse over the upper left slice, we can read the pointed pixel's Hounsfield scale value and the dose density value in gray units. The small console at the bottom of the screen is used to display messages to the user (errors, warnings, etc.).

2. Input settings (blue area—2): Check boxes allow the selection of all contours loaded from the active files. Radio buttons are used to set the visible range of the radiation dose. There are three possible options:

   - Show radiation as a heat map (a "warmer" color corresponds to a higher radiation dose)
   - Show radiation in a given range (the range is set by the sliders located below)
   - No radiation (the CT slice is displayed without the radiation scheme).

3. Calculation parameters (green area—3): The first set of radio buttons specifies whether the results are calculated for the whole body (all slices) or just for the current slice. The second set of radio buttons decides whether the calculations are applied for all pixels or just for pixels within the chosen contour. The last three radio buttons are used to select the settings related to the radiation. Calculation can be done for the chosen radiation range (the scope is set by sliders), for all the matrix of radiation (for all pixels included in the radiation area), or for pixels where the radiation is above zero.

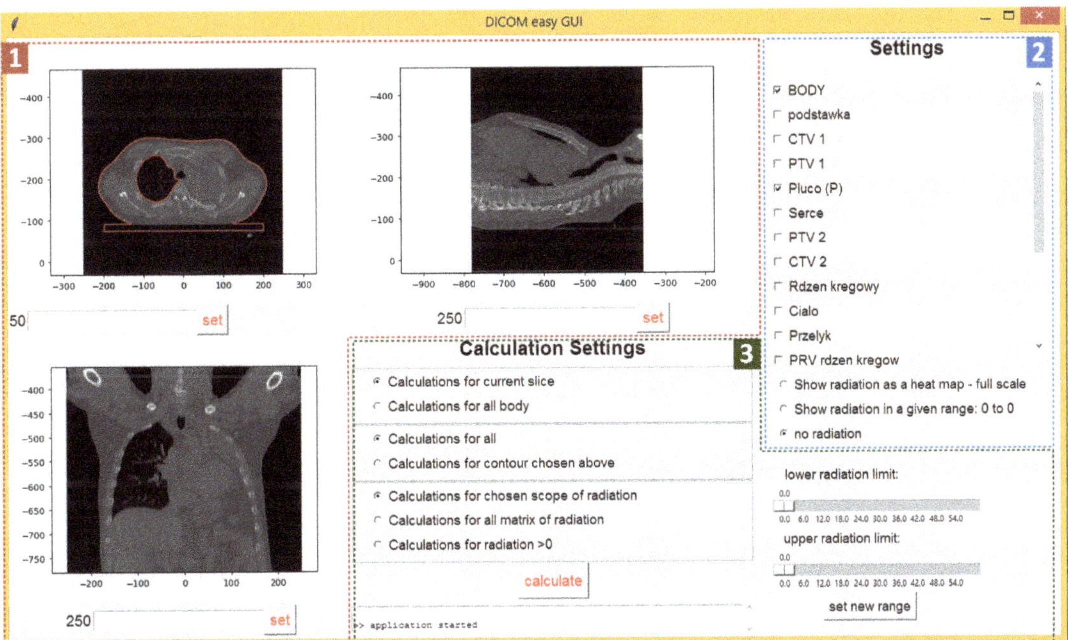

Figure 21. The main window of the system.

Clicking on the calculatebutton starts the calculations and generates the resulting window, presented in Figure 22. The reported results include:

- The image of the current slice with marked pixels selected for calculation,
- The number (indexes) of slices for which the calculation was done,
- The calculation parameters (radiation range, contour type),
- Min, max, median, mode, and average values in HU,
- The volume and area of the selected pixels.

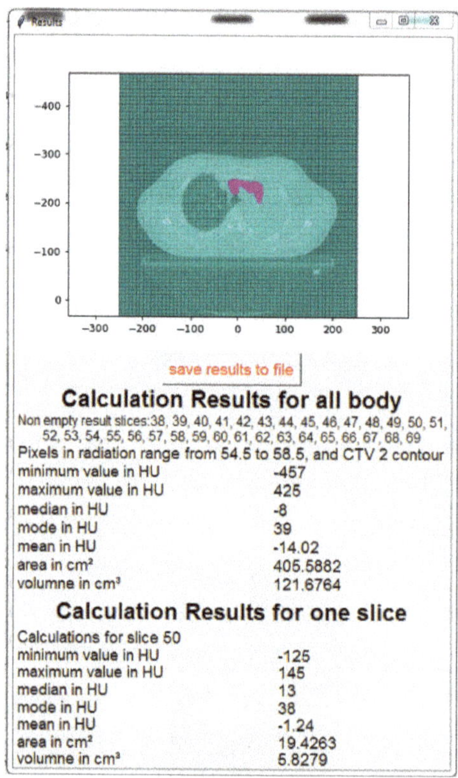

Figure 22. The result window with the calculation results.

When the whole body option is chosen, it is possible to scroll through CT slices. Then, the calculation results are updated on-the-fly. Finally, the results can be visualized (Figure 23) and stored in a file. The first picture, shown in Figure 23a, shows the CT slice with marked lung, esophagus, and spinal cord, which is an input slice for further calculation. Next, Figure 23b presents a CT slice with the dose density scheme, Figure 23c the lung contour with the selected dose range, and Figure 23d the region of calculation, which is a cross-section of the segmented lung and the area of the selected dose range.

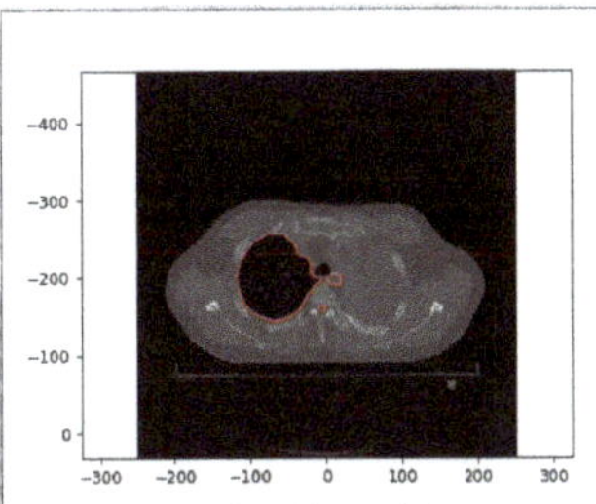

(**a**) CT slice with marked lung, esophagus, and spinal cord.

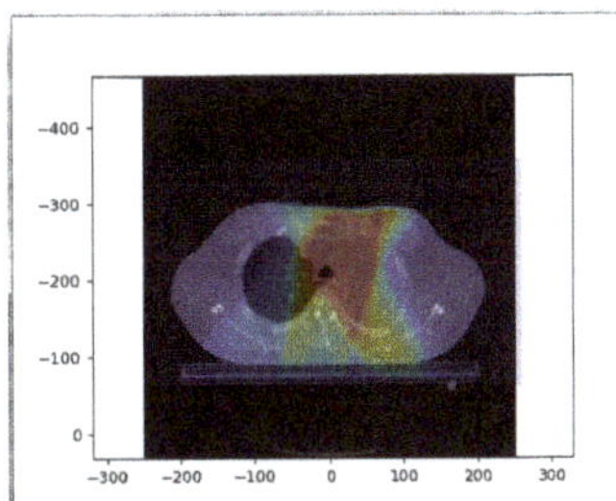

(**b**) CT slice with the dose density scheme.

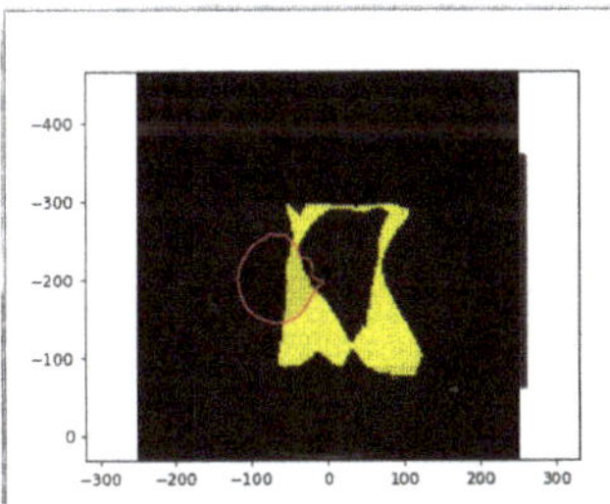

(**c**) Lung contour selected with the dose range.

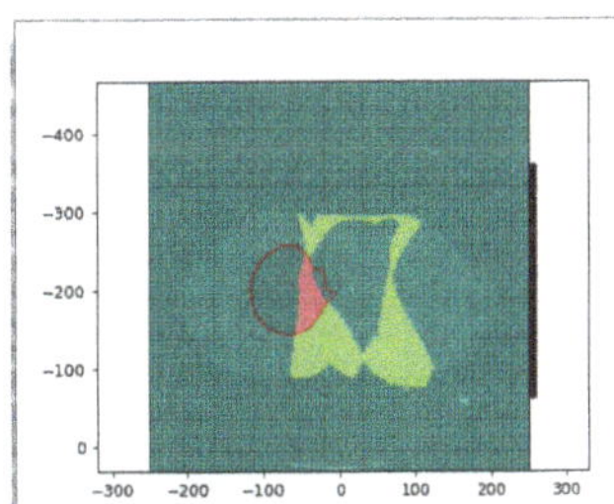

(**d**) Calculation results for a given range and selected contour. The area of interest us colored in pink.

Figure 23. The possible visualization of the CT slices.

6. Conclusions and Future Works

The paper presents a software tool that will be used for large-scale CT-based radiomics for radiation oncology specialists. This particular application is designed to analyze the correlation between the delivered dose and the density changes in pulmonary tissue, representing radiation-induced fibrosis. Currently, the tool is used for analyzing a big set of patients' results, together with the clinical data, focusing on the impact of "low doses"—between 0 and 5 Gy. We presume that even these lowest doses affect the lung function and morphology over time as fibrosis develops even if it is not visible with the naked eye. As an example, in Figure 24, you can see the initial pretreatment image of the tumor and the 18 month follow-up scan with complete remission achieved. There is also an easily noticeable area of RILF that corresponds to the high doses delivered (40–66 Gy). However, it is difficult to see any density changes in other areas. Analyzing the precise density using our software, we find that mean HU values of the lung tissue where the 1–5 Gy dose was delivered during RT increased from -821 HU to -772 HU (Figure 25). Thanks to this, we hope to prove the statistically significant impact of low doses on a large number of patients.

The results can be important in the era of modern conformal radiotherapy methods, which decrease high dose levels in patient's OARs at the cost of the, so-called, "low dose bath". We aim to assess clinically useful dose constraints for conventional lung RT.

Apart from a numerical measurement of density changes, we will work on using AI solutions to analyze and predict the patterns and evolution of fibrotic changes over time. We also want to develop radiobiological concerns on hypofractionated regimens and new modalities like particle therapy. There is also a novel ultra-high dose rate FLASH-RTtechnique [32] that can decrease the lung toxicity [33]; however, this method is still in early pre-clinical studies. What is more, precise fibrosis analysis is also crucial for patients treated with modern immunotherapy, especially combined with conventional chemotherapy and radiotherapy being one of the most common toxicities. We should also remember COVID-19 patients and distinguish different CT abnormalities.

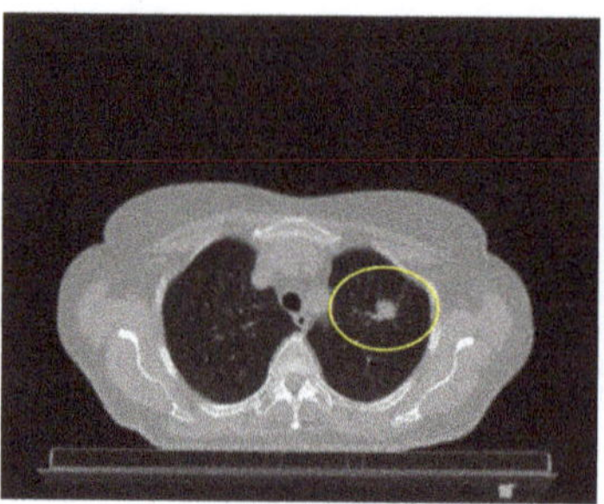

(**a**) CT scan before RT (tumor outlined).

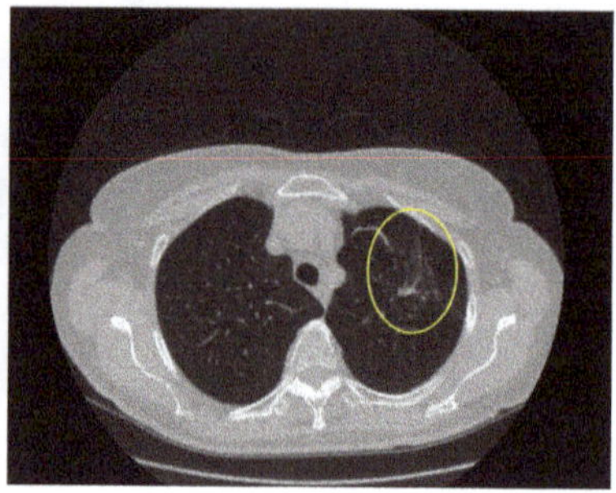

(**b**) CT scan 18 months after RT (fibrosis outlined).

Figure 24. CT images before and after radical radiotherapy (RT) of left lung cancer.

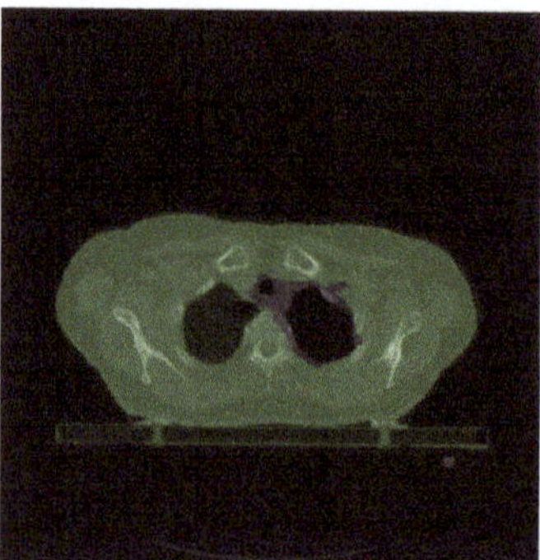

(**a**) Pre-RT CT scan; 1–5 Gy dose level covered in green.

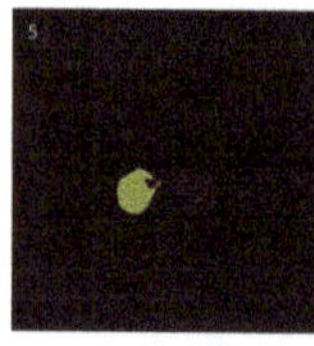
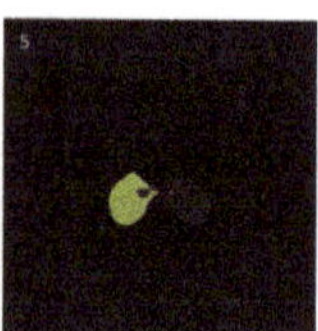

(**b**) Segmented lungs from the pre-RT scan with 1–5 Gy dose covered in yellow.

(**c**) Corresponding lung scan 18 months after RT.

Figure 25. CT images before and after radical RT of left lung cancer.

Author Contributions: Conceptualization: P.Ś. and M.K.; methodology: K.Ś., M.K., and B.B.; software: K.Ś. and B.B.; validation: K.Ś., M.K., B.B., and P.M.; formal analysis: S.W.; investigation: P.Ś., M.K., K.Ś., B.B., and S.W.; resources: P.Ś.; data curation: M.K. and P.Ś.; writing—original draft preparation: K.Ś., P.Ś., and M.K.; writing—review and editing: M.K., S.W., K.Ś., B.B., P.M., and P.Ś.; supervision: P.Ś.; funding acquisition: P.Ś. All authors read and agreed to the published version of the manuscript.

Funding: This research was funded by 0311/SBAD/0699 grants from Poznan University of Technology. The APC was funded by Poznan University of Technology.

Informed Consent Statement: Not applicable.

Acknowledgments: This work was partially supported by the 0311/SBAD/0699 grant from Poznan University of Technology, Institute of Computing Science.

Conflicts of Interest: The authors declare no conflict of interest.

References

1. Ginneken, B.V.; Romeny, B.M.T.H.; Viergever, M.A. Computer-aided diagnosis in chest radiography: A survey. *IEEE Trans. Med. Imaging* **2001**, *20*, 1228–1241. [CrossRef] [PubMed]
2. Wijsman, R.; Dankers, F.; Troost, E.G.; Hoffmann, A.L.; van der Heijden, E.H.; de Geus-Oei, L.F.; Bussink, J. Comparison of toxicity and outcome in advanced stage non-small cell lung cancer patients treated with intensity-modulated (chemo-) radiotherapy using IMRT or VMAT. *Radiother. Oncol.* **2017**, *122*, 295–299. [CrossRef]
3. Wennberg, B.; Gagliardi, G.; Sundbom, L.; Svane, G.; Lind, P. Early response of lung in breast cancer irradiation: Radiologic density changes measured by CT and symptomatic radiation pneumonitis. *Int. J. Radiat. Oncol. Biol. Phys.* **2002**, *52*, 1196–1206. [CrossRef]
4. Käsmann, L.; Dietrich, A.; Staab-Weijnitz, C.A.; Manapov, F.; Behr, J.; Rimner, A.; Jeremic, B.; Senan, S.; De Ruysscher, D.; Lauber, K.; et al. Radiation-induced lung toxicity–cellular and molecular mechanisms of pathogenesis, management, and literature review. *Radiat. Oncol.* **2020**, *15*, 214.
5. Warrick, J.H.; Bhalla, M.; Schabel, S.I.; Silver, R.M. High resolution computed tomography in early scleroderma lung disease. *J. Rheumatol.* **1991**, *18*, 1520–1528. [PubMed]
6. Bernchou, U.; Schytte, T.; Bertelsen, A.; Bentzen, S.M.; Hansen, O.; Brink, C. Time evolution of regional CT density changes in normal lung after IMRT for NSCLC. *Radiother. Oncol.* **2013**, *109*, 89–94. [CrossRef] [PubMed]
7. Mahmud, M.; Kaiser, M.S.; Hussain, A.; Vassanelli, S. Applications of Deep Learning and Reinforcement Learning to Biological Data. *IEEE Trans. Neural Netw. Learn. Syst.* **2018**, *29*, 2063–2079. [CrossRef] [PubMed]
8. Song, Q.; Zhao, L.; Luo, X.; Dou, X. Using deep learning for classification of lung nodules on computed tomography images. *J. Healthc. Eng.* **2017**, *2017*, 8314740.
9. González, G.; Ash, S.Y.; Vegas-Sánchez-Ferrero, G.; Onieva Onieva, J.; Rahaghi, F.N.; Ross, J.C.; Díaz, A.; San José Estépar, R.; Washko, G.R. Disease Staging and Prognosis in Smokers Using Deep Learning in Chest Computed Tomography. *Am. J. Respir. Crit. Care Med.* **2018**, *197*, 193–203. [CrossRef] [PubMed]
10. Kovács, A.; Palásti, P.; Veréb, D.; Bozsik, B.; Palkó, A.; Kincses, Z.T. The sensitivity and specificity of chest CT in the diagnosis of COVID-19. *Eur. Radiol.* **2020**. [CrossRef] [PubMed]
11. Zhou, L.; Li, Z.; Zhou, J.; Li, H.; Chen, Y.; Huang, Y.; Xie, D.; Zhao, L.; Fan, M.; Hashmi, S.; et al. A Rapid, Accurate and Machine-Agnostic Segmentation and Quantification Method for CT-Based COVID-19 Diagnosis. *IEEE Trans. Med. Imaging* **2020**, *39*, 2638–2652. [CrossRef]
12. Luo, N.; Zhang, H.; Zhou, Y.; Kong, Z.; Sun, W.; Huang, N.; Zhang, A. Utility of chest CT in diagnosis of COVID-19 pneumonia. *Diagn. Interv. Radiol.* **2020**, *26*, 437–442. [CrossRef] [PubMed]
13. Zuo, H. Contribution of CT Features in the Diagnosis of COVID-19. *Can. Respir. J.* **2020**, *2020*, 1237418. [CrossRef]
14. Shah, V.; Keniya, R.; Shridharani, A.; Punjabi, M.; Shah, J.; Mehendale, N. Diagnosis of COVID-19 using CT scan images and deep learning techniques. *Radiol. Imaging* **2020**. [CrossRef]
15. Cancer Image Archive Website. Available online: https://www.cancerimagingarchive.net/ (accessed on 10 September 2020).
16. Mansoor, A.; Bagci, U.; Foster, B.; Xu, Z.; Papadakis, G.Z.; Folio, L.R.; Udupa, J.K.; Mollura, D. Segmentation and Image Analysis of Abnormal Lungs at CT: Current Approaches, Challenges, and Future Trends. *Radiographics* **2015**, *35*, 1056–1076. [CrossRef] [PubMed]
17. Anjna, E.A.; Kaur, E. Review of Image Segmentation Technique. *Int. J. Adv. Res. Comput. Sci.* **2017**, *8*. [CrossRef]
18. Yuheng, S.; Hao, Y. Image Segmentation Algorithms Overview. *arXiv* **2017**, arXiv:1707.02051.
19. AucklandImageProcessing. Available online: https://www.cs.auckland.ac.nz/courses/compsci773s1c/lectures/ImageProcessing-html/topic3.htm (accessed on 22 May 2019).
20. Kaur, D.; Kaur, Y. Various Image Segmentation Techniques: A Review. *Int. J. Comput. Sci. Mob. Comput.* **2014**, *3*, 809–814.
21. Bala, A.; Kumar Sharma, A. Split and Merge: A Region Based Image Segmentation. *Int. J. Emerg. Res. Manag. Technol.* **2018**, *6*, 306. [CrossRef]
22. Rusu, A. Segmentation of Bone Structures in Magnetic Resonance Images (MRI) for Human Hand Skeletal Kinematics Modelling. Ph.D. Thesis, Institute of Robotics and Mechatronics, German Aerospace Center (DLR), Oberpfaffenhofen, Wessling, Germany, 2012.
23. Yellow Watershed Marked. Available online: http://www.cmm.mines-paristech.fr/~beucher/wtshed.html (accessed on 22 May 2019).
24. Shojaii, R.; Alirezaie, J.; Babyn, P. Automatic lung segmentation in CT images using watershed transform. In Proceedings of the IEEE International Conference on Image Processing 2005, Genova, Italy, 14 September 2005; Volume 2, p. II-1270, [CrossRef]
25. Elsayed, O.; Mahar, K.; Kholief, M.; Khater, H.A. Automatic detection of the pulmonary nodules from CT images. In Proceedings of the 2015 SAI Intelligent Systems Conference (IntelliSys), London, UK, 10–11 November 2015; pp. 742–746. [CrossRef]
26. Mistry, D.; Banerjee, A. Review: Image Registration. *Int. J. Graph. Image Process.* **2012**, *II*, 18–22.

27. Antoine Maintz, J.B.; Viergever, M.A. An Overview of Medical Image Registration Methods. Utrecht University Repository. 1998. Available online: http://dspace.library.uu.nl/handle/1874/18921 (accessed on 16 June 2019).
28. Cavoretto, R.; De Rossi, A.; Freda, R.; Qiao, H.; Venturino, E. Numerical Methods for Pulmonary Image Registration. *arXiv* **2017**, arXiv:1705.06147.
29. Lis, A. System Wspomagania Oceny Wpływu Rozkładu Dawki w Radioterapii Fotonowej na Późne Zmiany w Obrazie Radiologicznym Płuc. Master's Thesis, Poznan University of Technology, Poznań, Poland, 2018.
30. SimpleITK Registration Documentation. Available online: https://simpleitk.readthedocs.io/en/master/Documentation/docs/source/registrationOverview.html/ (accessed on 16 June 2019).
31. SimpleITK Registration Introduction. Available online: http://insightsoftwareconsortium.github.io/SimpleITK-Notebooks/Python\char'_html/60\char'_Registration_Introduction.html (accessed on 16 June 2019).
32. Wilson, J.D.; Hammond, E.M.; Higgins, G.S.; Petersson, K. Ultra-High Dose Rate (FLASH) Radiotherapy: Silver Bullet or Fool's Gold? *Front. Oncol.* **2020**, *9*, 1563. [CrossRef] [PubMed]
33. Favaudon, V.; Caplier, L.; Monceau, V.; Pouzoulet, F.; Sayarath, M.; Fouillade, C.; Poupon, M.; Brito, I.; Hupe, P.; Bourhis, J.; et al. Ultrahigh dose-rate FLASH irradiation increases the differential response between normal and tumor tissue in mice. *Sci. Transl. Med.* **2014**, *6*, 245–293. [CrossRef] [PubMed]

MDPI

St. Alban-Anlage 66

4052 Basel

Switzerland

Tel. +41 61 683 77 34

Fax +41 61 302 89 18

www.mdpi.com

Applied Sciences Editorial Office

E-mail: applsci@mdpi.com

www.mdpi.com/journal/applsci